Major Groups of Viruses

Group	Host	Morphology	Examples of Viruses
Class I Viruses: Double-Stranded DNA Genomes			
Myoviridae	Bacteria	Complex	T4
Siphoviridae	Bacteria	Complex	λ
Podoviridae	Bacteria	Complex	T7
Papovaviridae	Animal	Naked icosahedral	Polyomavirus, SV40
Adenoviridae	Animal	Naked icosahedral	Adenovirus
Herpesviridae	Animal	Enveloped icosahedral	Herpes simplex, varicella-zoster
Poxviridae	Animal	Complex	Smallpox, vaccinia
Hepadnaviridae	Animal	Enveloped icosahedral	Hepatitis B
Caulimoviruses	Plant	Naked icosahedral	Cauliflower mosaic
Class II Viruses: Single-Stranded DNA Genomes			
Microviridae	Bacteria	Naked icosahedral	φX174
Parvoviridae	Animal	Naked icosahedral	Parvovirus, adeno-associated virus
Geminiviruses	Plant	Fused-pair icosahedral	Maize streak
Class III Viruses: Double-Stranded RNA Genomes			
Reoviridae	Animal	Naked icosahedral	Reovirus, rotavirus
Class IV Viruses: Positive-Strand RNA Genomes			
Leviviridae	Bacteria	Naked icosahedral	MS2, Qβ
Picornaviridae	Animal	Naked icosahedral	Poliovirus, rhinovirus, hepatitis A, coxsackievirus
Togaviridae	Animal	Enveloped icosahedral	Sindbis
Coronaviridae	Animal	Enveloped helical	Murine hepatitis
Potyvirus	Plant	Naked helical	Potato Y
Tymovirus	Plant	Naked icosahedral	Turnip yellow mosaic
Tobamovirus	Plant	Naked Helical	Tobacco mosaic
Comovirus	Plant	Naked icosahedral	Cowpea mosaic
Class V Viruses: Negative-Strand RNA Genomes			
Rhabdoviridae	Animal and plant	Enveloped helical	Rabies, vesicular stomatitis
Paramyxoviridae	Animal	Enveloped helical	Mumps, measles, parainfluenza
Filoviridae	Animal	Enveloped helical	Ebola
Orthomyxoviridae	Animal	Enveloped helical	Influenza A, B
Bunyaviridae	Animal	Enveloped helical	Phlebovirus
Arenaviridae	Animal	Enveloped helical	Lassa
Class VI Viruses: Retroviruses			
Retroviridae	Animal	Enveloped icosahedral	Human immunodeficiency virus (HIV)

The Biology of Viruses

About the Cover

The cover is a stylized depiction of the beginning and ending stages in the reproduction of a human immunodeficiency virus. The temporal sequence of these events is read from right to left viewed as a whole. The front cover shows the initial stages of infection: A virion is illustrated attaching to specific receptors on the lymphocyte surface, then fusing its envelope with the plasma membrane and introducing its nucleocapsid into the cytoplasm. Uncoating then releases the virus's single-stranded RNA genome and the crucial enzyme reverse transcriptase. The double-stranded DNA molecule they produce is seen receding toward the nucleus where it will enter the nucleus and become inserted into the host cell genome. On the back cover, streams of new RNA genomes emerge from the nucleus to become foci for assembly of new nucleocapsids at the plasma membrane. Then new virions bud from the cell surface to complete the reproductive cycle of the virus.

This cover is the work of the visual artist Alan M. Clark. In addition to his scientific work, Clark has produced covers for the first edition of *The Biology of Viruses,* and for science fiction and fantasy books as well as art for computer games. Samples of his art can be seen on his website:

home.earthlink.net/~alanmclark/

The Biology of Viruses

Second Edition

Bruce A. Voyles

Patricia Armstrong Johnson Professor of Biological Chemistry
Professor of Biology
Grinnell College
Grinnell, Iowa

McGraw Hill

Boston   Burr Ridge, IL   Dubuque, IA   Madison, WI   New York   San Francisco   St. Louis
Bangkok   Bogotá   Caracas   Kuala Lumpur   Lisbon   London   Madrid   Mexico City
Milan   Montreal   New Delhi   Santiago   Seoul   Singapore   Sydney   Taipei   Toronto

McGraw-Hill Higher Education

A Division of The McGraw-Hill Companies

THE BIOLOGY OF VIRUSES, SECOND EDITION

Published by McGraw-Hill, a business unit of The McGraw-Hill Companies, Inc., 1221 Avenue of the Americas, New York, NY 10020. Copyright © 2002, 1993 by The McGraw-Hill Companies, Inc. All rights reserved. No part of this publication may be reproduced or distributed in any form or by any means, or stored in a database or retrieval system, without the prior written consent of The McGraw-Hill Companies, Inc., including, but not limited to, in any network or other electronic storage or transmission, or broadcast for distance learning.

Some ancillaries, including electronic and print components, may not be available to customers outside the United States.

This book is printed on acid-free paper.

1 2 3 4 5 6 7 8 9 0 QPF/QPF 0 9 8 7 6 5 4 3 2 1

ISBN 0–07–237031–9

Publisher: *James M. Smith*
Senior developmental editor: *Deborah Allen*
Associate marketing manager: *Tami Petsche*
Senior project manager: *Gloria G. Schiesl*
Senior production supervisor: *Sandy Ludovissy*
Coordinator of freelance design: *Michelle D. Whitaker*
Freelance cover designer: *Kelly Fassbinder/Imagine Design Studio*
Cover illustration: *Alan M. Clark*
Senior photo research coordinator: *Carrie K. Burger*
Photo research: *LouAnn K. Wilson*
Compositor: *Electronic Publishing Services Inc., NYC*
Typeface: *10/12 Times Roman*
Printer: *Quebecor World Fairfield, PA*

The credits section for this book begins on page 389 and is considered an extension of the copyright page.

Library of Congress Cataloging-in-Publication Data

Voyles, Bruce A.
    The biology of viruses / Bruce A. Voyles.—2nd ed.
        p.   cm.
    Includes bibliographical references and index.
    ISBN 0–07–237031–9
    1. Viruses. I. Title.

    QR360 .V69   2002
    579.2—dc21

                                        2001044142
                                        CIP

www.mhhe.com

CONTENTS

CHAPTER 2

Getting In: Attachment, Penetration, and Uncoating   35

CHAPTER 3

Expression and Replication of the Viral Genome in Prokaryotic Hosts   75

CHAPTER 4
Expression and Replication of the Viral Genome in Eukaryotic Hosts: The RNA Viruses 115

CHAPTER 5

Expression and Replication of the Viral Genome in Eukaryotic Hosts: The DNA Viruses 171

CHAPTER 6

Assembly, Maturation, and Release of Virions 201

CHAPTER 7
Effects of Viral Infection on Host Cells: Cytological and Inductive Effects   235

CHAPTER 8
Effects of Viral Infection on Host Cells:
Integrated Viruses and Persistent Infections   275

CHAPTER 9

Subviral Entities, Viral Evolution, and Viral Emergence   321

Both the first and second editions of this book have been driven by the same impetus. For more than 30 years I have been fascinated by the viruses, not just as disease-causing agents, but as organisms that have evolved a wonderful variety of solutions to the problems posed by their reliance on host cells for their reproduction. In the course of studying these organisms, virologists discovered many of the mechanisms by which prokaryotes and eukaryotes replicate and express their own genomes (the bidirectional replication of DNA and the splicing of RNA are just two examples). More recently, bacteriophages like lambda and M13 and eukaryotic viruses like baculovirus and the retroviruses have become versatile research tools of molecular biologists. Clearly, the viruses are for everyone, not just the virologist or clinician.

Probably like many authors, I was driven to write this book originally by frustration. When I wanted to share my fascination with the viruses with undergraduates, I could not find a suitable text. General microbiology texts relegate the viruses to a few examples covered in a handful of pages. Virology books range from everything-you-ever-wanted-to-know-plus-a-little-bit-extra encyclopedias to specialist books for medical students or plant pathologists that are narrowly focused on particular viruses or groups of viruses. While such books are wonderful things of their sort, I wanted an introduction to the viruses that illustrated the common features of their lifestyles rather than a description of their every detail; I wanted the forest, but could get only the trees. Hence, this book.

SCHEME OF THE BOOK

The core of this book is organized around the features of the reproductive cycle shared by all viruses. How are viruses structured? What strategies do viruses use to enter their host cells? How do they express and replicate their genomes? How do they produce new virions? How do host cells respond to viral infection? How might the viruses have evolved? Additional topics like how new viruses emerge as pathogens are also considered. The approach is to describe the general strategies that the viruses have evolved to meet particular situations, with illustrations taken from a variety of different viruses, rather than to present each taxonomic group of viruses in its entirety. For this reason, examples are taken across the spectrum of the bacterial, animal, and plant viruses. My hope is that this "theme and variation" approach will allow the nonexpert to appreciate the viruses as biological entities that share many common features. To my colleagues who may think this approach is too idiosyncratic for their courses, I can only say, "Try it! You'll like it!"

FEATURES

I have tried to create a book that is "user friendly" so that students will actually read it. To this end, I have incorporated a number of features into the book to make it easier to follow the development of the material and its presentation:

- To emphasize concepts, the material in each chapter is presented to the student in the form of questions that open each section. The text then provides the answers to the questions.

- Since many of the illustrations depict the organization or expression of genetic material, the book uses a set of standard conventions that allow the student to identify immediately the exact nature of the molecules being shown and their relationships to each other.

- Each chapter concludes with a summary and an annotated list of suggested readings from readily available sources. These include *Scientific American* articles, reviews of broad topics, some of the classic works in the field, as well as recent papers that are discussed in the chapter.

- Throughout the text there are frequent discussions on how virologists know what they know, so an appendix called *The Virologist's Toolkit* presents brief descriptions of the experimental techniques widely used in virology.

- The body of the text does not use a taxonomic approach, so a second appendix, *Characteristics of Selected Viruses,* provides brief summaries (keyed to the text) of the features and life cycles of the viral groups discussed in the book.

ACKNOWLEDGMENTS

Like Blanche Dubois, authors must frequently "rely on the kindness of strangers" in the preparation of their books. I am especially grateful to a fine panel of reviewers who read either the original manuscript or the second edition manuscript carefully and gave me invaluable assistance both by their brickbats and their bouquets:

Prakash Bhuta, *Eastern Washington University*
Jeffrey Byrd, *St. Mary's College of Maryland*
Kathy Dunn, *Boston College*
Linda Fisher, *University of Michigan–Dearborn*
Darrell Galloway, *Ohio State University*
Jon Lowrance, *Lipscomb University*
Mark McBride, *University of Wisconsin–Milwaukee*
Robert Miller, *Oklahoma State University*
Kevin Oshima, *New Mexico State University*
Susan Payne, *The University of Texas at Arlington*
Marie Pizzorno, *Bucknell University*
Laraine Powers, *East Tennessee State University*
Theresa Singleton, *Delaware State University*
Paul Wanda, *Southern Illinois University–Edwardsville*
David Westenberg, *University of Missouri–Rolla*

Any errors remaining in the text are the result of my own cussedness. Other "kind strangers" include the scientists who permitted me to use original photographs or illustrations from their papers in this book. The credit line for each illustration in the credits section is only a token of my thanks for their courtesy and assistance.

Authors also require the kindness of nonstrangers. Special thanks go to Carolyn Bosse of Grinnell College's Biology Department for her help in preparing photographs, mailing requests for permissions, and for patiently listening to my moanings and groanings. Great thanks are also due to my editors at McGraw-Hill, especially my good friend Deborah Allen who started me down this path originally and has overseen creation of the second edition.

Finally, thanks beyond measure to my wife, Martha, and my children, Paul and Erin. Their loving support made it possible.

DEDICATION

I would like to dedicate this edition of *The Biology of Viruses* to Patricia Armstrong Johnson and to her son, Harold B. Johnson, Jr., who endowed in her memory the professorship that I have the honor to hold. Dr. Johnson, a historian who taught at the University of Virginia, developed an intense interest in biological chemistry and genetics. This interest led him to endow a professorship to support the study of these disciplines and to contribute to the immense promise they offer for the betterment of mankind in the twenty-first century. It is my hope that this book will advance the objectives that Dr. Johnson has so generously supported.

Bruce A. Voyles
Patricia Armstrong Johnson Professor of Biological Chemistry

Chapter 1

Viruses and Host Cells

CHAPTER OUTLINE

- What is a virus?
- Where are viruses found in nature?
- What are the physical properties of viruses?
- What structural features of cells are important in their role as viral hosts?
- What genetic features of cells are important in the virus–host cell interaction?
- How does a virus reproduce?
- What are the basic techniques used to study the virus–host cell interaction?
- How are viruses named?
- Summary: What is a virus?

Why study viruses? In just the two decades since its identification, the human immunodeficiency virus (HIV) has killed more than 18,000,000 people and twice that number have been infected. Sub-Saharan Africa is being decimated by the explosive spread of the virus. While antiviral therapies have improved the quality and length of life of HIV-infected persons, no cure and no vaccine to prevent infection are available. Ebola virus has killed far, far fewer people than has HIV, but in a manner so horrific that Ebola has become a household name and the subject of best-selling books and movies. Newspapers and the evening news shows report frightening new virus outbreaks in America and elsewhere—West Nile virus in the Northeast, the Sin Nombre hantavirus in the Four Corners area, Whitewater Arroyo virus in California, Nipah virus in Malaysia, and Hendra virus in Australia. After hundreds of people are stricken, Great Britain destroys hundreds of thousands of cattle to prevent the spread of Mad Cow disease, but new cases appear in France and Italy nonetheless. Doctors fear that transplantation of pig tissues into patients may introduce a pig retrovirus into the human population. Viruses appear to cause several types of cancer, including cervical carcinoma, Kaposi sarcoma, endemic adult T-cell leukemia, and Burkett's lymphoma. Evidence accumulates to suggest links between diseases like multiple sclerosis, diabetes, or obesity and viral infections. On the positive side of the equation, viruses hold promise as delivery systems for gene therapies to treat or eliminate genetic diseases. Viruses and the diseases they cause, as well as the hopes they may offer, are shaping the future of humankind.

WHAT IS A VIRUS?

Virus. The word itself is unadorned Latin for "poison," dramatic evidence of the long association of these entities with the medical woes of humankind. As the germ theory of disease became established as a central paradigm of medical science at the end of the nineteenth century, a great blossoming of medical microbiology occurred. The causative agents of numerous diseases were isolated, observed under the microscope, cultured on artificial media, and demonstrated to cause their particular pathologies. The causative agents of some diseases, however, could not be studied in that fashion. Among these were the viruses. Since they did not "follow the rules," these mysterious entities came to be described in terms that were all essentially negative in their construction and connotations.

First, viruses were *small.* Unlike bacteria, they could not be seen under the light microscope. More significantly, they were so small they could pass through the filters used to sterilize solutions by removing bacteria and other contaminants. This quality of smallness is not a defining characteristic, however, since it is not unique to the viruses. Some bacteria like the mycoplasma are also small enough to pass through the filters used for bacterial sterilization.

Second, *viruses could not be cultivated on artificial media* in the same fashion as other organisms like bacteria. This characteristic is also not the exclusive property of the viruses, since a number of bacteria cannot be cultivated on artificial media. Despite more than a century of searching, no artificial medium has yet been found to support the growth of the bacterium that causes syphilis, for example, so it still must be cultured within the tissues of a host organism or in conjunction with animal cells in *in vitro* culture systems.

Viruses are even more demanding than the syphilis organism, however. They must be cultivated not only in a host organism, but within a host *cell.* Viruses are *intracellular parasites* that require the metabolic activities of a host cell to support their growth. This, too, is not a characteristic unique to the viruses. Two groups of bacteria, the rickettsia and the chlamydia, are also intracellular parasites and require specific metabolites from their host cells for growth.

What, then, are viruses? We see from these examples that viruses cannot be unambiguously defined by negatives, that is, by how they fail to fit the characteristics of other organisms like the bacteria. Their small size and their requirement for a host cell to support their growth are not unique. What characteristics do define the viruses? To address that question we shall begin by considering several sets of hypothetical experimental data in order to develop a positive picture of the virus and how it differs from the cells that may be its hosts. These will be "generic" data that will enable us to consider the properties of a "typical" bacterial virus and a "typical" bacterial host cell such as *Escherichia coli.*

The first experiment involves a biochemical analysis of our generic bacterial virus and its host cell. The results of this analysis are presented in Table 1.1. The differences are striking. The virus contains only a single type of nucleic acid (DNA in this case), while the host cell contains both DNA and RNA. The remainder of the virus is protein or glycoprotein. It contains none of the lipids, glycolipids, simple sugars, polysaccharides, nucleotides like ATP

Table 1.1 | Comparison (in Percentages) of the Biochemical Composition of a Bacterial Virus and a Bacterium

Component	Host Cell	Virus
DNA	3	40
RNA	21	0
Protein and glycoprotein	55	60
Lipid	12	0
Polysaccharide	5	0
Small molecules and ions	4	0

and ADP, free amino acids, and other small molecules present in its bacterial host cell. Clearly this bacterial virus is fundamentally simpler than its host cell at the biochemical level.

The second experiment involves determining growth curves for our "typical" bacterial virus and bacterium. The growth curve for a bacterium like *E. coli* is shown in Figure 1.1, *A*. One bacterium was introduced into a suitable nutrient broth and allowed to grow. At closely spaced intervals, an exceptionally able scientist carefully determined the exact number of organisms present and plotted those numbers using the conventional semilogarithmic plot, which relates the number of organisms to time. The data fall on a straight line on this semilogarithmic plot, which means that growth in the culture is *exponential;* that is, a geometric progression resulted

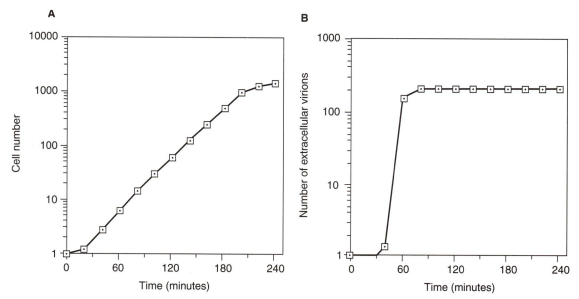

FIGURE 1.1 Growth curves for a "typical" bacterium and a "typical" bacterial virus. **A,** Growth of a bacterium. The number of cells in this culture doubles about every 20 minutes until exhaustion of the medium causes the rate to slow. **B,** Growth of a bacterial virus. The virus put into the culture at the beginning of the experiment disappears and no extracellular viruses are detected until an abrupt increase occurs at about 60 minutes. There is no further increase after that sharp rise.

as that one cell produced two, those two produced four, those four produced eight, and so forth. Eventually there is no further increase in cell number and growth appears to cease in the culture. If our scientist continues this experiment by setting the *E. coli* culture aside to sit on a shelf for 1 year and at the end of that time examines a bacterium taken from it, she will not be able to detect any metabolic activity of any kind in the cell. Furthermore, when a bacterium from that aged culture is transferred into fresh medium, nothing happens; the bacterium appears no longer able to initiate growth.

The graph in Figure 1.1, *B,* shows a similar experiment performed with a "typical" bacterial virus. The pattern is completely different. When the virus is introduced to a culture containing suitable host cells (since we have already established that a host is necessary for growth of the virus), our scientist initially sees no increase whatsoever in the number of viruses. In fact, when the scientist searches very, very diligently within that culture, she is not even able to find the virus she initially introduced. It seems to have disappeared even from within its host cell! Then, suddenly, there is a dramatic, almost instantaneous, increase in the number of viruses in the culture. One virus appears to have given rise to 150 new viruses all at once rather than by means of a geometric progression. If no additional host cells are available, there is no further change in the number of viruses. Addition of new host cells, however, causes the cycle to repeat itself. When this experiment is continued in parallel with the first, with the culture being allowed to sit for a year untouched, and then one of the viruses is examined carefully, our scientist again detects no metabolic activity, just as with the bacterium. When a virus that shows no metabolic activity is introduced into a culture containing live host cells, however, the same pattern of growth occurs as in the first instance. Clearly the absence of metabolic activity means something quite different in a virus and a bacterium.

What can we conclude from comparison of these experiments? First, it is obvious that the nature of growth of the bacterial virus is fundamentally different from the nature of growth of the bacterium itself. The bacterium appears to multiply by a process of cell division since its numbers increase exponentially. The virus does not appear to multiply by cell division since its growth is not exponential, but rather shows a pattern of plateaus and very sharp rises. Secondly, the lack of any detectable metabolic activity in a bacterial cell and a bacterial virus indicates fundamentally different states. Since the bacterium cannot grow again, it appears to have changed from a living state to a nonliving state. Although the virus would also appear to be in that same nonliving state, it nonetheless could initiate growth when introduced into the appropriate conditions. Although questions about the nature of life itself properly belong to the realm of metaphysics, it is clear that viruses and host cells differ in some fundamental fashion in this context as well.

WHERE ARE VIRUSES FOUND IN NATURE?

The short answer to this question is *everywhere.* It appears that wherever cellular life occurs, viruses also occur. We are all familiar with some of the viruses that cause human disease, but viruses infect all of the kingdom *Animalia,* including both vertebrates and invertebrates like insects. The kingdom *Planta* also provides hosts for viruses. Since plant viruses may cause significant agricultural losses, they are the objects of intensive research. Although they are only just being discovered and studied in detail, viruses have been described in both filamentous fungi and yeast, in algae of all types, and in protozoans. Seawater teems with viruses that infect phytoplankton and zooplankton. And finally, there are the bacterial viruses whose study has produced a wealth of information about the basic processes of molecular genetics and the viruses of the *Archaea* that are being discovered.

WHAT ARE THE PHYSICAL PROPERTIES OF VIRUSES?

Our experiments on a "typical" bacterial virus and its host cell suggested that viruses are very different in structure from the cells that host them. We noted that the virus that was the subject of those experiments contained only DNA, while the host cell had both DNA and RNA. In many other viruses the single nucleic acid type present is *RNA,* not DNA. The only other biochemical constituents of the simple bacterial virus in our example were proteins and glycoproteins. The host cell, in contrast, contained proteins and glycoproteins, lipids, and glycolipids, simple sugars and polysaccharides, as well as a vast array of small biochemical molecules like nucleotides (especially ATP and ADP) and amino acids. Clearly the virus must possess a much simpler architecture than its host cell. Equally clearly, since the sole nucleic acid present can be RNA rather than DNA, viruses can function with different genetic systems than their host cells.

Size

We have noted that viruses are small. Figure 1.2 illustrates just how small they are in comparison to bacteria and eukaryotic cells. Most virions fall into the range between 30 and 100 nm in size. As in most generalizations, however, there are many exceptions, such as the poxviruses whose oval virions are between 200 and 400 nm long and the filoviruses like Ebola whose virions may be up to 1000 nm long. Most bacteria are about 100 times larger than the average virus, while eukaryotic cells are another order of magnitude larger yet.

The Genome

The genome of host cells occurs only in the form of double-stranded DNA. In prokaryotic cells the single molecule of the genome is arranged as a closed circle, which obviously has neither beginning nor end. In eukaryotic cells the multiple molecules of the genome are linear rather than circular and have characteristic repeated sequences of nucleotides at their ends, or telomers.

In contrast to the regularity seen in the prokaryotes and eukaryotes, diversity is the rule in the viruses, almost as if nature were playing with all the possibilities for arrangements of nucleic acids. Figure 1.3 illustrates some of these possibilities. DNA genomes can be single-stranded as well as double-stranded, and each of these can form either a linear or circular molecule. More unusual forms, such as linear double-stranded DNA with nicks at various points in the chain, or double-stranded DNA with closed ends, occur as well. The size of viral DNAs ranges over about two orders of magnitude, from masses of approximately 1.2×10^6 daltons to 2×10^8 daltons. That corresponds to a range of about 3000 bases to 375,000 base pairs. RNA viruses are less adventuresome since all have linear genomes, but these may be either single- or double-stranded and there may be only one molecule or several making up the entire genome. A major variation within the single-stranded RNA genomes is whether the molecule can serve directly as messenger RNA or is the complement of the viral mRNAs. The size range of viral RNAs is much narrower than that of viral DNAs, being less than a single order of magnitude: about 1.2×10^6 daltons to about 7×10^6 daltons, or about 3000 to 30,000 bases.

As we also noted in the introductory experiment, the virus as a structural entity seemed to disappear immediately after its interaction with a new host cell and then seemed to reappear some time later in the culture medium. Given this striking characteristic of the virus "lifestyle," it is often useful to distinguish between the *virus,* an entity with genetic if not physical continuity, and the **virion** or **virus particle,** the physical entity that occurs extracellularly.

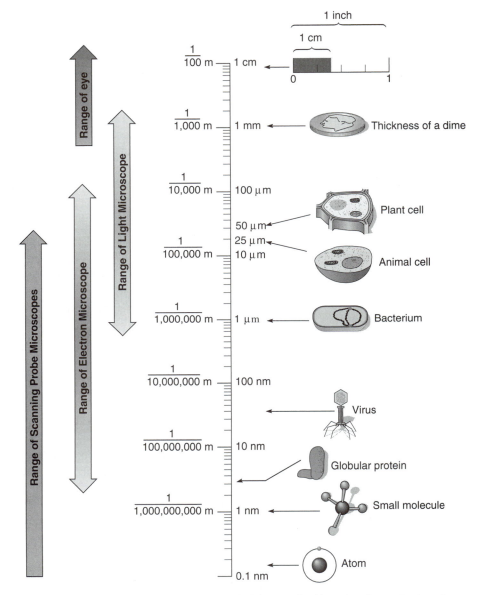

FIGURE 1.2 Viruses are far smaller than their host cells. Note that the vertical scale on this figure is logarithmic to accommodate a vast range of sizes depicted.

Virion Architecture

The virion always contains, at minimum, both nucleic acid and protein. What is the arrangement of these constituents within the virion? Treatment of intact virions with nucleases, which would degrade any exposed nucleic acid, does not destroy their ability to infect cells or reproduce themselves. In contrast, treatment of intact virions with proteases, which would degrade

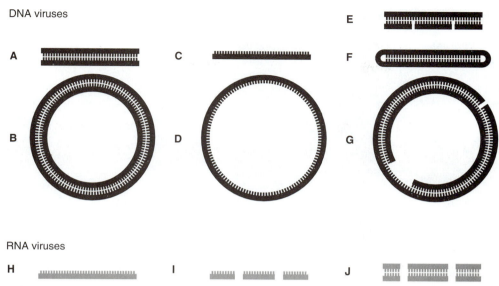

DNA viruses

A

B

C

D

E

F

G

RNA viruses

H

I

J

FIGURE 1.3 The organization of viral nucleic acids. In addition to the double-stranded molecules seen in eukaryotic and prokaryotic cells that are linear **A,** or circular **B,** respectively, viral DNAs can be single-stranded linear **C,** or circular **D.** Other unusual forms of viral DNA are double-stranded linear with breaks in the phosphodiester backbone in one strand **E,** linear double-stranded with closed ends **F,** or circular with large gaps in each strand **G.** Unlike their host cells, viruses can also use RNA as their genetic material. All RNA viral genomes are linear, but they may be one single-stranded molecule **H,** segmented single-stranded molecules **I,** or segmented double-stranded molecules **J.** The convention for depicting nucleic acids is shown on the inside of the front cover.

exposed proteins, may have that result. Since the nucleic acid of the virion appears to be protected from degradation but its proteins are not, these data indicate that the basic arrangement of a virion is with the nucleic acid contained within a protective coat of proteins.

The electron microscope supports this reasoning. Figure 1.4 shows photomicrographs of the virions of two typical viruses: a plant virus (tobacco mosaic virus) and an animal virus (adenovirus). These photos illustrate that the architecture of virions, regardless of their host, is usually based on two simple themes: the *helix* and the *sphere.* They also reveal that the helix and sphere are composed of smaller subunits in regular arrangements. It would thus appear that the virion is built from nucleic acid surrounded and protected by a layer of protein molecules organized in either a helix or sphere. The term **nucleocapsid** ("capsid" is from the Latin word *capsa* meaning "box") is given to this nucleic acid–protein complex, while **capsid** alone refers to the protein shell empty of nucleic acid. The morphological units visible in photomicrographs are called **capsomers** ("mer" is a suffix derived from the Greek word *meros* meaning "part" and usually refers to a subunit or building block of a larger structure). The capsomers, in turn, are generally formed by the association of a definite number of individual proteins, often termed the **structural subunits** or **protomers** (*proto* is Greek for "first").

Crick and Watson proposed in 1956 that the regular arrangement of the protein in the nucleocapsid could be explained by making some simple assumptions about the nature of viruses.

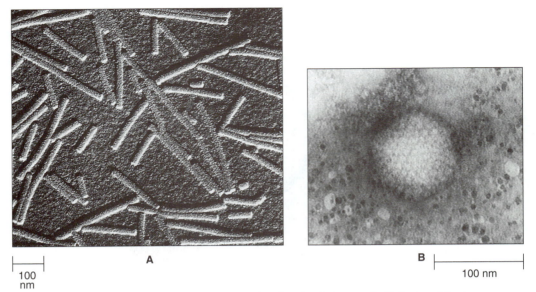

A

B

100
nm

100 nm

FIGURE 1.4 Transmission photomicrographs of stained viruses. **A,** Helical architecture. Plant viruses like the tobacco mosaic virus shown here have the form of rigid rods, while animal viruses have the form of a flexible helix. **B,** Spherical architecture (adenovirus). Although the term *spherical* is applied to this arrangement, the structure is actually an icosahedron.

They reasoned that the protein capsid would have to be relatively large compared to the nucleic acid it contained in order to ensure adequate protection of its cargo. They further hypothesized that a virus would find it easier to force its host cell to synthesize a large number of copies of a few different small proteins rather than to synthesize only one or two copies of very large proteins, which could enfold the nucleic acid. These small proteins would be able to interact with each other and the nucleic acid in only a very limited number of ways if they use the same symmetrical unit arrangement repeatedly. The result is either helical or spherical virions.

Their reasoning has proven correct. As we shall see, viruses usually contain only a small number of genes and therefore can code for only a limited number of proteins. In addition to capsid proteins, there are usually nonstructural proteins necessary for the virus to reproduce as well, so only a small number of small structural proteins are encoded.

Although the term "spherical" is applied to nucleocapsids, the actual arrangement is usually in the form of an **icosahedron,** a three-dimensional structure with 20 faces in the shape of equilateral triangles (Figure 1.5). Each of the 12 corners or vertices of the icosahedron is the intersection of five triangular faces. This structure requires minimal energy to assemble from its subunits and is therefore the most efficient design for a capsid enclosing a spherical space. In small virions (Figure 1.5, *A*) each of the triangular faces is constructed of three structural subunits or protomers, so a total of 60 subunits is used for the whole. Larger virions (Figure 1.5, *B*) are built using multiples of three structural proteins in each capsomer (and multiples of 60 subunits for the whole virion), rather than relying on synthesizing larger

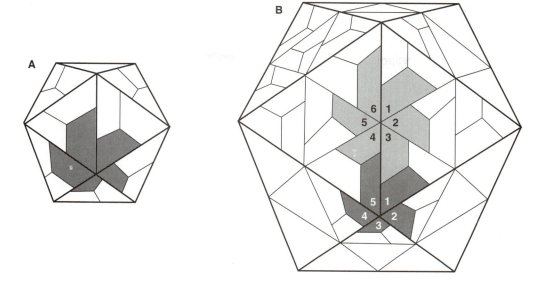

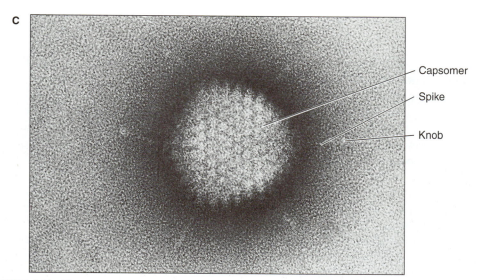

FIGURE 1.5 The arrangement of structural subunits in viruses with icosahedral symmetry. The simplest icosahedral capsid **(A)** has three identical structural subunits in each triangular face. Each vertex is surrounded by five structural subunits *(shaded)*. In larger virions **(B)** each face is formed from multiple clusters of three structural subunits (four small triangular groupings of three subunits = 12 subunits/face = 240 subunits/capsid in this example). Note that each vertex is still surrounded by five structural subunits *(dark gray)* while the edge between adjacent triangular faces consists of six structural subunits *(light gray)*. Spikes may be attached to some of the vertices of the capsid in some viruses **(C).** Adenovirus has an icosahedral virion with spikes at each vertex.

FIGURE 1.6 The arrangement of structural subunits in viruses with helical symmetry. The identical subunits of helical capsids form bonds both to the subunits before and behind themselves in the ribbon and to subunits adjacent to themselves in the helix.

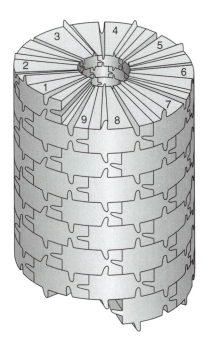

individual proteins. In these capsids the structural proteins of each face of the icosahedron are usually arranged so that five subunit proteins surround each vertex (and are therefore called **pentamers** or **pentons,** from the Greek word for "five"), while the subunits of adjoining triangular faces form groups of six (and are therefore called **hexamers** or **hexons,** for the Greek word for "six"). In most bacterial and plant viruses, the same proteins can function in either location, so penton or hexon is a "distinction without a difference." Some animal viruses, however, have special proteins in the penton position that are different from those in the hexon position. These penton proteins are frequently the attachment points for long projections from the nucleocapsid called **spikes,** as shown in Figure 1.5, *C.*

The other basic structure of the nucleocapsid appears in photomicrographs as a hollow tube or cylinder (Figure 1.6). The capsomers are actually arranged in the form of a ribbon of protein wound into a helix rather than a cylinder, however, since this helical arrangement permits all proteins to interact equivalently with each other and with the nucleic acid molecule within the nucleocapsid. Helical nucleocapsids may be rigid, as in many plant viruses, or long and flexible, as with many animal viruses.

In many bacterial viruses the virion has both helical and icosahedral components (Figure 1.7). In the simplest of these, represented by the bacterial virus λ (lambda), the viral nucleic acid is enclosed by an icosahedral "head" to which is attached a helical "tail" section. More complex bacterial viruses such as T4 elongate the icosahedral head and add a variety of spikes and fibers to the helical tail to form a virion that bears a striking resemblance to Apollo mission lunar landers (or perhaps it is actually the other way around!).

The nucleocapsids of many animal viruses, as well as a rare few plant and bacterial ones, are surrounded by an **envelope** or membrane derived from its host cell. The presence of an envelope means that the virus particle contains lipids in addition to proteins and nucleic acids.

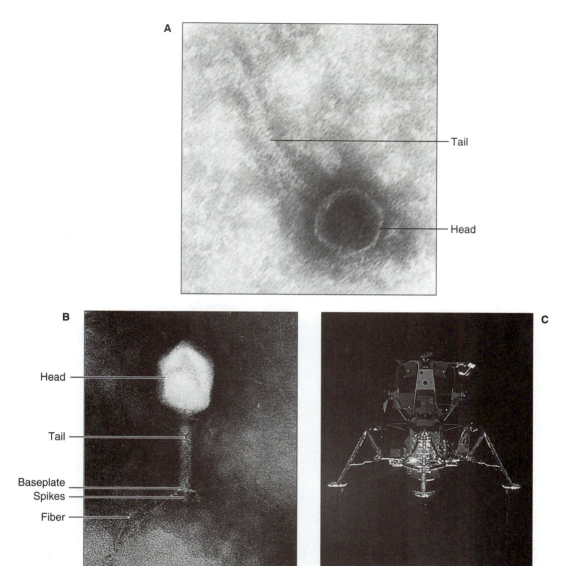

FIGURE 1.7 Virions of bacterial viruses. λ virus **A,** has an icosahedral "head" with a helical "tail" attached to it. T4 virus **B,** adds a variety of plates, spikes, and fibers to this basic structure. T4 resembles the lunar lander **C.**

As Figure 1.8 shows, enveloped viruses may have either helical or icosahedral nucleocapsids. The envelope membrane frequently contains projections, again called spikes, which are virus-specific proteins inserted into the basic membrane system of the host cell. Enveloped viruses are sometimes termed **ether-sensitive viruses** since the envelope is destroyed when membrane lipids are extracted from the envelope membrane by organic solvents like ether, rendering the treated virus unable to infect new host cells. Nonenveloped, or **naked viruses,**

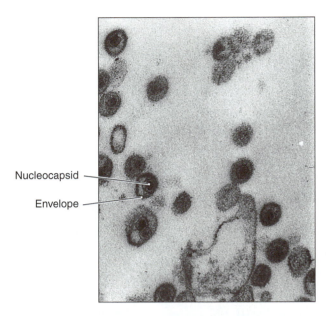

Nucleocapsid

Envelope

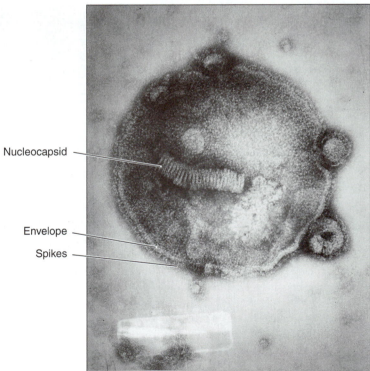

Nucleocapsid

Envelope

Spikes

FIGURE 1.8 Enveloped virions. The membranes of the envelope can surround either an icosahedral nucleocapsid (**A,** mouse mammary tumor virus) or a helical nucleocapsid (**B,** influenza A virus). Note the blunt projections or spikes on the surface of the influenza virion envelope.

are not affected by organic solvents. They contain no lipids and their protein nucleocapsids are hydrophilic on their surfaces, so they do not interact with nonpolar solvents.

These are the basic themes and variations of architecture in the viruses. In the course of this book we shall examine a number of different viruses in detail to investigate the relationships between the structure of a virus and how it functions as an intracellular parasite.

WHAT STRUCTURAL FEATURES OF CELLS ARE IMPORTANT IN THEIR ROLE AS VIRAL HOSTS?

The first event that must occur in the interaction of a virus with its host cell is the attachment of the virus to that cell. Since the cell surface and any structures associated with it obviously play a crucial role in this process, we shall review the architecture of bacterial, animal, and plant cell surfaces. As we shall see, the marked differences in cell surface organizations among these groups produce significant differences in the strategies used by their respective viruses both in entering and leaving host cells.

Animal Cells

Animal cells interact with their environments directly through their plasma membranes. Figure 1.9 shows the **fluid-mosaic membrane model** proposed by Singer and Nicolson. According to this model, phospholipids are arranged with their hydrophobic fatty acid tails toward the center of the membrane and their hydrophilic phosphate and other charged groups facing the outer and inner surfaces of the membrane. Proteins are associated with this lipid

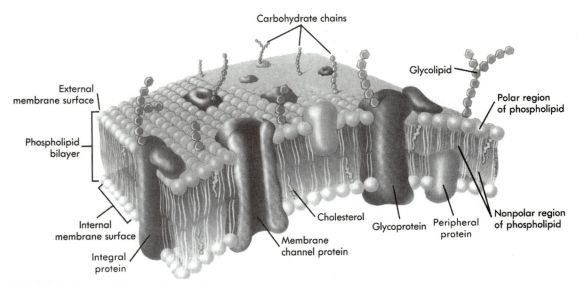

FIGURE 1.9 The fluid-mosaic membrane model. Proteins and glycoproteins may either span the membrane lipid bilayer *(integral proteins)* or "float" in the upper or lower surface of it *(peripheral proteins)*. Some of the membrane phospholipids may also have carbohydrate chains attached to them, forming glycolipids.

bilayer in two fashions. Some membrane proteins span the entire bimolecular leaflet of phospholipids and can thus serve as pores conferring selective permeability or as transducers of signals across the membrane. The membrane must be completely disrupted by detergents in order to release these proteins, so they are termed **integral** or *intrinsic* proteins. Other proteins "float" embedded in only one side of the lipid bilayer and can be released from the membrane by gentle treatment with chelating agents or by an osmotic shock created by changes in ionic strength of the surrounding solution. These are termed **peripheral** or *extrinsic* proteins. The amount of protein relative to lipid in animal cell membranes differs widely, depending on the role a given membrane plays in a cell with a particular function.

In addition to lipids and proteins, the cell surface membranes of animal cells contain a large number of carbohydrates. In most cases, short, branched chains of sugars and aminosugars are attached to proteins to form glycoproteins. Some carbohydrates are attached to phospholipids to form glycolipids. The carbohydrate chains of both glycoproteins and glycolipids play major roles in a wide variety of recognition and/or binding reactions at the cell surface, including those involved in cell-cell adhesion and transduction of signals across the plasma membrane.

Plant Cells

Plant cells are bounded by a plasma membrane with the same structural features as the membranes found in animal cells. External to the plasma membrane, however, is rigid, nonliving **cell wall** that serves to protect the cell from osmotic shocks and to provide mechanical support to the entire organism. This plant cell wall is made up of long fibrils of cellulose, a polysaccharide built of glucose molecules, enmeshed in a gel-like matrix of hemicellulose, pectin, lignin, and proteins. This means that the metabolically active cell surface is covered by a thick, protective layer of inert cell wall.

Although each plant cell is surrounded by its own cell wall, individual cells in a multicellular organism do not live in splendid isolation. Their cell walls are pierced in numerous locations by channels called **plasmodesmata** (singular: *plasmodesma,* from the Greek words meaning "formed" plus "band"), which allow fusion of the plasma membranes of adjoining cells (Figure 1.10. Elements of the endoplasmic reticulum called *desmotubules* are often seen on either side of the opening formed by the plasmodesma and may actually cross the cell wall through the channel. Plasmodesmata therefore allow the formation of a continuous cytoplasmic connection from cell to cell throughout an entire plant tissue and can permit the rapid transfer of materials in a plant.

Prokaryotic Cells

A rigid cell wall also bounds all prokaryotic cells, with the exception of the mycoplasma. As with plant cells, the bacterial cell wall provides the mechanical protection necessary for the survival of bacterial cells that are often subjected to extreme changes in their osmotic environment. It also gives the various groups of bacteria their characteristic shapes as rods, spheres, or spirals. The structure of the prokaryotic cell wall is far more complex than that of the plant cell wall, however, almost as if the bacteria were making up for their lack of internal structural complexity by elaborating their surface structures.

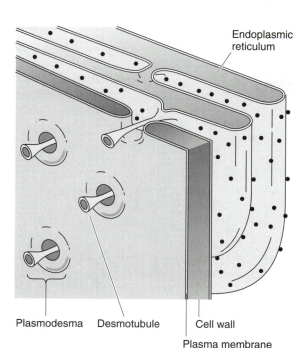

Plasmodesma   Desmotubule   Cell wall

Plasma membrane

FIGURE 1.10   Plant cell walls. Between the cell surface membranes of adjacent cells in a plant is a thick cell wall. Plasmodesmata, or channels through the cell wall, create cytoplasmic connections between the cells. The endoplasmic reticula of adjoining cells may also be connected via desmotubules, which pass through plasmodesmata.

From the very earliest days of microbiology, a procedure called the Gram stain allowed scientists to distinguish two distinct kinds of prokaryotic cells. Electron microscopic studies and biochemical analyses have demonstrated that the differential staining of gram-positive and gram-negative cells reflects profound differences in the architecture of their cell walls. These differences mean that the recognition events between virions and bacterial cells involve different structures in gram-positive and gram-negative cells.

The Gram-Positive Cell Wall.   The gram-positive cell wall is a relatively simple structure. In photomicrographs the gram-positive cell wall appears as a thick, amorphous-looking layer surrounding the cell. As Figure 1.11 illustrates, this layer is, in fact, a single gigantic molecule called **peptidoglycan** ("peptido" = amino acid chain, "glycan" = sugar chain) composed of long chains of alternating N-acetylglucosamine and N-acetylmuramic acid ("glycan") cross-linked at intervals by chains of four amino acids ("peptido"). The cell wall of a gram-positive organism, with as many as 40 layers of peptidoglycan cross-linked in two dimensions, creates a formidable barrier surrounding the cell. It can, however, be broken down by enzymes such as the lysozyme found in secretions like tears and saliva.

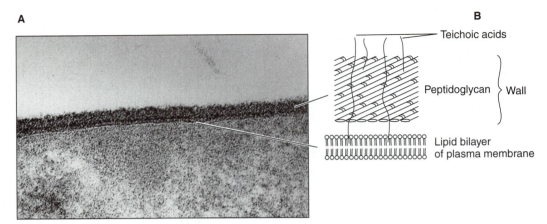

A

B

Teichoic acids

Peptidoglycan } Wall

Lipid bilayer
of plasma membrane

FIGURE 1.11 The gram-positive bacterial cell wall. **A,** Thin section photomicrograph of *B. subtilis* surface. External to the cytoplasmic membrane is a thick wall *(arrow).* **B,** Structure of the gram-positive cell wall. The wall is a three-dimensional network of cross-linked peptidoglycan molecules. Long teichoic acid molecules anchored either to peptidoglycan or to the plasma membrane extend well beyond the peptidoglycan layer.

Since all gram-positive organisms share this basic architecture, the peptidoglycan layer does not allow viruses to identify particular species or strains of bacteria as potential hosts. Gram-positive cells do, however, have a way of bearing "badges of identity" in their environment. These badges of identity are a class of molecules called **teichoic acids.** Teichoic acids are based on long chains of either glycerol or ribitol (the alcohol of ribose) linked together by phosphodiester bonds. A wide variety of different amino acids, sugars, or amino sugars may be attached to the alcohols to generate many different types of teichoic acids. The additions, together with the phosphate groups linking the alcohols, create molecules with large net negative charges on the cell surface.

The Gram-Negative Cell Wall. Photomicrographs show that **gram-negative** cells have a far more complex cell wall than gram-positive cells. As Figure 1.12 illustrates, gram-negative cells have a peptidoglycan layer similar in construction to that in gram-positive cells but with two significant differences. First, the peptidoglycan layer is much thinner (usually only one or two layers thick) than that in gram-positive cells, and second, fewer of the amino acid chains are involved in cross-linking the glycan chains. The result is a looser overall construction.

The most important difference between gram-positive cells and gram-negative cells, however, lies outside the peptidoglycan layer in the gram-negative bacteria. In addition to their normal plasma membrane, gram-negative cells possess an outer "membrane" with a distinctive composition. This outer membrane is a true lipid bilayer containing phospholipids and proteins similar to those of plasma membranes. In addition, however, the gram-negative cell outer membrane contains a unique compound called **lipopolysaccharide (LPS)** that plays a major role in viral and bacterial cell interactions.

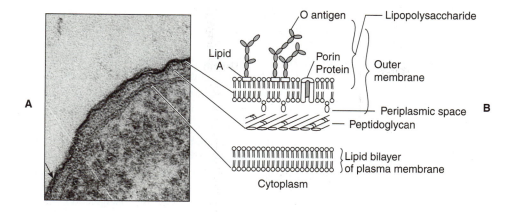

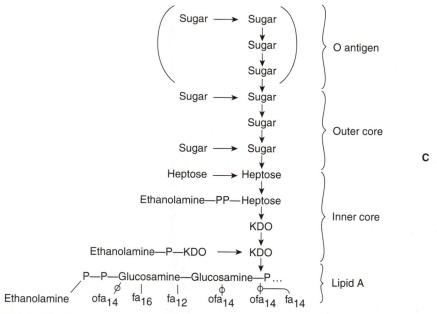

FIGURE 1.12 The gram-negative bacterial cell wall. **A,** Thin-section photomicrograph of *E. coli* surface. The distinctive feature of the gram-negative cell wall is a second membrane system external to a thin layer of peptidoglycan *(arrow)*. **B,** Structure of the gram-negative cell wall. The external side of the lipid bilayer of the outer membrane contains molecules of lipopolysaccharide. **C,** The complex structure of lipopolysaccharide includes lipid A (fa = fatty acid), a long chain of sugars called the core (KDO = keto-deoxy-octonate), and a repeating set of sugars called the O antigen. Proteins involved with the gram-negative cell wall include porins embedded in the outer membrane and transport proteins in the periplasmic space between that membrane and the peptidoglycan layer.

Lipopolysaccharide, also diagrammed in Figure 1.12, is an extremely complex molecule, but it can be subdivided into three subsections. The first of these, called **lipid A,** anchors LPS in the outer membrane through its long chain fatty acids attached to a pair of *N*-acetyl-glucosamines. One position on one *N*-acetyl-glucosamine, however, carries a large, complex carbohydrate chain that sticks out from the membrane and into the environment surrounding the cells rather than being embedded in the lipid bilayer. The sugar chain that is attached directly to the *N*-acetyl-glucosamine of lipid A is called the **core polysaccharide.** This highly branched structure usually includes a number of somewhat "oddball" monosaccharides with seven or eight carbons rather than the usual five or six. Although the core polysaccharide is similar in most gram-negative organisms, numerous mutant forms occur. One of the most widely used laboratory strains of *E. coli* is such a mutant; *E. coli* B lacks the more distal portion of the usual *E. coli* core polysaccharide.

Attached to the distal end of the core polysaccharide is a specific repeating sequence of sugars termed the **O antigen** or O side chain. The O antigen unit sequences are short, usually consisting of four or five carbohydrates, but the entire structure contains dozens of units. The entire molecule is therefore very long and protrudes a great distance from the cell surface. O antigens show enormous variability, even within the same taxonomic species, so they probably play major roles in recognition events that occur at the cell surface among the gram-negative organisms. Numerous mutant forms of O antigens also occur, including the laboratory strain *E. coli* K12, which lacks an O antigen entirely (as does *E. coli* B since it is missing the part of the core polysaccharide to which O antigen is attached).

In addition to lipopolysaccharide, numerous proteins are also found in the outer membrane. Many of these are integral proteins called "porins," which serve as transport channels for ions and other small nutrient molecules. Other proteins involved with transport functions occur in the periplasmic space between the outer membrane and the peptidoglycan layer. The core and O antigen sugars and the porin and periplasmic proteins provide a great variety of possible recognition molecules for virus-bacterial cell interactions.

Appendages

In addition to the cell wall, bacteria may have other structures that can allow viruses to attach to them and introduce their nucleic acids into the cell. One of these structures is the **flagellum** (plural: *flagella,* Latin for "whip") which allows a bacterium to move. Flagella are built of a single protein called flagellin arranged in the form of a hexagonal array that produces a 12 to 20 nm diameter filament with a hole down its middle. The flagellin proteins of different species or strains of bacteria have different amino acid sequences, so the flagellum serves as an important distinguishing feature among bacteria.

The second structures are very fine tubules. The **pilus** (plural: *pili,* Latin for "hair") allows the passage of genetic information from one cell to another in the process of conjugation. Similar structures are termed *fimbria* when they occur on bacteria that attach themselves to surfaces such as the lining of the respiratory or gastrointestinal tract. Each tubule is a hollow helix formed by a single type of protein called pilin, which is attached to the plasma membrane and passes out through the bacterial cell wall. The diameter of the filament is only about 3 to 10 nm.

WHAT GENETIC FEATURES OF CELLS ARE IMPORTANT IN THE VIRUS–HOST CELL INTERACTION?

Once a virus has attached to a cell surface and introduced its genetic material into a suitable host cell, the organization of the host cell's genome and the steps by which that genome becomes expressed play crucial roles in determining the pattern of the virus–host cell interaction. Both prokaryotic and eukaryotic cells carry out the same processes of the **Central Dogma of Molecular Biology:** DNA is *replicated* to make more DNA; DNA is also *transcribed* to make mRNAs, which are then *translated* to make proteins. Differences between prokaryotic and eukaryotic cells in the details of the organization of their genomes and how they carry out the processes of the Central Dogma are particularly significant in viral replication strategies. We shall therefore review these differences, which are summarized in Table 1.2.

The Organization of the Host Cell Genome

All of the genetic information required to meet the normal structural and metabolic needs of a prokaryotic cell is carried on a single circular molecule of double-stranded DNA. For convenience, this molecule is often called a "chromosome" even though it does not have the architecture and associated proteins of the eukaryotic chromosome. Since the prokaryotic chromosome contains only a single copy of each protein-coding gene, the genome is considered haploid. The "information density" in the prokaryotic chromosome is very high, since nearly all its bases are in either coding or regulatory sequences.

The chromosome is usually not the entire genetic complement of the prokaryotic cell, since nearly every bacterium also harbors one or more small, independently replicating, circular DNA molecules called **plasmids.** In addition to the genes necessary for their own replication and transfer to new host cells, many plasmids carry genetic information that may confer some selective advantage on the host cell under particular environmental conditions. These plasmid-borne genes may code for enzymes that degrade particular antibiotics, for example, or allow the host cell to utilize an unusual nutrient.

Table 1.2 | Comparison of Genomic Organization and Expression of Prokaryotic and Eukaryotic Cells

Prokaryotic Cell	Eukaryotic Cell
1 DNA molecule	>1 DNA molecule
Circular DNA	Linear DNA
Haploid	Diploid
Operons	No operons
Polycistronic mRNAs	Monocistronic mRNAs
No posttranscriptional modifications	Capping, tailing, and splicing of transcript
No compartmentalization	Compartmentalization
Continuous cell cycle	May exit cycle to G_0

The placement of particular genes within the prokaryotic chromosome is not random. In most instances the genes coding for the enzymes in a particular biochemical pathway are physically grouped together as a structural unit whose expression is governed by a single set of controlling elements. This arrangement of structural genes and their controlling elements is termed an **operon.** All the structural genes in an operon are transcribed as a single continuous messenger RNA molecule, termed a **polycistronic mRNA** (*poly* is Greek for "many," plus *cistron,* a coined term for "gene" based on a particular type of genetic test). Within the polycistronic mRNA, the coding information for each protein is identified by its own start and stop codons, so each is translated separately to produce individual proteins.

The organization of genetic information in the genomes of eukaryotic cells is fundamentally different from that in prokaryotic cells. First, the eukaryotic cell's genome is distributed on two or more double-stranded DNA molecules rather than being contained within the single molecule characteristic of prokaryotic cells. Secondly, the individual DNA molecules (or chromosomes) are linear rather than circular and are associated with a number of proteins, particularly histones. Eukaryotic cells can have either a single set of chromosomes (haploid) or two sets of each (diploid). The eukaryotic cells commonly encountered as virus host cells are diploid.

Eukaryotic genomes do not have operons. This means that the genes for a particular biochemical pathway are not grouped together as a genetic unit located at a given site on a particular chromosome, but rather are distributed out among sites that may even be on different chromosomes. As a consequence, each individual gene is regulated by its own controlling elements. It is transcribed into its own **monocistronic mRNA** (*mono* from the Greek word for "one"), which is then translated to produce its particular protein.

Posttranscriptional Modification of mRNA

As indicated previously, nearly all the bases in the prokaryotic genome have either a coding or regulatory function. This means that the processes of the Central Dogma have a "what you see is what you get" quality: a gene's sequence of bases in the DNA is reproduced (in its complement) in its entirety in the corresponding mRNA, which is, in turn, translated directly to yield a protein. In eukaryotic cells, this is decidedly not the case. Interspersed among the coding portions of a particular gene called **exons** (from "expressed" information) are intervening sequences called **introns,** which do not carry information relating to the sequence of amino acids in the protein ultimately produced. The process of transcription, which forms an RNA molecule complementary to all the bases in the gene, therefore only begins the formation of the ultimate mRNA in eukaryotes. Before it can be translated, the transcript must undergo a series of **posttranscriptional modifications.**

The first posttranscriptional modification made in eukaryotes is the addition of a "cap" in the form of a 7-methyl-guanosine to the 5′ end of the transcript. This addition is made "backwards" with a 5′ to 5′ linkage, so the transcript no longer has a free 5′ end at all. The second addition is a "**poly-A tail**" or chain of adenosines to the 3′ in the molecule. Following the addition of the cap and tail, "**splicing**" reactions remove the non-coding introns from the transcript and link together the remaining coding exon sequences. Only after the cap and tail have been added and splicing has been completed can the resulting molecule be considered true messenger RNA.

Internal Compartmentalization

Even after posttranscriptional modification, eukaryotic messenger RNA is not immediately available for translation. A hallmark of eukaryotic cells is the sequestering of different metabolic activities in different internal compartments of the cell. These compartments are either in the form of membrane systems or membrane-bound organelles. For the processes of the Central Dogma, DNA resides in the membrane-bound nucleus; both replication and transcription occur within that organelle. The transcript, after its modifications, must be transported from that compartment into the cytoplasm in order for the next step in gene expression to occur. Translation itself can occur either with ribosomes free in the cytoplasm or in association with another compartment of the cell, the endoplasmic reticulum. The messages translated in association with endoplasmic reticulum contain information for proteins that either will become associated with cellular membranes or will be exported from the cell to function in an extracellular environment.

Prokaryotic cells generally do not have internal compartmentalization of their metabolic activities. The plasma membrane, being the only membrane system in prokaryotic cells, carries out the functions of the endoplasmic reticulum as well as other metabolic activities. Since prokaryotic cells do not need to modify their transcripts, and since they lack any internal compartmentalization, the processes of the Central Dogma are carried out in a continuous fashion. As soon as the initiation signals for translation have been transcribed into mRNA, ribosomes attach to the message and translation begins. Transcription and translation are thus tightly coupled reactions in prokaryotes.

The Cell Cycle

The **cell cycle** is the series of stages through which a cell passes as it grows and divides. The prokaryotic cell is, by its very nature, a generalist. Each individual cell must be capable of carrying on all the metabolic activities associated with that particular species and be able to adapt to the environments in which that cell may find itself. It must also be able to reproduce itself when the environment is favorable. This means that prokaryotic cells are always capable of progressing through the stages of the cell cycle.

Eukaryotic cells that are parts of multicellular organisms, in contrast, have the opportunity (or necessity) to become specialists. In multicellular organisms, individual cells may become differentiated into a particular cell type that carries out only a very limited number of metabolic functions. For example, during the process of development, a mammalian cell in one location of the developing fetus may differentiate to become muscle cell, while a cell in another location may differentiate into a neuron with a totally different function. A very important consequence of this differentiation process is that many cells in multicellular organisms do not divide after they have differentiated. Neurons, for example, never divide after they become differentiated into neurons. The cell cycle in eukaryotes is divided into four parts:

1. G_1 ("gap" 1) in which the cell enlarges and prepares to synthesize its genome;
2. S (synthesis) in which DNA synthesis occurs;
3. G_2 ("gap" 2) in which the cell enlarges further and prepares to divide; and,
4. M (mitosis) in which cell division occurs.

When cells differentiate they exit the cell cycle during the G_1 stage and enter a condition called G_0. Since viruses depend on their host cell to provide some or all of the components needed for the expression and replication of their genomes, the stage of the cell cycle can be a crucial determinant in the outcome of a virus's infection of a cell.

HOW DOES A VIRUS REPRODUCE?

In answering the question "What is a virus?" at the beginning of this chapter, we considered a hypothetical experiment describing the growth of a bacterial virus. This experiment was actually a generic version of a classic experiment called the **one-step growth experiment,** which demonstrates the important features of the interaction of a virus and its host cells. The process of viral replication that occurs in the one-step growth experiment can be divided into six stages, as illustrated in Figure 1.13:

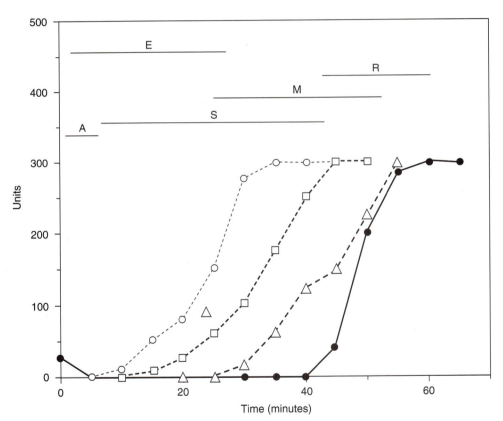

FIGURE 1.13 The events underlying the one-step growth experiment. (●) = extracellular virions, (○) = nucleic acids, (□) = proteins, (△) = intracellular virions, (A) = attachment period, (S) = synthesis period, (M) = maturation period, (R) = release period, (E) = eclipse period.

1. **Attachment** of the virion to the host cell by means of specific recognition events at the cell surface. This stage is also sometimes termed *adsorption* (from Latin *ad* = "to" and *sorbere* = "to suck").

2. **Penetration** and **uncoating** introduce the virus genome into the host cell. When this process occurs, the virus as a complete structural entity disappears. The period between penetration and the formation of new virions is therefore sometimes called the **eclipse** period. Attachment, penetration, and uncoating take about 5 minutes for the viruses shown in Figure 1.13.

3. **Synthesis of viral proteins and nucleic acids** begins the actual process of replication of the virus. Viral proteins and enzymes synthesized at this time usually aid in mobilizing the host cell's resources for viral synthesis and permit the efficient expression of the viral genome.

4. **Synthesis of viral structural proteins** may occur concurrently with synthesis of viral nucleic acids, but usually lags somewhat behind it. Both new viral genomes and structural proteins accumulate within the host cell during these two stages. The synthesis of viral nucleic acids and protein molecules occurs between about 5 minutes and 40 minutes for the virus illustrated in Figure 1.13

5. **Maturation** or *assembly* of new virions begins at about 22 minutes of infection in the example viruses as genomes and structural proteins become available. This process of assembling new virions from component parts is a distinguishing feature of the viruses.

6. **Release** of the new virions occurs in different fashions depending on the nature of the host cell (does it have a cell wall?) and of the virion itself (does it have an envelope?). The release may be sudden or gradual.

We shall consider a variety of strategies that different viruses have evolved to carry out these steps within the constraints imposed by the organization and mode of expression of the genomes of their host cells.

WHAT ARE THE BASIC TECHNIQUES USED TO STUDY THE VIRUS–HOST CELL INTERACTION?

A variety of different assays for viruses have been developed based on the characteristics of particular virus–host cell interactions. The first of these assays to be developed, called a **plaque assay,** was for bacterial viruses. In this procedure, diagrammed in Figure 1.14, *A,* multiplying host bacteria and a small number of viruses (often from a dilution series) are mixed together in a tube containing "soft" or low-concentration agar. This virus/host cell/agar mixture is poured onto a plate of normal bacterial agar and allowed to solidify. Because the concentration of agar in the overlying layer is low, each virus can diffuse through the liquid in the agar to reach a host cell and infect it. After the virus reproduces in that host cell, it lyses its host to release more viruses, which can in turn diffuse into the immediate area surrounding the original infected cell and infect the bacteria they encounter. After several cycles of infection, release, diffusion, and reinfection, a hole is created in what would otherwise be a continuous sheet or "lawn" of bacteria in the agar overlay. These holes are termed *plaques* from the French term for "to plate." In order to determine the number of viruses originally introduced into the tube, one only needs to count the number of plaques. The concentration of viruses in a solution, the *titer,* may be determined by making a dilution series and assaying each tube. In addition, since different viruses can produce different plaque morphologies (Figure 1.14, *B*), the plaque assay is often used to evaluate the purity of virus preparations or to screen for genetic mutants within a population.

Animal Virus Techniques

When the bacterial viruses began to be widely studied, virologists already had at their disposal standard techniques for handling the bacterial host cells. This was not at all the case for scientists interested in studying plant or animal viruses, however. Animal viruses were first studied outside diseased organisms using developing embryos as hosts, particularly chicken embryos. A suspension of virus particles is injected through a hole punched in the egg shell and delivered to one or more of the membrane systems associated with the developing chick

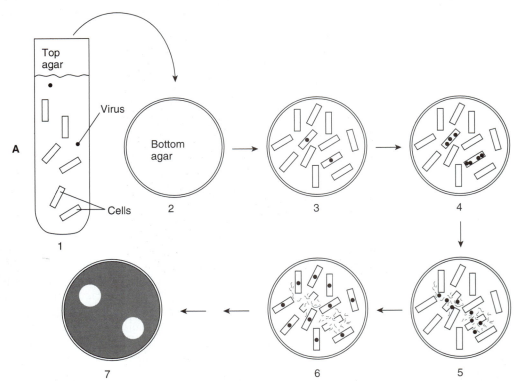

FIGURE 1.14 The plaque assay for bacterial viruses. **A,** Formation of plaques. Viruses and host cells are mixed together in a tube of low-density agar *(top agar) (1)* and then poured onto a layer of regular agar *(2)*. Each virus infects a single host cell *(3)*, multiplies in that cell *(4)*, and then lyses it *(5)*, releasing the new virions. The newly released virions infect neighboring cells *(6)* to begin successive cycles of infection, multiplication, release, diffusion to neighboring cells, and reinfection. The result is that each original virus eventually produces a clear area (plaque) *(7)* in a continuous sheet (lawn) of bacteria. *(Continued.)*

embryo. The viruses infect the cells of those membrane systems or the embryo itself, reproduce, and are released into the various fluids associated with the developing embryo. While this type of embryo culture allows the propagation of viruses, it is of very limited usefulness in studying the various metabolic and genetic properties associated with particular viruses. Studies of that nature required the development of tissue culture techniques.

Tissue culture, as the name suggests, first developed when explants of tissues were removed from an organism to be grown *in vitro* (Latin for "in glass"), that is, in a petri dish and bathed in a suitable nutrient solution. Tiny pieces of liver, for example, can be maintained

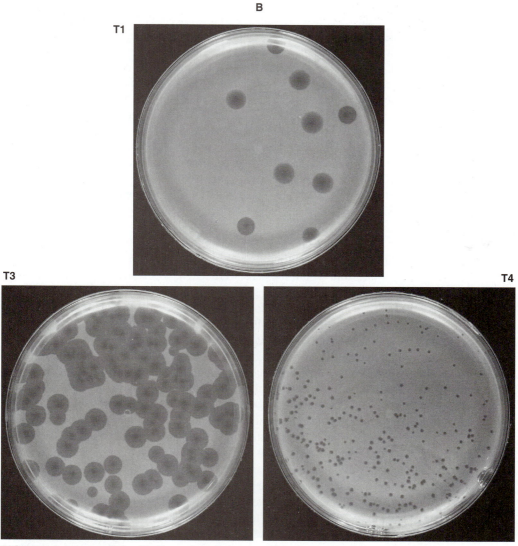

FIGURE 1.14, cont'd. **B,** Examples of plaques. The viruses T1, T3, and T4 produce distinctive types of plaques when grown in the same host *(E. coli).*

in a medium containing the appropriate salts, vitamins, amino acids, and a supplement of either serum or lactalbumin hydrolysate. It was shown in 1949 that such tissue cultures could support the growth of animal viruses.

In the 1950s, techniques were developed that would allow the tissue blocks to be disassociated into individual cells, which could then be cultured in a manner analogous to that used for bacteria. Chicken or mouse embryos, for example, are first minced into small pieces and then digested with enzymes such as collagenase or trypsin in order to produce individual cells. These are placed into petri dishes containing the appropriate nutrient solutions so that the cells can grow attached to the dish surface. Such **cell cultures** are termed *primary cultures* because the cells are taken directly from an animal and placed in culture.

Primary cultures have both positive and negative aspects for culturing viruses. The positive aspect of such cultures is that the cells are as much as possible like those in the actual host organism. They are not adapted to culture and have not undergone significant genetic changes as a result of their artificial cultivation. The negative aspect of primary cultures is that cells must be taken from experimental animals each time a new culture is necessary. This is necessary because the cells in most primary cultures do not continue to divide *in vitro*. Such cultures can be subdivided (or transferred) only a limited number of times. Other cultures, particularly those derived from fetal tissues, may retain most of the features of normal primary cell cultures but also have the ability to multiply for a number of generations in culture. These cell cultures, termed *diploid cell lines,* do eventually die out, however.

In some cultures a so-called *immortality event* may occur. The immortality event usually involves significant changes in the chromosomal complement of the cells; diploid cells may become subtetraploid, for example. These mutations permit the cells to adapt very well to the artificial environment in which they are being cultivated. Such cells, instead of dying out after a limited number of generations, become a *continuous cell line* that may continue to divide for a very long time. These cells can therefore be treated in many respects like bacterial cultures. The first continuous cell line, called HeLa from the first letters of the name of the patient from whom the cells were taken, was started in 1952 from a cervical carcinoma. Continuous cell lines are generally identified by laboratory abbreviations: BHK for baby hamster kidney cells; 3T3 for a line of mouse cells that are transferred every 3 days at a dilution of one to three; and MCF-7 for a human mammary epithelial cell line developed at the Michigan Cancer Foundation.

Animal cells in tissue culture serve as hosts for a great variety of assays to quantify viruses. The way in which an assay is conducted in a culture depends on the nature of the particular virus's growth process. A number of animal viruses cause their host cells to burst as the virus is released. These viruses can be assayed in a tissue culture equivalent of the plaque assay, as shown in Figure 1.15. Many other animal viruses do not destroy their host cells in the process of their replication and therefore cannot be assayed by looking for plaques in a tissue culture. These viruses may often be assayed by examining the host cells for characteristic morphological and metabolic changes that are associated with virus infection. For example, some viruses cause changes in the cell surface that result in infected cells being able to bind red blood cells. It is therefore possible to assay for these viruses by determining the number of cells in an infected culture to which red blood cells will bind. Other viruses may be assayed by detection of viral proteins by techniques like immunofluorescent staining or viral nucleic acids by methods like nucleic acid hybridization, as described in the Virologist's Toolkit in Appendix A.

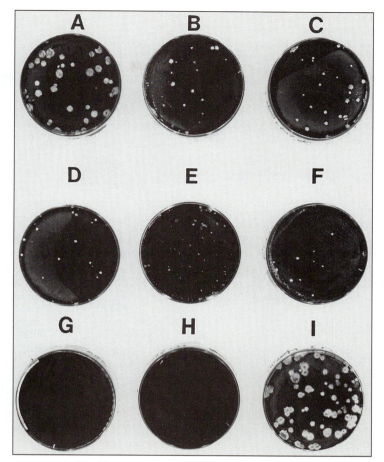

FIGURE 1.15 Plaque assays for animal viruses. Destruction of cultured cells (HeLa) by poliovirus produces clear plaques against the background of stained host cells. Plate A contains plaques produced with wild-type poliovirus; plates B through I contain plaques produced by polioviruses with different mutations in one of the virion capsid proteins.

Plant Virus Techniques

Plant viruses are difficult to assay because of the methods plant viruses use to enter their host cells. Plant cells are surrounded by thick cellulose cell walls, which present a formidable physical barrier to entry of a virus. Most plant viruses therefore require mechanical weakening of the plant cell wall before they can enter. The most common type of assay for plant viruses, the plant leaf "local lesion" assay, requires that the plant leaf be damaged by a mechanical process such as rubbing with carborundum and applying the virus to the abraded surface. After a suitable incubation time, lesions characteristic of a virus and plant appear on the damaged leaf, as shown in Figure 1.16. These types of assays are difficult to quantify but do allow propagation of the virus. More recently, there has been an explosion in

FIGURE 1.16 Plant leaf local lesion assays. Leaves were abraded and inoculated with different virus strains, each of which produces a characteristic lesion that appears as a white area in these photographs.

the development of tissue culture methods in plant science, so it is reasonable to expect that plant virology will soon benefit from this new technology.

HOW ARE VIRUSES NAMED?

Names are important. Virtually every religion or culture begins its descriptions of creation with acts of naming. The same is true of virology. Historically the bacterial viruses have been given names that reflect the conventions of the laboratory in which they were discovered, so they include both letters (Roman, Greek, or both) as well as numbers such as T4, MS2, λ, and ϕX174 (the ϕ [phi] in ϕX174 is from the "abbreviation" for "phage" used by virologist cognoscenti). Plant and animal viruses have been named for the diseases they cause (tobacco mosaic virus, influenza virus, human immunodeficiency virus); the person discovering the virus (Rous sarcoma virus, Epstein-Barr virus); or the location where the virus was first described, isolated, or where its disease occurred (Sendai virus, coxsackievirus, St. Louis encephalitis virus, West Nile virus, Ebola virus).

In the biological sciences, names are usually part of formal taxonomies. Well-designed taxonomies are especially useful and important because they present not only the names of organisms but also the evolutionary relationships between those organisms. A taxonomic designation is, therefore, a shorthand way of conveying a great deal of information in a compact fashion. Unfortunately, the evolutionary relationships between the viruses are presently understood only at a rudimentary level. For this reason there are separate taxonomies based on the kingdom of the host cell rather than a single, inclusive taxonomy for all viruses.

The degree of formal organization in these viral taxonomies differs considerably. The animal virologists within the International Committee on Taxonomy of Viruses have proposed a system based on several levels of the classical Linnean scheme. The highest level of taxon is the *family,* whose names end in the suffix *-viridae.* Families represent clusters of viruses with apparently common evolutionary origins based on their types of nucleic acids, the number and type of their capsomers, and so forth. Families are subdivided into *genera* (singular: *genus*), which are identified by the suffix *-virus.* Genera are delineated by the particular host cells infected, immunological characteristics of the virions, and so forth. The lowest level of taxon, the *species,* has not yet been formally defined, but it will probably be equivalent to the present vernacular usage of the term *virus.* For example, the taxonomic designation of an age-old scourge that has been eliminated by the good efforts of a united humankind is as follows: Family: Poxviridae, Genus: *Orthopoxvirus,* Species: *smallpox virus.*

Thus far the bacterial viruses have been grouped only into families based on structural characteristics of the virions. The viruses with complex virions have been placed into three different families based on the size of the head and the length and nature of the tail, for example. The viruses within each family are merely named rather than grouped into genera.

The plant virologists also do not classify the objects of their study into conventional taxonomic categories. Instead, they have created clusters of viruses that are related by their shapes, types of nucleic acids, modes of transmission and host, and so forth. These groups are named for a prototype virus, such as the cucumoviruses named for cucumber mosaic virus, or the potyviruses (with a long O sound, please) named for the potato Y virus.

The Baltimore Classification

The plant virologists demonstrate that formal taxonomies based on presumed evolutionary relationships are not the only organizational schemes that can be applied to groups of organisms. This book will emphasize the interaction of viruses with their host cells. An organizing scheme that is particularly useful in that context is that proposed by David Baltimore. All viruses, regardless of the nature of their genomes, must use messenger RNA as the template for the synthesis of proteins. The **Baltimore classification** is based upon the relationship between the viral genome and the messenger RNA used for translation during expression of the viral genome.

As shown in Figure 1.17, the central feature of the Baltimore classification is *mRNA,* which is defined as a **positive-strand** molecule by the conventions of molecular biology. Other molecules of nucleic acid that have the same sequence as that mRNA (with thymines replacing uracils in DNA, of course) are also designated as positive strands. Nucleic acid molecules that have sequences that are the *complement* of mRNA are designated as **negative strands.** The six classes of the Baltimore classification are organized as follows:

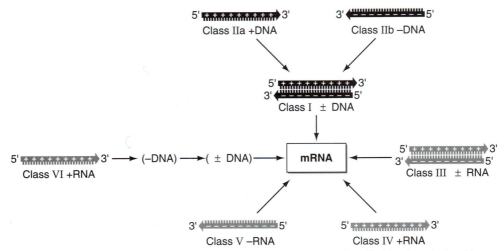

FIGURE 1.17 The Baltimore classification. Viral genomes are classified by the relationship between the mRNA, which is used to synthesize viral proteins, and the viral genome. Positive or " + " nucleic acids have the same sequence as that mRNA; negative or " – " nucleic acids have sequences that are complementary to that mRNA. "±" denotes double-stranded nucleic acids. The viruses of all but Class IV make mRNA by transcription from a negative-strand template molecule. The representations for positive- and negative-strand DNA and for positive- and negative-strand RNA used in this figure will be the standard forms used throughout this book.

- *Class I viruses:* The genome is double-stranded DNA, so mRNA is synthesized in the normal fashion using negative-strand DNA as a template.
- *Class II viruses:* The genome is single-stranded DNA. At the time this scheme was proposed, only positive-strand Class II viruses were known, but negative-strand viruses have since been found. They are now designated Class IIa and Class IIb, respectively. As Figure 1.17 indicates, synthesis of mRNA by these viruses involves a double-stranded DNA intermediate.
- *Class III viruses:* The genome is double-stranded RNA, one strand of which is therefore equivalent to mRNA.
- *Class IV viruses:* The genome is single-stranded RNA that can serve as mRNA directly, so these are positive-strand viruses. There are two subclasses (IVa and IVb) based on differences in mechanisms for expression and replication of the genome in these viruses.
- *Class V viruses:* The genome is single-stranded RNA that cannot serve directly as mRNA, but is instead the template for the synthesis of viral mRNA. Since the genome is complementary to the viral mRNAs, these are negative-strand viruses. Class Va and Vb viruses are also distinguished by differences in the mechanisms used in the expression and replication of their genomes.
- *Class VI viruses:* The genome is positive-strand RNA but its expression and replication require synthesis of a double-stranded DNA molecule. Class VI viruses are also termed *retroviruses* or RNA-tumor viruses.

The major virus groups of the Baltimore classification are presented in Table 1.3. The characteristics of the groups are given more fully in Appendix B, Characteristics of Selected Viruses.

Table 1.3 | Major Groups of Viruses

Group	Host	Morphology	Examples of Viruses
Class I Viruses: Double-Stranded DNA Genomes			
Myoviridae	Bacteria	Complex	T4
Siphoviridae	Bacteria	Complex	λ
Podoviridae	Bacteria	Complex	T7
Papovaviridae	Animal	Naked icosahedral	Polyomavirus, SV40
Adenoviridae	Animal	Naked icosahedral	Adenovirus
Herpesviridae	Animal	Enveloped icosahedral	Herpes simplex, varicella-zoster
Poxviridae	Animal	Complex	Smallpox, vaccinia
Baculoviridae	Insect	Enveloped helical	Nuclear polyhedrosis
Hepadnaviridae	Animal	Enveloped icosahedral	Hepatitis B
Caulimoviruses	Plant	Naked icosahedral	Cauliflower mosaic
Class II Viruses: Single-Stranded DNA Genomes			
Microviridae	Bacteria	Naked icosahedral	φX174
Parvoviridae	Animal	Naked icosahedral	Parvovirus, adeno-associated virus
Geminiviruses	Plant	Fused-pair icosahedral	Maize streak
Class III Viruses: Double-Stranded RNA Genomes			
Reoviridae	Animal	Naked icosahedral	Reovirus, rotavirus
Class IV Viruses: Positive-Strand RNA Genomes			
Leviviridae	Bacteria	Naked icosahedral	MS2, Qβ
Picornaviridae	Animal	Naked icosahedral	Poliovirus, rhinovirus, hepatitis A, coxsackievirus
Caliciviridae	Animal	Naked icosahedral	Norwalk
Togaviridae	Animal	Enveloped icosahedral	Sindbis
Flaviviridae	Animal	Enveloped icosahedral	Yellow Fever, Hepatitis C
Coronaviridae	Animal	Enveloped helical	Murine hepatitis
Potyvirus	Plant	Naked helical	Potato Y
Tymovirus	Plant	Naked icosahedral	Turnip yellow mosaic
Tobamovirus	Plant	Naked helical	Tobacco mosaic
Comovirus	Plant	Naked icosahedral	Cowpea mosaic
Class V Viruses: Negative-Strand RNA Genomes			
Rhabdoviridae	Animal and plant	Enveloped helical	Rabies, vesicular stomatitis
Paramyxoviridae	Animal	Enveloped helical	Mumps, measles, parainfluenza
Orthomyxoviridae	Animal	Enveloped helical	Influenza A, B
Bunyaviridae	Animal	Enveloped helical	Phlebovirus, Hantaan
Arenaviridae	Animal	Enveloped helical	Lassa
Filoviridae	Animal	Enveloped helical	Ebola
Class VI Viruses: Retroviruses			
Retroviridae	Animal	Enveloped icosahedral	Human immunodeficiency virus (HIV)

SUMMARY: WHAT IS A VIRUS?

We now have begun to discover the answer to the question that opened this chapter: "What is a virus?" Viruses are entities that are biochemically much simpler than either prokaryotic or eukaryotic cells. They have no independent metabolic activity of their own but depend upon a host cell for the metabolites and organelles required to express and replicate their genomes and to generate new viruses. Viruses reproduce in a fundamentally different fashion from their host cells, using a process of assembly rather than cell division.

Chapters 2 through 6 will explore the events that occur during each stage of the viral life cycle. They will examine the many ways viruses interact with their host cells by looking in detail at the strategies different viruses have evolved to reproduce themselves in hosts with fundamentally different genetic organizations and modes of expression. Chapters 7 and 8 discuss the effects viruses have on their host cells, as well as situations in which viruses and host cells have evolved to coexist. Chapter 9 discusses a variety of subviral entities and theories about the evolution of viruses.

SUGGESTED READINGS

Journals

1. "The structure of small viruses" by F.H.C. Crick and J.D. Watson, *Nature* 177:473–475 (1956), presents their hypotheses about the architecture of viruses.

2. "The assembly of a virus" by P.J. Butler and A. Klug, *Scientific American* vol 239, number 5:62–69 (1978), and "The structure of poliovirus" by J.M. Hogle, M. Chow and D.J. Filman, *Scientific American* vol 256, number 3:42–49 (1987) contain wonderful illustrations of a helical virus (tobacco mosaic) and icosahedral virus (polio), respectively.

3. "Common cold viruses" by M.G. Rossmann, et al., *Trends Biochem Sci* 12:313–318 (1987), is a very readable discussion of the structure of a human rhinovirus and its relationship to the virus's interaction with its host cells and immune system.

4. "Three-dimensional model of human rhinovirus type 14" by C.J. Hurst, W. Benton, and J.M. Enneking, *Trends Biochem Sci* 12:460 (1987), presents a pattern of the virus Rossman discusses that can be photocopied, cut out, the indicated structural subunits colored-coded, and then assembled into an icosahedral virion model; it's wonderful fun and instructive!

5. "Viral taxonomy for the nonvirologist" by R.E.F. Matthews, *Ann Rev Microbiol* 39:451–474 (1985), presents the various families of viruses, discusses the shortcomings of the present taxonomic categories, and reviews the evolutionary relationships between groups of viruses.

6. "Expression of animal virus genomes" by D. Baltimore, *Bacteriol Rev* 35:235–241 (1971), presents his classification of viruses based on their modes of gene replication and expression.

Books

1. *Fields Virology,* 3rd edition, edited by B.N. Fields, D.M. Knipe, and P.M. Howley, Lippincott-Raven, Philadelphia, PA, 1995, is a two-volume, 3000-page compendium about the animal viruses and the human diseases they cause.

2. *Principles of Virology: Molecular Biology, Pathogenesis, and Control,* edited by S.J. Flint, et al., American Society for Microbiology Press, Washington, DC, 2000, focuses on the molecular aspects of viral reproduction and pathogenesis.

3. *Plant Virology,* 3rd edition, by R.E.F. Matthews, Academic Press, San Diego, CA, 1991, describes numerous plant viruses and the diseases they cause.

Popular Accounts About Emerging Viruses

1. *A Dancing Matrix: How Science Confronts Emerging Viruses,* by Robin M. Henig, Vintage Books, New York, 1994.

2. *The Coming Plague: Newly Emerging Diseases in a World Out of Balance,* by Laurie Garrett, Penguin Books, New York, 1994.

3. *Virus Hunting,* by Robert Gallo, New Republic Book of Basic Books, New York, 1991.

4. *The Hot Zone,* by Richard Preston, Random House, New York, 1994.

Chapter 2

Getting In: Attachment, Penetration, and Uncoating

CHAPTER OUTLINE

- Why does a virus infect only certain cells and not others?
- What types of interactions occur between the virion and the cell surface during attachment?
- How do bacteriophages recognize their host cells?
- How do bacteriophages introduce their nucleic acids into host cells?
- How do animal viruses recognize their host cells?
- How do animal viruses enter their host cells and uncoat their nucleic acids?
- How do plant viruses infect their host cells?
- Summary

The reproduction of any virus requires that it enter a host cell to use that cell's metabolic systems to synthesize the components of the new virions. The first events of the virus replication cycle, therefore, involve the recognition of a suitable host cell by a virion, the attachment of the virion to that cell, and then the introduction of the viral genome into the host cell. The manner in which these processes occur depends on both the architecture of the virion and the nature of the structures surrounding the target host cell. In this chapter we shall examine the processes of attachment, penetration, and uncoating in bacterial, animal, and plant viruses.

WHY DOES A VIRUS INFECT ONLY CERTAIN CELLS AND NOT OTHERS?

Everyone has observed that different viruses seem to infect different populations of cells. Their characteristic diseases suggest that measles virus has involved cells of the skin, while mumps virus has infected those of the salivary glands in the throat. What is less obvious to the casual observer is that measles and mumps viruses are both members of the same group (the paramyxoviruses) and both enter the body by the same route (initial infection of the respiratory tract epithelium). Why then do they eventually infect different **target tissues,** the cells in which they produce their characteristic diseases?

A similar picture emerges when **host ranges,** or the types of different cells or organisms that a particular virus can infect, are examined. Host ranges can be broad enough to cross phylum boundaries (many togaviruses can infect both insect and mammalian cells) or be limited to cells of a single taxonomic class (poliovirus will infect only primate cells). Other host ranges are so narrow that only certain strains of a single species, or a certain cell type within a particular organism, can be infected. The bacteriophage T4 can infect only a few strains of the single species *E. coli,* for example, while human immunodeficiency virus (HIV, the AIDS virus) infects primarily human helper T lymphocytes or antigen-presenting cells like macrophages or dendritic cells. In an extreme example, some RNA bacteriophages can infect just certain strains of bacteria within the same species and these bacteria must be expressing a particular gene carried on a single type of plasmid.

What can be responsible for the *specificity* that appears to be a characteristic of viral interactions with their host cells? The specificity of a virus for its host cell might be the result of events at two different stages of the viral replication cycle: the attachment of the virion to the surface of its host cell, or the expression of the viral genome within the host cell. Of these, attachment to the cell surface is more likely for reasons of evolutionary economy. If a virus attempts to attach to an inappropriate cell, but cannot do so (and thus does not enter it), the virus may be able to try again on a different cell. If the virus actually enters a cell that is incapable of supporting its replication, however, all is lost. The virus is effectively inactivated since it cannot produce progeny viruses and cannot leave the cell to try again on a new potential host. Evolutionary pressures would therefore be expected to lead to specificity as a virus/cell surface phenomenon.

Experimental evidence supports this hypothesis. First, virtually all cells in organisms within a particular kingdom are capable of carrying out the metabolic activities of the Central Dogma of Molecular Biology in the same fashion. If a virus can reproduce itself in one type of cell, human hepatocytes (liver cells) for example, then one might reasonably expect that it should be able to reproduce in hepatocytes from other species as well. But this is usually not the case. Poliovirus isolated from nature can readily infect lymphoid cells of humans but cannot infect mouse cells of the same type with the same metabolic processes. If the RNA genome of a poliovirus is artificially introduced into a mouse cell, however, the virus can use the mouse cellular synthetic machinery to produce new virions (which are, in turn, able to infect human cells but unable to reinfect mouse cells). This clearly indicates that the failure of poliovirus to infect mouse cells is not a result of some defect in the ability of a mouse cell to support all the biosynthetic processes necessary to replicate the virus. The basis of selectivity must lie elsewhere.

Although all the cells in organisms within a particular kingdom carry out the basic activities of the Central Dogma in the same fashion, these cells do not share the same surface structures. As was described in Chapter 1, different bacteria may have different biochemical groups on their cell walls. Likewise, different animal cells may have markedly different proteins, glycoproteins, and glycolipids on their surfaces. Many of these cell surface structures have evolved to mediate specific cell surface recognition events, such as those involved with nutrient uptake, transduction of signals across the plasma membrane, or binding to other cells. Other molecules on the surfaces of animal cells, the histocompatibility antigens, play very important roles in labeling those cells as "self" during the cellular interactions of the organism's immune system. This vast array of different cell surface structures makes them natural candidates for involvement in the attachment of viruses to their host cell surfaces. A virus's host range would be the result of its ability to bind only to cells bearing particular cell surface proteins, glycoproteins, or glycolipids.

Several types of evidence support this conclusion. One of the most direct is genetic studies on the process of infection of *E. coli* by bacteriophages. When a plaque assay on *E. coli* is performed with a virus such as T3, a colony of bacteria occasionally arises within a plaque (Figure 2.1). The cells from that colony are resistant to infection by T3. Similarly, bacteria resistant to infection by the bacteriophage T4 can be isolated from T4 plaques.

Careful genetic analysis of bacteria taken from T4 plaques reveals that they have undergone a mutation that changes the structure of their cell wall lipopolysaccharide molecules. This suggests that T4 binds to the original lipopolysaccharide molecules but is unable to recognize and/or bind to the mutant form. Occasionally a T4 will arise in the viral population that is able to infect the mutant *E. coli*. These virions are found to contain changes in the

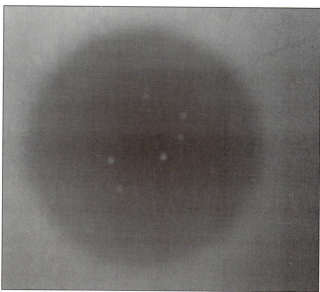

FIGURE 2.1 T3 plaque containing eight colonies that arose from individual cells and are resistant to T3 infection.

structure of their tail fibers. Clearly, reciprocal mutations in the host cell surface and the virus binding structure have restored the usual host-range relationship. Thus, it appears that virus–cell surface interactions are responsible for determining whether a virus is able to infect a particular cell and whether host ranges are the result of coevolution of those structures.

Recombinant DNA technology has permitted a direct demonstration of the relationship between a particular cell surface structure and the ability of an animal virus to infect its target cells. Human immunodeficiency virus (HIV) has been found to infect lymphoid or brain cells bearing a surface glycoprotein called CD4. When the human gene for this structure is introduced into other types of cells, CD4 molecules may become expressed on their cell surfaces. HIV is then able to infect the cells that have been "engineered" to have the appropriate cell surface receptors.

WHAT TYPES OF INTERACTIONS OCCUR BETWEEN THE VIRION AND THE CELL SURFACE DURING ATTACHMENT?

The binding of a virion to its host cell is probably mediated by *electrostatic* interactions between proteins or glycoproteins on the virion capsid or envelope and host cell surface proteins, glycoproteins, or glycolipids. Amino acids with charged side chains can bind to other charged amino acids or to the carboxyl groups on the *N*-acetyl-sugars (especially *N*-acetyl neuraminic acid, also called sialic acid), which are commonly found at or near the ends of the carbohydrate chains in glycoproteins or glycolipids. The formation of such bonds is highly dependent on two factors. The virion and cell surface molecules that are to interact must be able to "fit together" properly. This means they must each have the correct three-dimensional shape. In addition, the chemical groups on each of them that will form the electrostatic bonds must have the correct ionization. Both the three-dimensional configuration and the ionization of proteins are influenced by the pH and ionic strength of the medium, as well as the presence of particular ions.

pH, ionic strength, and the presence of particular ions have all been found to influence the attachment of virions to cell surfaces, compelling evidence for the role of electrostatic interactions in the process. Table 2.1 presents a sample of the conditions that have been found necessary to produce optimal binding of particular virions to their host cells under *in vitro* assay conditions. pH requirements are usually fairly narrow for a given virus, but the pH optima for different viruses range widely. This probably reflects the need to have the cor-

Table 2.1 | Environmental Influences on the Attachment of Virions to Their Host Cells

Virion and Host Cell	Condition Needed
Adenovirus/erythrocytes	pH 5.5–8.7
Coxsackie B4/HeLa	pH 3.0–3.5
T4/*E. coli* B	0.5% NaCl
λ/*E. coli*	Mg^{++}
T5/*E. coli*	Ca^{++}
Rhinovirus/HeLa	Mg^{++} or Ca^{++}

rect ionization on different sets of interacting amino acids, sugars, or both. For example, a physiological pH around 7.0 to 7.4 is thought to create a positively charged arginine on an adenovirus virion, which can bind to a negatively charged carboxyl group on an erythrocyte surface membrane. Divalent cations like Ca^{++} and Mg^{++} may aid in attachment either by providing bridges between two negatively charged groups or by influencing the three-dimensional conformation of the interacting proteins.

The attachment of a virion to its target cell obviously depends on a fortuitous meeting of the virion (moved by Brownian collisions) with the appropriate receptor molecule on the cell surface. Many factors can influence the efficiency of attachment. An important factor is the density of receptor molecules, which can vary between 5×10^2 and 5×10^5 per cell. The absolute concentrations of both the virions and the host cells also play a significant role in determining the rate of attachment. If virions and host cells are mixed together and the number of virions that have attached to host cells are determined at various time intervals, the rate of attachment is $dA/dt = kVH$, where V and H are the concentrations of free virions and host cells respectively, and k is the attachment rate constant (Figure 2.2). The values of

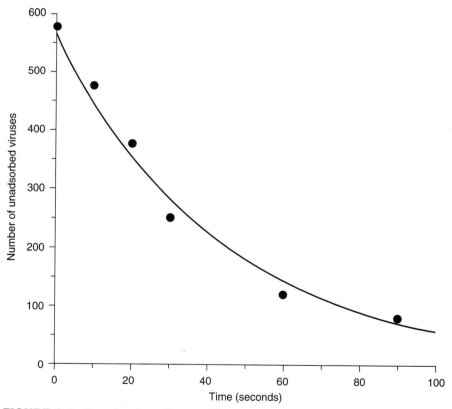

FIGURE 2.2   The rate of attachment of T4 to *E. coli* B. Virus and host cells at equal concentrations were mixed at 37°, aliquots removed at 10-second intervals, centrifuged, and the number of unadsorbed viruses remaining in the supernatant determined by plaque assays.

k for viruses attaching to animal cells measured in tissue cultures vary tremendously, from a level that represents diffusion-limited conditions to a rate five orders of magnitude slower.

The relationship $dA/dt = kVH$ has important consequences to an experimenter interested in studying the events that occur at a particular time after a virus begins to infect a cell. It is possible to make "synchronized" infected cultures by mixing together virions and host cells for a brief period and then diluting the mixture. This dilution step prevents further attachment by reducing the probability that a virion will encounter a host cell and bind to it. For example, a 100-fold dilution reduces the concentrations of both virions and host cells by 10^2 and consequently the attachment rate by a factor of $10^2 \times 10^2 = 10^4$.

Similar types of experiments have demonstrated that a *single* virus is sufficient to infect a cell. Figure 2.3 presents the results of an experiment relating the concentration of T4 viri-

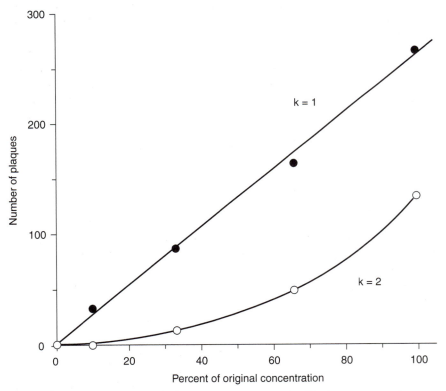

FIGURE 2.3 Comparison of a dose-response curve for T4 plaque assayed on *E. coli* B and a theoretical curve for production of plaques by double infections. The theoretical curve is based on the Poisson distribution: $P(k) = (e^{-m}m^k)/k!$ in which $P(k)$ is the number of cells infected by *k* viral particles and *m* is the multiplicity of infection or average number of viruses per infected cell. Under conditions of the plaque assay the number of viruses added is extremely small compared to the number of potential host cells, so *m* is extremely small. Therefore, the number of cells doubly infected (k = 2) is extremely small and the equation reduces to $P(2) \approx m^2/2$. The theoretical dose-response curve for k = 2 is parabolic rather than linear.

ons to the number of plaques produced on *E. coli*. The experimental dose-response curve is linear, indicating that one virion produces one plaque. If a higher **multiplicity of infection** (the average number of viruses that entered each infected cell) were required to produce a plaque (for example, two virions infecting the same cell concurrently), the curve would be parabolic rather than linear. The linear nature of the dose-response curve does *not* indicate, however, that every virus that initially binds to a host cell actually infects that host cell. For most virus–host cell combinations, many virions first become *reversibly* attached and can be released from the cell surface if the host cell is transferred into fresh medium. After this initial reversible phase of attachment, alterations in the virion and/or host cell surface may lead to *irreversible* binding and progression to the next stage, introduction of the viral genome into the host cell.

The term **efficiency of plating (EOP)** is the proportion of virions that actually produce infected cells. For example, T4 usually has an EOP approaching 1, meaning that virtually every virion introduced into a culture of host cells will infect a cell, while poliovirus has an EOP of only about 0.001 to 0.02 even under the most favorable conditions. Several factors influence the EOP of a particular virus/host cell combination. The first, and usually the most important, is the efficiency with which both reversible and irreversible binding occurs, a factor that varies widely from virus to virus. Other factors that contribute to the EOP value relate to the entry of the virus into the cytoplasm of its host cell and the preparation of the nucleocapsid for expression of the viral genome.

HOW DO BACTERIOPHAGES RECOGNIZE THEIR HOST CELLS?

Genetic studies have provided a wealth of information about the natures of virus attachment proteins and the cell surface receptors they recognize. Isolation of mutant cells that are resistant to infection can allow dissection of the interaction between viruses and their host cells, while recombinant DNA techniques permit direct demonstration of the identity of various components.

The cell receptors for a number of bacteriophages have been elucidated, some of which are listed in Table 2.2. In gram-negative organisms (see Figure 1.12), lipopolysaccharide

Table 2.2 | Cell Surface Receptors for Bacteriophages

Virus	Receptor
T4, T3	LPS core polysaccharide
T2	OmpF porin protein
T1, T5	TonA ferrichrome transport protein
T6	Tsx nucleoside transport protein
λ	LamB maltose transport protein
χ	Flagellum *(Salmonella)*
f1, MS2	F pilus
SP-50	Teichoic acid *(Bacillus)*

Host cells are *E. coli* unless otherwise noted.

(LPS) provides a variety of binding sites, especially in sugars of the O antigen and the core polysaccharide. Proteins of the outer envelope may also be recognition sites. In gram-positive cells, teichoic acids provide the necessary specificity (see Figure 1.11). Appendages such as flagella and pili are also sites of attachment for bacteriophages.

T4 Attachment to *E. coli*

The bacteriophage T4 has an extremely complicated virion, with an icosahedral head and a helical tail to which are attached a baseplate and six spikes and tail fibers (Figure 2.4). In all bacteriophages with complex architectures, attachment occurs via the components of the tail sec-

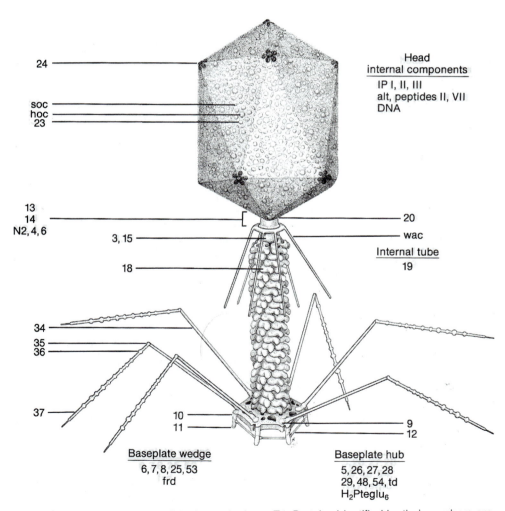

FIGURE 2.4 The structure of the bacteriophage T4. Proteins identified by their numbers are involved in the processes of attachment or penetration by this virus.

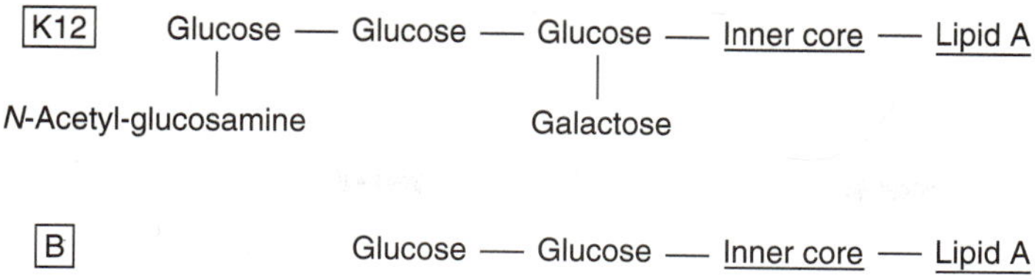

K12   Glucose —— Glucose —— Glucose —— Inner core —— Lipid A

$\quad\quad\quad\quad\quad\quad\quad$ |$\quad\quad\quad\quad\quad\quad\quad\quad\quad$ |

N-Acetyl-glucosamine$\quad\quad\quad\quad\quad\quad$ Galactose

B$\quad\quad\quad\quad\quad\quad\quad\quad$ Glucose —— Glucose —— Inner core —— Lipid A

FIGURE 2.5 The structures of lipopolysaccharide in *E. coli* B and *E. coli* K12. The structures of the inner core polysaccharide and lipid A were presented in Figure 1.12. Both *E. coli* B and *E. coli* K12 have lost distal portions of the wild-type lipopolysaccharide molecules.

tion. The attachment of T4 to *E. coli* B has both reversible and irreversible components, which genetic and biochemical studies have shown occur at different sites on the cell surface.

Reversible attachment of T4 involves an interaction between the distal portion of the virion tail fibers and the polysaccharides of the outer membrane lipopolysaccharide. Early experiments demonstrated that T4 tail fibers would bind specifically to the LPS extracted from *E. coli* B, and that this binding was inhibited by the addition of specific monosaccharides or disaccharides. Biochemical analysis of resistance mutants has localized the specific binding site to the end of the very truncated outer core polysaccharide of *E. coli* B (Figure 2.5). T4 can also attach to cells of the strain *E. coli* K12 whose outer core polysaccharide chain contains three more sugars than that of *E. coli* B. The presence of those sugars distal to the glucose binding site thus does not block binding completely, although it may reduce its efficiency. T4 is also able to recognize a second binding site in *E. coli* K12, an outer membrane protein, the porin OmpC. As the data in Table 2.3 demonstrate, binding to LPS and binding to OmpC appear to be independent of each other since the effects of mutations removing both of them are additive.

Table 2.3 | Relationship of Polysaccharide and Protein Receptors to Efficiency of Plating of T4 on *E. coli* K12

Strain	Lipopolysaccharide Structure	OmpC Porin Protein	Efficiency of Plating
Wild type	Lipid A-KDO-Hep-Hep-Glu-Glu-Glu-GluNAc $\quad\quad\quad\quad\quad\quad\quad$ | $\quad$ | $\quad\quad\quad\quad\quad\quad$ Hep Gal $\quad\quad\quad\quad\quad\quad$ Hep $\quad\quad\quad\quad\quad\quad$ Hep	Present	1.0
		Absent	10^{-3}
Mutant 1	Lipid A-KDO-Hep-Hep-Glu	Present	1.2
		Absent	0.7
Mutant 2	Lipid A-KDO	Present	10^{-3}
		Absent	$<10^{-7}$

Efficiency of plating values is normalized using that of the wild-type strain with OmpC porin present as the standard value = 1.

The lipopolysaccharide structures are KDO = ketodeoxyoctonic acid, Hep = heptose, Glu = glucose, Gal = galactose, GluNAc = N-acetyl-glucosamine).

The T4 virion has six tail fibers and the concentration of LPS on the cell surface is high, so even though the binding between any one tail fiber and the cell is weak, after the first tail fiber binds, the rate of binding of the others increases, so all six will find their targets and produce an effective collective binding. When all six fibers are bound to the cell wall, a conformational change is transmitted to the virion's baseplate to trigger penetration, the second phase of the attachment process. The virion has been said to "browse" about on the cell surface (actually, it is probably bounced around by Brownian vibrations of the cell) until a suitable site for attachment of the baseplate to the surface is found. These sites may be locations of cell wall synthesis where the plasma membrane and the outer membrane adhere to one another. It is likely, although not yet demonstrated, that a specific protein in the membrane is the site that leads to irreversible binding. Studies with phage mutants have determined that irreversible virion binding occurs by means of the protein (P12) that forms the connections between the baseplate spikes, and not by the spikes themselves. Irreversible binding positions the virion for the next phase of the replication cycle, introduction of the viral DNA into the host.

Other Bacteriophages

The bacteriophage λ has both an icosahedral head and a helical tail but lacks the complex of baseplate and tail fibers seen in T4 (Figure 2.6). λ does, however, have a single tail fiber (product of the J gene) located at the tip of the λ virion tail that is responsible for specific recognition of its *E. coli* host. The J protein fiber recognizes an integral protein (the *lam*B

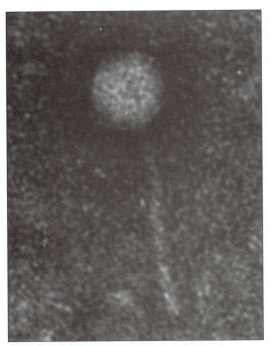

FIGURE 2.6 Structure of the bacteriophage lambda (λ). The tail of λ is longer than that of T4 but lacks a baseplate, spikes, and tail fibers.

protein) in the outer membrane of the *E. coli* host cell, which is responsible for specific transport of the sugar maltose. Since the *lam*B gene is inducible, that is, its synthesis is increased when maltose is available and glucose is absent, the ability of λ to infect its host is governed in some measure by the nutritional status of that host. Mg^{++} ions are also required for efficient binding, so the J protein/*lam*B protein interaction is probably electrostatic.

λ is but one example of the many bacteriophages that have evolved to use specific transport proteins as their binding sites. The bacteriophage T6 receptor is a protein responsible for transport of nucleosides in *E. coli,* while both the bacteriophages T1 and T5 use a ferrichrome transport protein as their cellular receptor.

Other bacteriophages are sometimes given the unfortunate identification "male-specific." Virtually all RNA bacterial viruses bind only to the so-called "male" pilus encoded by the conjugative F plasmid in *E. coli,* or its equivalent in other bacteria. (Cells with the F plasmid are sometimes called "male" since they donate DNA to recipient cells during conjugation, but since this is emphatically *not* a form of sexual reproduction, the term is both inaccurate and misleading. The correct term for such cells is F^{+}.) Infection by one of these viruses has come to be used as a "diagnostic test" for F^{+} cells and also as a type of "biological stain" for visualizing pili in photomicrographs. The icosahedral viruses MS2, R17, Qβ, and f2 are examples of F^{+}-specific RNA bacteriophages (Figure 2.7). Several

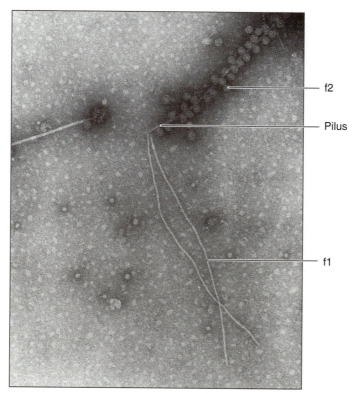

FIGURE 2.7 Virions of f2 (spherical RNA virus) and f1 (helical DNA virus) attached to F pili of *E. coli.*

helical DNA viruses (M13, f1, and fd) also are F^+ specific. Since binding to the pilus allows direct access to the plasma membrane and interior of the host cell, these viruses appear not to need the complex spikes and other attachment devices found in bacteriophages that bind to the cell wall. The F^+-specific viruses have only one (or a few) molecules of an *adsorption protein* (the A protein) that mediates both binding of the virion to the pilus and the introduction of the genome into its host cell.

HOW DO BACTERIOPHAGES INTRODUCE THEIR NUCLEIC ACIDS INTO HOST CELLS?

Attachment is only the first step in a continuous process resulting in the viral genome entering the host cell. **Penetration** and **uncoating** are the terms used to identify the next two steps in which the viral nucleic acid moves from the virion into the host cell (penetration) and, if necessary, is made ready for expression (uncoating). The thickness and biochemical complexity of the bacterial cell wall make penetration a formidable task for viruses that attach to that structure. The direct access to the plasma membrane afforded by flagella and pili makes the process easier for viruses attaching to them.

Penetration by the Coliphages T2 and T4

T2 and T4 are very similar-appearing **coliphages** (viruses that infect *E. coli*), differing only in minor details such as the absence of a collar around the tail piece where it attaches to the head of the virion. The classic experiments of Albert Hershey and Martha Chase were the first demonstration that bacteriophage nucleic acid entered a host cell, but the capsid proteins of the virion did not. As Figure 2.8 illustrates, T2 bacteriophages were labeled with ^{32}P in their DNA and ^{35}S in their proteins, allowed to attach to host *E. coli* cells, and then subjected to violent agitation in a Waring blender. This treatment ensured that any connections between virus components and the cell surface were disrupted. Hershey and Chase found that the bulk of ^{32}P-labeled DNA entered the *E. coli* cells, while the ^{35}S-labeled proteins of the capsid were left on the outside of the hosts.

How does the DNA get from the virion head into the cytoplasm of the host cell? Photomicrographs (Figure 2.9) clearly illustrate that penetration involves a marked change in the architecture of these viruses, especially in the overall configuration of the sheath proteins surrounding the tail tube of the virion. It is traditional to say that these images indicate that the sheath "contracts" to force the tail tube through the cell wall and into contact with the plasma membrane, leading to the "injection" of viral DNA through the tube and across the plasma membrane.

This hypodermic syringe analogy is misleading, however. The sheath does not contract; it *reorganizes*. In its extended configuration the sheath is made of 144 copies of a single protein (gp18) arranged in groups of six to form 24 rings around the tail tube (see Figure 2.4). The vertical distance between the ring subunits is 4.1 nm. The overall structure surrounding the tail tube is a helix with a twist angle of 17° (that is, each protein is shifted 17° to the right of the one below it). The process of penetration is initiated by conformational changes in tail fibers bound to the cell wall. These, in turn, trigger changes

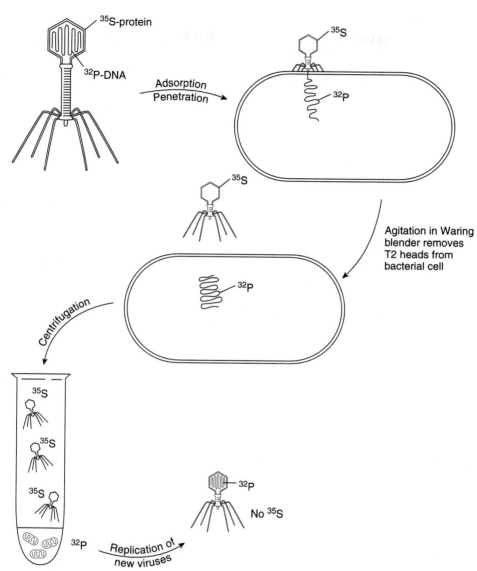

FIGURE 2.8 The Hershey-Chase experiment. Bacteriophages with ^{35}S-labeled proteins and ^{32}P-labeled DNA were allowed to interact briefly with host cells, and then the solution was violently agitated in a Waring blender. After centrifugation, most of the ^{35}S-labeled proteins were found in the supernatant, while the ^{32}P-labeled DNA molecules were pelleted along with the host cells.

in the arrangement of proteins in the baseplate (Figure 2.10, *A*). In its initial state the proteins of the baseplate are tightly arranged in a hexagonal pattern around the tail tube. Triggering causes the baseplate proteins to become expanded (the "star" conformation)

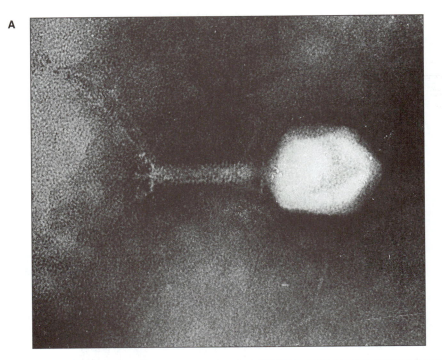

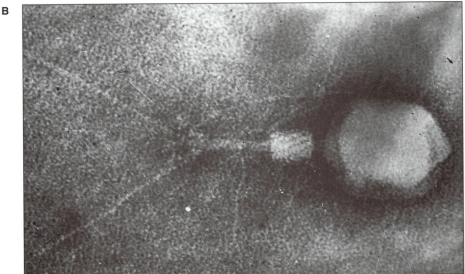

FIGURE 2.9 Comparison of the architecture of the bacteriophage T4 before **(A)** and after **(B)** penetration. The sheath surrounding the tail tube appears shorter and thicker after penetration, while the head appears empty.

and, as a result, the connections of the bottom layer of sheath proteins to the tail tube are weakened. As Figure 2.10, *B,* illustrates, this permits the sheath proteins to slide down in succession to form 12 new rings of 12 proteins each. This new shorter helix has a twist angle

A

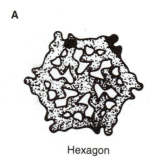

Hexagon

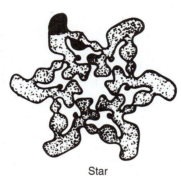

Star

FIGURE 2.10 Changes in the tail of T4 during penetration. **A,** Initiation of penetration involves a rearrangement of the tightly organized proteins of the baseplate *(hexagon)* into a looser configuration *(star)*. The tail tube that occupies the center hole in both configurations has been omitted from these drawings. *(Continued.)*

of 32°, with the rings packed only 1.5 nm apart. As a result of this rearrangement of sheath proteins, the tail tube slides through the cell wall and into contact with the plasma membrane.

Contact with plasma membrane triggers ejection of T4 DNA from the virion head through the tail tube. Genetic studies indicate that the DNA may have a **pilot protein** bound to its end that provides assistance in transferring the DNA through the tube and in initiating DNA replication within the *E. coli* host cell. (It has been noted that Hershey and Chase were very lucky that this protein does not have significant amounts of sulfur-containing amino acids!) The tail tube does not actually pierce the plasma membrane, but rather interacts with it to form an opening or pore from the tube into the cytoplasm through which the DNA can pass.

Other Bacteriophages

Bacteriophages such as T4 have evolved very complex structures, and with them, complex mechanisms for penetration. Less elaborate virions such as λ have tails that are not contractile but do serve to deliver the virus DNA to the plasma membrane. Movement of DNA through the tube from the λ virion head is triggered by contact of the distal end of the tail tube with a plasma membrane integral protein (the pts protein) that is part of the

B

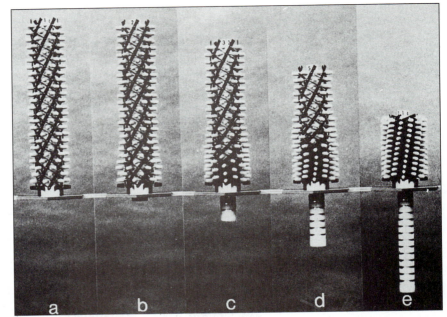

FIGURE 2.10, cont'd. **B,** A model of the rearrangement of sheath proteins during penetration, converting 24 narrow rings into 12 wider rings and lowering the tail tube through the cell wall.

maltose phosphotransport system in *E. coli*. Recall that the λ cell surface receptor for attachment is a maltose binding protein in the *E. coli* outer membrane, so the recognition events for both attachment and penetration by λ involve components of the same host sugar transport system.

The mechanism of penetration by RNA bacteriophages that attach to pili rather than to the cell wall is uncertain. It is clear from Hershey-Chase-type experiments that the capsid proteins of RNA bacteriophages remain outside the host cell, while the genome and its associated *A protein* enter the cytoplasm. This is not the case with F⁺-specific DNA bacteriophages. Both the genomic DNA and the capsid appear to enter the cytoplasm of the host cell. This has led to the suggestion that penetration by F⁺-specific DNA bacteriophages may be the result of retraction of the pilus into the host cell.

HOW DO ANIMAL VIRUSES RECOGNIZE THEIR HOST CELLS?

The animal cell surface presents a vast smorgasbord of proteins, glycoproteins, and glycolipids from which a virus can select a receptor. Some viral "gourmands" sample from nearly every possibility available and thus exhibit wide host ranges. Viral "gourmets," on the other hand, select their individual receptors with discrimination and thus have very narrow host ranges. Regardless of the nature of the virus–host cell combination of receptors, the interactions are electrostatic in nature.

Attachment by Naked Viruses

Unless they have spikes like adenovirus, all unenveloped animal viruses attach to their host cells through interactions directly between their capsids and the host cell receptor molecules. The nature of these interactions is best understood for the picornaviruses, the small (*pico* in Greek) RNA viruses whose members include the enteroviruses (polioviruses, coxsackieviruses [named for Coxsackie, New York] and the echoviruses [*e*nteric, *c*ytopathic, *h*uman, *o*rphan {associated with no particular disease}]), the rhinoviruses (*rhino* means "nose" in Greek so these are the cold viruses), the aphthoviruses (*aphtho* is the Greek for the disease "thrush," which is characterized by small ulcers; the disease now called thrush is caused by a yeast, while the aphthoviruses cause diseases like hoof-and-mouth disease—so much for etymology!), and the cardioviruses (*cardio* means "heart" in Greek).

Early cell culture studies indicated that host cell receptors could be divided into groups based on the picornaviruses that bound to them. One method for demonstrating this involves determining the susceptibility of cell surface structures to degradation by enzymes that cleave at different molecular sites. The data summarized in Table 2.4 show that the architecture of receptors even within the enterovirus subgroup of the picornaviruses can be different. Treating the target host cells with the enzyme trypsin, for example, degrades the receptors for poliovirus and rhinovirus, but not those for the related echoviruses and coxsackie B viruses. Direct competition studies of the ability of one virus to interfere with the binding of another virus have refined these studies. Saturating a cell surface with any of the polioviruses, for example, does not prevent the attachment of human rhinoviruses to the same cells. This indicates that the poliovirus and rhinovirus receptor sites are distinct from each other, even though they are both degraded by the same enzyme.

More recently monoclonal antibodies directed against host cell surface molecules have been used to discriminate among receptors. A **monoclonal antibody** preparation results when a single ("mono") antibody-producing cell is made to proliferate (that is, to form a "clone" of cells) *in vitro*. Since all the cells in the clone are identical, all the antibody molecules

Table 2.4 | Differences in Picornavirus Host Cell Surface Receptors Demonstrated by Enzymatic Digestion

Virus	Enzyme		
	Trypsin	**Chymotrypsin**	**Neuraminidase**
Poliovirus	Removed	Present	Present
Rhinovirus	Removed	Present	Present
Echovirus	Present	—	—
Coxsackie B virus	Present	Removed	Present
Cardiovirus	Present	Removed	Removed

Host cells were treated with enzyme immediately before the indicated viruses were added. If the enzyme removed the cell surface receptor, the virus was unable to infect the treated cell.

produced are identical and have exactly the same specificity or ability to bind to a particular small biochemical structure. Competition studies show that monoclonal antibodies against the poliovirus receptor do not interfere with the binding of rhinovirus to the same cell surface, and conversely, monoclonal antibodies against the rhinovirus receptor do not block poliovirus attachment.

Recombinant DNA technology has permitted the molecular characterization of the cell receptors for both poliovirus and human rhinovirus. Animal cells that were incapable of binding these viruses can be made to take up and incorporate molecules of human DNA in a process called **transfection.** Some cells acquire the information for the viral receptor and therefore have that new structure expressed on their cell surfaces. Clones of these cells are identified first by monoclonal antibodies, and that identification is confirmed by infection studies. This process has made it possible to isolate and then sequence the DNAs encoding both the poliovirus and rhinovirus receptors. These sequences reveal that both receptors are members of the *immunoglobulin supergene family* of proteins. These integral membrane glycoproteins are involved in a variety of recognition and binding activities relating to the immune system or to cellular communication. The rhinovirus receptor was found to be the intercellular adhesion molecule-1 (ICAM-1), originally identified by its ability to bind to a molecule on lymphocyte cell surfaces. The cellular function of the poliovirus receptor is not yet known.

Since rhinovirus is very small, the conformations of the capsid proteins within its virions have been determined by X-ray crystallography. These studies indicate that the three structural subunits that make up each capsomer (see Chapter 1) are arranged in a manner that leaves a "canyon" running along the capsomer (Figure 2.11). When five capsomers are arranged around a vertex, the canyon circles that point. The amino acids within that canyon are highly conserved among the various rhinoviruses, while those on the capsomer surfaces are much more variable. Several lines of evidence support the hypothesis that the amino acids within the canyon are involved in receptor binding. First, the size of the canyon in human rhinovirus is suitable for a binding site for ICAM-1, known to be the rhinovirus receptor. Second, chemically induced changes or mutations in those amino acids, as well as drugs that alter the shape of the canyon, all interfere with binding of the rhinovirus to its receptors. Finally, antibodies raised against the sequence of conserved amino acids in the canyon also prevent infection of target cells.

Other picornaviruses also locate their binding sites in a cavity on the virion surface. In the cardioviruses the cavity is in the form of pits rather than canyons between structural subunits. Another group of picornaviruses, the aphthoviruses, take the contrary approach, however, placing their virion binding sites on loops of protein exposed on the virion surface.

Another group of nonenveloped animal viruses do the aphthoviruses one better. The adenoviruses have a **spike,** or a long, slender fiber with a knob at its end, attached to 12 vertices of its capsid (see Figure 1.3). Human adenoviruses have been classified into subgroups based on the ability of the viruses to cause either rat or rhesus monkey erythrocytes to clump together. This reaction is termed **hemagglutination** (*heme* is Greek for "blood" and *agglutino* is Latin for "to glue to"). Many naked and enveloped viruses can cause hemagglutination. Since each virus may interact with only particular types of erythrocytes, hemagglutination tests are extremely useful in the laboratory. In nature, however, binding to erythrocytes would represent a catastrophe for the viruses involved. Erythrocytes do not carry out any of the processes of the Central Dogma, so such cells are totally unable to support viral reproduction.

FIGURE 2.11 Model of the formation of a "canyon" binding site by the structural subunits of the rhinovirus capsomers. A canyon encircles each vertex of the capsid.

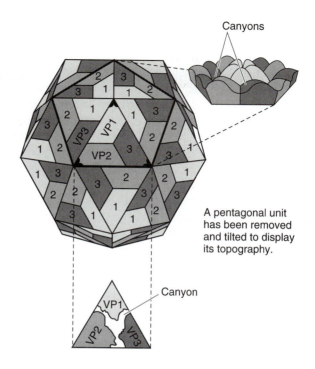

Canyons

A pentagonal unit has been removed and tilted to display its topography.

Canyon

The adenovirus spike is a rigid triple helix constructed from three identical fiber proteins. The C-terminal amino acids of each fiber form globular shapes that together create a knob with a deep indentation on its exterior surface. Different subgroups of adenovirus attach to different cell surface structures. A number of subgroups use a molecule called *car* (*c*oxsackievirus and *a*denovirus *r*eceptor, since both types of virus use the same molecule) that is a member of the immunoglobulin superfamily. Another subgroup uses a histocompatibility class I molecule that is a member of the same superfamily. Yet another subgroup uses one of the integrin molecules as its receptor. The viruses that recognize a given cell surface receptor have a distinctive set of amino acids in the knob indentation, suggesting that it is the site of the attachment reaction.

Adenoviruses are relatively inefficient in attaching to and penetrating their host cells. In fact, attachment of the knob to its receptor, while necessary, is insufficient to allow the virus to enter its host cell. This observation, which has been made for a number of different viruses, indicates that a second reaction is needed to trigger the process of penetration. We shall return to this topic shortly.

Attachment by Enveloped Viruses

Many groups of animal viruses have a membranous envelope surrounding their nucleocapsids. These envelopes are derived from host cell membranes (usually the plasma membrane), but contain viral proteins or glycoproteins as well. The viral proteins that are involved in virion binding to host cell receptors often form projections or spikes from the surface of the envelope (see Figure 1.7). The spike proteins of different viruses appear to share several common structural themes. In contrast, the nature of the host cell receptors recognized by different enveloped viruses is quite varied.

As their names indicate, both the orthomyxoviruses (*ortho* is Greek for "straight," *myxo* is Greek for "mucus," a substance that contains complex mucopolysaccharides) and the paramyxoviruses (*para* is Greek for "beyond" or "beside" since members of this group were included in the orthomyxoviruses until their molecular differences were discovered) use carbohydrates as their host cell recognition molecules. Two glycoproteins in the virion envelope spikes of influenza A virus (an orthomyxovirus) have been extensively investigated. One, termed HA for its hemagglutinin activity, is involved in recognition of host cells. Variability in the amino acid sequence of this glycoprotein plays a major role in the development of new strains of the virus that can produce epidemics. The other spike glycoprotein, called NA for its neuraminidase activity, is not involved in cell surface attachment but possibly has roles in the assembly and release of new virions.

Unlike the adenovirus knob, influenza A virus hemagglutinin has two peptide subunits called HA1 and HA2. The genes for HA1 and HA2 have been sequenced and the three-dimensional structure of the entire HA molecule determined by X-ray crystallography. The entire HA complex is shown in Figure 2.12. HA1 (328 amino acids long) forms a large globule, part of the stalk down to the small globule at the envelope surface, and then passes through the envelope to anchor the entire molecule. HA2 (221 amino acids long) forms part of both the stalk and the small globule. The entire HA spike on the envelope is a trimer composed of three HA1/HA2 molecules. Each of the three large globules contains a small pocket or binding site

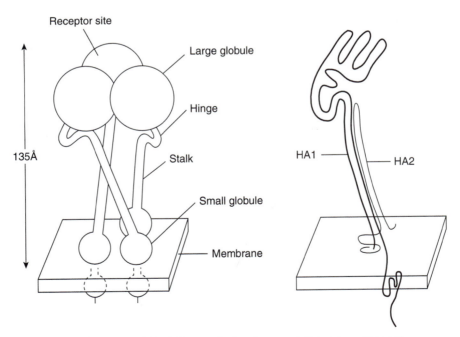

FIGURE 2.12 Model of the influenza A virus hemagglutinin trimer. One HA molecule is drawn at right to show the relationship of the HA1 and HA2 peptides.

for *N*-acetyl-neuraminic acid, a molecule that occurs in numerous cell surface glycoproteins and glycolipids on a wide variety of cell types. As was the case with rhinovirus binding sites described previously, the amino acids (both polar and nonpolar) in the actual influenza binding site are strongly conserved, while the amino acids surrounding the pocket show considerable genetic variability.

Sendai virus (a paramyxovirus) envelope spikes also recognize a set of *N*-acetyl-neuraminic acid–containing glycolipids in the host cell plasma membrane. This can be shown by treating host cells with the enzyme neuraminidase to remove the terminal neuraminic acids from the target glycolipids. This enzymatic treatment blocks the virion recognition event. Binding can be restored by replacing the plasma membrane glycolipids with exogenous neuraminic acid-containing glycolipids. The spike glycoprotein responsible for binding (called HN in Sendai virus since it has both hemagglutinin and neuraminidase activities) appears to form dimers and tetramers on the envelope surface in a fashion similar to the formation of trimers of HA proteins by influenza virus.

Although *N*-acetyl-neuraminic acid is frequently involved in attachment of animal viruses, many other molecules may be recognized. The herpesviruses herpes simplex 1 and 2 viruses (also called human herpes viruses 1 and 2), for example, have icosahedral nucleocapsids surrounded by envelopes with numerous small spikes containing viral glycoproteins. Inhibition studies indicate that attachment of herpes simplex virus to its host cells is a two-step process that begins with attachment of the virion spikes to molecules of the extracellular matrix rather than to the cell surface itself. Spike glycoproteins termed gB and gC bind initially to heparan sulfate, a long, sulfate-containing polysaccharide covalently attached to a protein core, which is prevalent in connective tissues in particular. Following their low-affinity attachment to heparan sulfate, the viruses attach to the cell surface through other spike glycoproteins to initiate the process of penetration.

The rhabdoviruses (*rhabdo* is Greek for "rod") generally have a bullet shape that is produced by an envelope surrounding a helical nucleocapsid wrapped into a spiral that tapers at one end but not the other (Figure 2.13). The envelope is covered by prominent spikes

FIGURE 2.13 Photomicrograph of negatively stained vesicular stomatitis virus (rhabdovirus) showing prominent spikes on envelope.

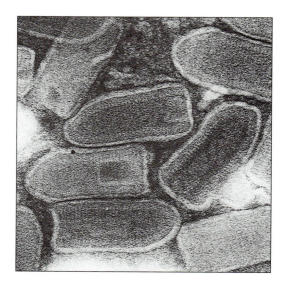

that are the product of a single viral gene (the G gene). Rhabdoviruses have extraordinary host ranges. Some can infect not only species in different classes in the same kingdom (mammals and insects, for example), but also members of entirely different kingdoms (insects and plants). The ability to attach to such vastly different cell types suggests that the virus recognizes a surface constituent common to many different cells.

Vesicular stomatitis virus (VSV) can infect both insects and mammals. VSV's attachment to its cell surface receptor may be inhibited by treating the target cells with phospholipase C, but not by digestion with either protease or neuraminidase. This indicates that the VSV receptor is probably a phospholipid rather than a protein or glycoprotein. Competition studies with different phospholipids showed that only phosphatidylserine totally inhibits VSV infection of mammalian cells, so this common membrane constituent may be the host cell receptor molecule.

Rabies virus, another rhabdovirus, couples a very broad range of species it can infect (although they are all mammals) with a striking sequestering process, which results in the virus entering first the peripheral nerves and then the central nervous system. This neurotropism (*trope* is Greek for "a turn") is likely a result of the surface receptors used by rabies virus to enter target cells. Several elegant studies have demonstrated that a major rabies surface receptor in nerve cells is the nicotinic acetylcholine receptor that mammalian neurons express at neuromuscular junctions. Binding of rabies virus to neuromuscular junctions can be inhibited by neurotoxins known to have high affinities for these acetylcholine receptors. In addition, the neurotoxin amino acid sequences known to bind to the acetylcholine receptor show great similarity to the sequence of the binding region in the envelope spike protein (the G protein) of rabies virus. Cells lacking the acetylcholine receptor can also be infected by rabies virus, so the G protein must also be able to recognize other receptors. These are possibly phospholipids or glycolipids, since chloroform-methanol-soluble cell membrane fractions will bind rabies virus. The relationship, if any, between these binding activities is not clear.

Many other viruses also use specialized cell surface structures of particular cell types as their receptors. Epstein-Barr virus (EBV) is a herpesvirus associated with infectious mononucleosis and several forms of human cancer (Burkitt's lymphoma and nasopharyngeal carcinoma). The target cells of EBV are B lymphocytes. Several studies using monoclonal antibodies determined that EBV recognizes a large B-lymphocyte surface glycoprotein normally involved in binding a cleavage fragment of the third complement component (C3d).

Another virus that infects cells of the immune system is the human immunodeficiency virus. The receptor for the HIV retrovirus is a molecule designated CD4, which occurs on the surface of helper T lymphocytes and is involved in those cells' recognition of major histocompatibility antigens on B lymphocytes or cytotoxic T lymphocytes. Although HIV was originally identified as a T-lymphotropic virus, many other types of cells may be infected after the initial introduction of the virus. CD4 is also expressed on the surface of macrophages and monocytes, which are also components of the immune system, as well as glial cells in the central nervous system.

The receptor role of CD4 has been demonstrated in a variety of ways. Monoclonal antibodies directed against CD4 prevent HIV infection of T lymphocytes. In addition, transfection of human epithelial cells with the CD4 gene leads to expression of the CD4 protein on their surfaces and makes them susceptible to infection by HIV. When a similar

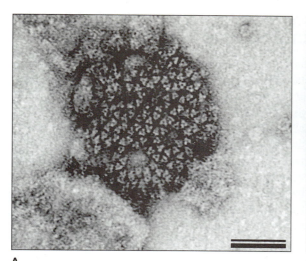
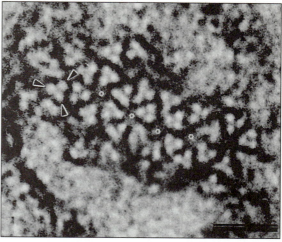

A   B

FIGURE 2.14 Spike proteins of retroviruses form trimeric complexes on the virion surface. Negative staining clearly shows this three-fold organization in the envelope glycoprotein of human foamy virus particles (**A;** the bar represents 50 nm). The trimeric spike proteins are arranged predominantly into rings of six spikes (**B;** the bar represents 25 nm).

transfection is done with mouse cells, however, HIV can bind to the cells expressing CD4 but cannot enter those cells. This finding indicates that a second reaction is necessary for penetration to occur.

The HIV envelope glycoprotein, termed SU for its surface location, appears to be responsible for binding to CD4. This large molecule is not anchored directly to the envelope membrane but is held in place by numerous interactions with a smaller integral membrane glycoprotein called TM, for transmembrane. Trimers of SU-TM molecules form the spikes that protrude from the viral envelope. The SU binding site for CD4 appears to be in a protected pocket that is not exposed to the action of antibodies. The same general trimeric structure for envelope glycoproteins appears in other retroviruses, as Figure 2.14 illustrates.

Table 2.5 summarizes the virus–host cell receptor relationships discussed in this section.

Table 2.5 | Cell Surface Receptors for Attachment of Animal Viruses

Virus Family	Virus	Receptor
Picornavirus	Human rhinovirus	Intercellular adhesion molecule I (ICAM-I)
Orthomyxovirus	Influenza A virus	N-acetyl-neuraminic acid
Paramyxovirus	Sendai virus	N-acetyl-neuraminic acid in glycolipids
Herpesvirus	Herpes simplex virus	Heparan sulfate
	Epstein-Barr virus	B-lymphocyte C3d receptor
Rhabdovirus	Vesicular stomatitis virus	Phosphatidylserine
	Rabies virus	Neuron acetylcholine receptor
Retrovirus	Human immunodeficiency virus	CD4

HOW DO ANIMAL VIRUSES ENTER THEIR HOST CELLS AND UNCOAT THEIR NUCLEIC ACIDS?

The surfaces of animal cells are dynamic structures in continual flux. Attachment of a virus to this dynamic surface triggers events that lead to the entry of the virus into the cell (*penetration*) and introduction of the viral genome in a suitable form to the cellular compartment where it is replicated (*uncoating*). The bulk of early evidence regarding the processes of penetration and uncoating in animal cells has been developed from electron microscopic studies. While such work has been quite valuable, it must be interpreted with care. The efficiency of plating can be extremely low in animal virus systems. It is thus possible that the virions visualized as being associated with the plasma membrane or within cytoplasmic vesicles are actually "dead ends" rather than examples of productive infection.

As has been mentioned several times already, in many viruses attachment to the cell surface appears to be a necessary, but not sufficient, event to produce an infection. After the initial binding step has brought the virion into contact with the cell surface, a second specific binding reaction to a **coreceptor** molecule is needed to induce the processes of penetration and uncoating. In several cases a virus or closely related group of viruses that share the same attachment receptor display different cell tropisms based on recognition of different coreceptors.

General Mechanisms of Penetration and Uncoating

Animal viruses use two general mechanisms for penetration and uncoating of their virions: direct fusion of viral envelopes with the cytoplasmic membrane of a host cell and virus-induced endocytosis. These mechanisms are based on the normal processes an animal cell uses to move materials from the external environment of the cell into its cytoplasm. Fluids, proteins, and small particles can enter a cell through **endocytosis** (*endo* is Greek for "within" and *cytos* is Greek for "a closed vessel"), a process that involves internalization of the plasma membrane in the form of a vesicle enclosing the material to be taken in. Nonspecific uptake of small droplets of fluid (and whatever that fluid contains) is termed *pinocytosis*. In contrast, *receptor-mediated endocytosis* transports specific molecules or particles that have been bound to cell surface receptors. These receptors are often located near specialized areas of the membrane called *coated pits*. In both pinocytosis and receptor-mediated endocytosis, the plasma membrane forms a depression that gradually deepens until a vesicle is created by self-fusion of the membrane at the cell surface. Proton pumps in the vesicle membrane then transport hydrogen ions into the vesicle, lowering its pH. The endocytic vesicle may also fuse its membranes with those of a *lysosome,* a vesicle containing a variety of degradative enzymes that are active only at the low pH levels found within the hybrid vesicle.

Animal viruses take advantage of the ability of membranes both to fuse with one another and to form vesicles within the cytoplasm. Enveloped viruses may penetrate their host cell and begin the process of uncoating by binding to plasma membrane receptor molecules, followed by direct fusion of their envelope membrane with the target host cell plasma membrane (Figure 2.15, *A*). As the envelope and host plasma membrane become merged, the viral nucleocapsid is delivered into the cytoplasm of the host. Alternatively, enveloped viruses may penetrate a host cell by means of endocytosis. The virus envelope binding to specific cell surface membrane receptors causes endocytosis in that region of the plasma membrane (Figure 2.15, *B*). When the pH within the vesicle drops, the virus envelope may then fuse with the vesicle membrane to deliver the nucleocapsid to the cytoplasm.

Naked viruses obviously cannot enter host cells by means of direct membrane fusions, so they penetrate host cells only by endocytosis. The term *viropexis* (*pexis* is Greek for "fixation") has been applied to endocytosis of naked viruses induced by their binding (fixation) to the cell surface (Figure 2.15, *C*). Earlier studies had suggested that some naked viruses like poliovirus entered their host cells through a process called "direct" penetration in which the virion attached to the plasma membrane and was degraded as the nucleocapsid was delivered into the cytoplasm in some fashion. Present evidence does not support this mechanism, however.

Related to the process of penetration is uncoating, which may be broadly considered to mean all the events that occur between entry of a virion into its host cell and the beginning of expression of the viral genome. Uncoating may include removal or rearrangement of capsid proteins and loss of spikes to allow the genome access to the cellular machinery for

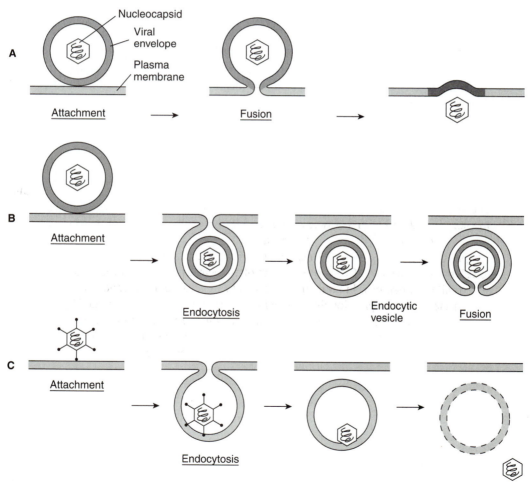

FIGURE 2.15 Models of penetration by animal viruses. **A,** Entry by direct fusion of envelope and plasma membrane. **B,** Entry by receptor-mediated endocytosis, followed by fusion of envelope and vesicle membranes. **C,** Entry of naked virus by receptor-mediated endocytosis followed by breakup of the vesicle to release the viral nucleocapsid.

the activities of the Central Dogma. These events frequently occur as a direct result of the process that allowed the virus to enter its host cell, although additional steps may be required.

Although it is not formally a part of uncoating, transport of the genome to the site where it is expressed (often the nucleus) can logically be included as a component of this stage of the viral replication cycle.

Fusion Proteins

The merging of two different membranes does not occur spontaneously, but as the result of the activity of specific viral **fusion proteins** that mediate the process. The genes for numerous fusion proteins of both DNA and RNA viruses have been cloned and sequenced, permitting identification of several structural and functional themes. Fusion proteins are all integral membrane proteins that have most of their amino acids exposed to the exterior of the envelope. They are usually glycoproteins, and many have fatty acids attached to them as well, although neither of these features is essential to their function. They are usually synthesized from a single mRNA as a large precursor peptide that undergoes posttranslational cleavage into two subunit peptides held together by disulfide bonds. The two-subunit fusion proteins function, in turn, as part of larger structures in the membrane that are usually trimers or tetramers of the fusion protein.

Most fusion proteins contain a **fusion peptide,** a highly conserved series of hydrophobic amino acids flanked by charged amino acids that are thought to be involved directly in the fusion process. Computer models of the possible three-dimensional structure of the fusion peptide region of various viral fusion proteins suggest that the largest, most hydrophobic residues of the fusion peptide are arrayed on the same side of a helix, while the smaller, less hydrophobic amino acids (frequently alanines and glycines) are on the opposite side of the helix (Figure 2.16. The arrangement of the fusion peptide region of the fusion protein may be pH-dependent in some viruses, making the fusion process itself pH-dependent.

The fusion peptide is always part of the same subunit peptide that contains the hydrophobic transmembrane anchoring segment for the fusion protein. This suggests that fusion is

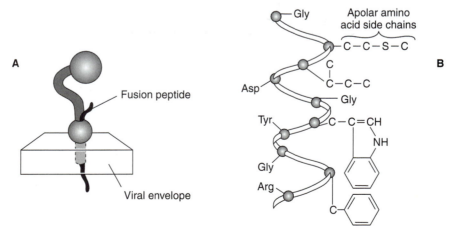

FIGURE 2.16 Fusion peptides. **A,** Fusion peptides are usually near the surface of the viral envelope. **B,** The arrangement of amino acids in fusion peptides of many viruses appears to be an α-helix. The amino acids with large hydrophobic side chains are clustered on one side of the helix. Part of the sequence for the influenza A virus fusion peptide is shown. *(Continued.)*

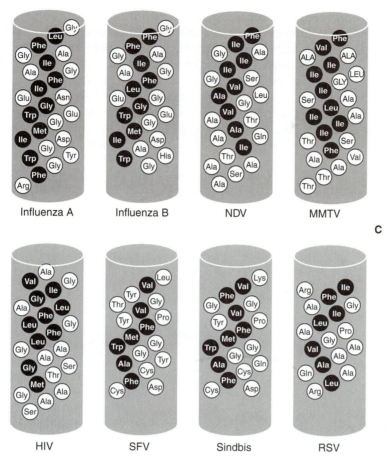

FIGURE 2.16, cont'd. **C,** Sequences of other fusion peptides with their clusters indicated by black circles. NDV = Newcastle disease virus, MMTV = mouse mammary tumor virus, HIV = human immunodeficiency virus, SFV = Semliki forest virus, RSV = Rous sarcoma virus.

facilitated by hydrophobic interactions occurring simultaneously with both the viral envelope and the host cell membrane.

Although the actual mechanism of fusion is not known precisely, various models of the process suggest that when the fusion peptides of oligomeric binding proteins are inserted into the cell membrane, conformation changes occur that draw together the viral envelope and the host membrane. This, in turn, leads to the formation of a hemifusion intermediate in which the outer phospholipid leaflets of the two membranes merge. As the binding proteins undergo further conformational changes, the inner leaflets fuse to link the viral envelope and the cell membrane.

Penetration by Enveloped Viruses: Fusion with the Plasma Membrane

As was mentioned previously, enveloped viruses may penetrate their host cells either by direct fusion with the plasma membrane or by receptor-mediated endocytosis and later fusion of the virion envelope with the vesicle membranes. Electron microscopy is one way of distinguishing between these possibilities, but it is often difficult to determine which events result in productive infections and which do not. Another method to examine the mechanism of entry by an enveloped virus is to determine whether newly infected cells bear envelope glycoproteins on their surfaces. When specific antibodies against viral envelope glycoproteins were applied to the surface of cells immediately after infection with Sendai virus (paramyxovirus), vesicular stomatitis virus (VSV) (rhabdovirus), or Sindbis virus (togavirus), only the Sendai-infected cells bound the antibody. This meant that Sendai entered its host by fusion, leaving its envelope as part of the cell surface membrane, while VSV and Sindbis entered by an engulfment process that left no envelope residue on the surface.

Paramyxoviruses, herpesviruses, and retroviruses enter their hosts by direct fusion of their envelopes with plasma membranes. In these viruses, attachment and penetration are separate activities mediated by different envelope proteins. For example, Sendai virus attaches to its host cell membrane via its HN spike proteins (Figure 2.17, A), but a separate F protein is responsible for fusion of the envelope and plasma membranes. The Sendai virus F protein, like all other viral fusion proteins studied, is an oligomer (a tetramer in this case) built of sets of two subunit peptides (F1 and F2) joined by a disulfide bond. The F1 peptide contains the distinctive stretch of hydrophobic amino acids characteristic of a fusion peptide. The membrane fusion reactions of Sendai virus do not depend on a pH change or any viral reproduction. In fact, Sendai virions that have been inactivated by ultraviolet irradiation so that they cannot reproduce are widely used in research to cause the fusion of adjacent cells in cell cultures. After fusion of the Sendai viral envelope and a host cell membrane into a single structure, the Sendai nucleocapsid is free in the cytoplasm to begin the synthesis of virus-specific molecules (Figure 2.17, B).

Penetration by herpes simplex virus appears to be more complex than in Sendai virus since it involves a number of different viral proteins as well as two different host cell molecules. Herpes simplex virus has eight glycoproteins on its envelope surface, two of which appear to play roles in attachment (gB and gC). Genetic studies indicate that gB and a second glycoprotein, gD, mediate membrane fusion reactions. Mutant viruses lacking either gB or gD are able to attach to cell surfaces, but fusion of the viral envelope with the plasma membrane does not occur. This indicates that penetration is a multistep process. Evidence that productive infection is through direct fusion with the plasma membrane rather than by endocytosis comes from other studies on the function of gD. Although direct fusion with the plasma membrane does not occur in gD mutants, these viruses can be taken into the cell by endocytosis. This route of entry is a dead end, however, since fusion of the viral envelope and the endocytic vesicle membranes does not occur. Instead, the virion within the vesicle is completely degraded rather than uncoated.

Penetration requires that the viral glycoprotein gD bind to one of several different cell surface proteins. Both human herpes simplex type I and type II viruses can use a protein termed HveA (herpesvirus entry mediator A) as this coreceptor. HveA has been characterized as a member of the tumor necrosis factor receptor family. Although it was the first coreceptor identified, HveA only allows these viruses to enter human lymphoid but not other cell types, so it cannot be the coreceptor that is used when these viruses infect the epithelial and neuronal cells that are their target tissues. Other coreceptors that appear to be related to the poliovirus receptor have been identified for herpes simplex viruses. Transfection experiments that have introduced these coreceptor molecules into Chinese hamster ovary cells that express heparan sulfate but do not support penetration by herpes simplex viruses suggest that

A

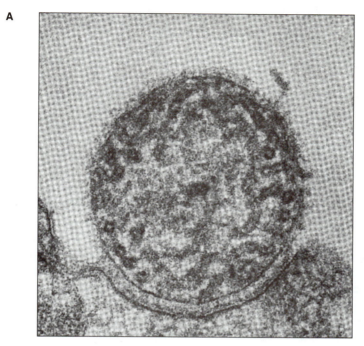

B

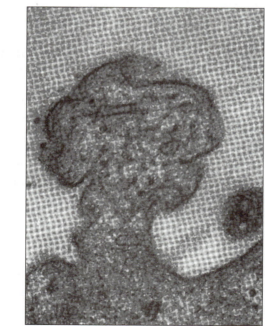

FIGURE 2.17 Sendai virus (paramyxovirus) attachment and penetration by direct fusion with the plasma membrane of its host cell. **A,** The virus has become attached to the plasma membrane. **B,** The viral envelope has fused with the plasma membrane so that the helical nucleocapsid is released directly into the cytoplasm.

HveC (or PRR-1 for poliovirus receptor-related protein 1) is the strongest candidate for allowing infection of epithelial and nervous system cells.

After the herpesvirus envelope fuses with the cytoplasmic membrane in the usual fashion, two proteins are released from the virion and play roles in shifting the cell's biosynthetic machinery over to the production of viral products. The nucleocapsid is transported to the nucleus, where its DNA contents are released through a pore into the nucleoplasm.

Retroviruses are the third group of viruses in which penetration occurs by fusion of their envelopes with the plasma membrane. With the exception of the lentiviruses like HIV, retrovirus binding of a single type of cell surface receptor by the viral envelope spike glycoprotein SU is sufficient to cause the conformation change necessary in TM to embed its fusion peptide into the plasma membrane and induce fusion. A variety of different receptors have been identified in various species, including phosphate and amino acid transport proteins and members of the tumor necrosis factor receptor and immunoglobulin supergene families.

In contrast to other retroviruses, penetration by HIV requires attachment to both a receptor and a coreceptor to induce penetration. The initial binding of SU to CD4 appears to cause a conformational change in SU that creates a new binding site for a coreceptor molecule. Binding of SU to the coreceptor in turn induces a conformational change in TM that allows a fusion peptide to interact with the plasma membrane. Identification of the HIV coreceptors provided the answer to a puzzling feature of HIV infection, namely that some strains of the virus are T-lymphotropic, others strains are macrophage-tropic, and yet others are dual-tropic. As Table 2.6 indicates, the HIV coreceptors are members of a group of receptors on the surfaces of immune system cells that bind small signaling molecules released by other cell types. Since these signaling molecules cause the recipient cells to move toward the signaling cell, they are termed *chemotactic cytokines,* or *chemokines* for short (*taxis* [adjective form, *tactic*] means "drawing up in rank and file," while *kine* comes from the word for "motion"). Different strains of HIV show different tropisms because they use different chemokine receptors as their coreceptors.

Although discovery of the HIV coreceptors solved one puzzle, the story is still not told completely. Many studies are underway to determine the influence of coreceptors on both susceptibility to infection by HIV and the progression of HIV-caused disease. For example, study of a group of Caucasians who remained uninfected despite numerous exposures to HIV found that group members were homozygous for a deletion mutation in the gene encoding CCR-5, the macrophage-tropic virus coreceptor. The absence of the coreceptor is probably not the whole answer to their resistance, however, since these persons have also not been infected by T-lymphotropic viral strains for which they do have the coreceptor. Mutations in chemokine receptors also appear to influence the course of AIDS development.

Table 2.6 | Major Coreceptors for HIV Infection of CD4-Bearing Cells

Tropism	Coreceptor	Natural Ligands of Coreceptor
Macrophage	CCR-5	Rantes
		Macrophage inflammatory proteins 1α and 1β
T lymphocyte	CXCR-4	Stromal-derived factor 1
Dual-tropic	CCR-2b	Rantes
		Monocyte chemoattractants 1, 2, 3, or 4

CXCR-4 is the major receptor for some strains of HIV-2. Some strains of macrophage-tropic HIV use chemokine receptors other than CCR-5. More than a dozen "minor" coreceptors have been identified.

Changes in CCR-2 have been associated with a slower progression of AIDS. Mutations in the promoter of CCR-5 or in CX3CR1, a chemokine receptor found in high levels in brain cells, appear to accelerate the course of disease.

Figure 2.18 summarizes the three different types of virus–cell surface molecule interactions that occur during attachment and penetration of enveloped viruses.

A One viral attachment molecule binds to one host cell receptor molecule (influenza A virus).

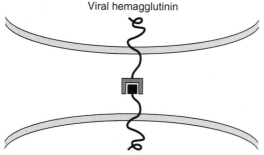

B One viral attachment molecule binds to a receptor and a coreceptor (HIV).

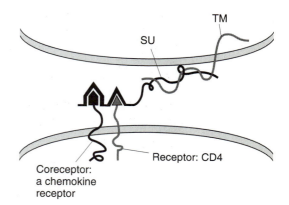

C Two viral attachment molecules: One attaches to a receptor and the other to a coreceptor (herpes simplex virus).

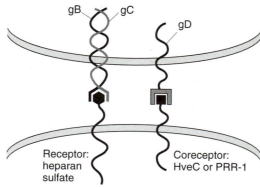

FIGURE 2.18 Models of three different types of interactions between viruses and host cells during attachment and penetration. **A**, Influenza A virus has a single attachment molecule, hemagglutinin, that both attaches the virion to the cell surface and induces penetration. **B**, HIV also has a single molecule, SU, that functions in both attachment and penetration. It first binds to a CD4 receptor molecule on the host cell surface. This binding causes a conformational change in SU that allows a new binding reaction to occur between SU and a coreceptor that is one of a variety of cell surface chemokine receptors. Binding of the coreceptor in turn causes a fusion peptide in TM to induce fusion of the viral and cellular membranes. **C**, In herpes simplex virus, attachment is by means of the binding of an attachment complex of gB and gC proteins to heparan sulfate on the cell surface. A second viral protein, gD, then binds to a coreceptor, HveC or PRR-1, to initiate penetration.

Penetration by Enveloped Viruses: Endocytosis

All orthomyxoviruses, togaviruses, and rhabdoviruses penetrate their host cells by endocytosis of the virion followed by fusion of the virion envelope membrane with that of the endocytic vesicle. A number of enveloped viruses, including Semliki forest virus (togavirus), make use of specialized regions of the plasma membrane called *clathrin-coated pits* in this process (Figure 2.19). These pits normally invaginate to form "coated vesicles" during pinocytic turnover of the membrane, so the endocytic vesicles containing the virions are also "coated" when they first form. The clathrin coat is removed from the vesicle before the process of uncoating the virion occurs by fusion of the viral envelope and the vesicle membrane.

Fusion of the envelope and vesicle membranes is pH-dependent, occurring in the pH range of 5 to 6.5 that is characteristic of the contents of endocytic vesicles. Although the details of the triggering process are known for only a few viruses, it appears that the lower pH causes one or more conformational changes in the fusion proteins of the envelope. Those conformational changes may expose the hydrophobic fusion peptide sequences that at neutral pH had been held in the interior of the fusion protein. They may also cause a clustering or a bending of the subunit proteins that make up the entire fusion protein.

Influenza A virus (an orthomyxovirus) provides the best understood example of penetration by endocytosis, since the molecular structure of its spike hemagglutinin glycoprotein is known in detail. The HA molecule is responsible both for attachment and penetration, in contrast to the separate molecules used by viruses that enter their host cells by direct fusion of the envelope and plasma membranes. Binding of the virion to host cell *N*-acetyl-neuraminic acid-containing receptors induces endocytosis, forming an endocytic vesicle with the influenza virion attached to its inner surface. Fusion of this vesicle with an acidic lysosome lowers the pH of the vesicle to around 5, triggering several conformational changes in the HA glycoprotein (Figure 2.20). Studies using antibodies directed against influenza A HA fusion peptide indicate that first the three HA stalks change shape near the viral membrane surface to make their fusion peptides swing out from a groove in the stalk. The fusion peptides are then available to interact with vesicle membranes. This is followed by bending at the hinge region of the stem that causes the globular heads of the trimeric HA molecule to move apart from each other. Both of these conformational changes appear to be necessary for fusion to occur, since mutants in which the heads are locked together by disulfide bonds are unable to fuse, even though their fusion peptides were exposed. Fusion of the envelope and vesicle membranes releases the eight segments of the influenza genome directly into the cytoplasm. Transport of these segments into the nucleus for expression appears to be an active process that is not dependent on microfilaments, microtubules, or intermediate filaments.

The pH-induced changes in influenza HA are irreversible, but those of vesicular stomatitis virus (a rhabdovirus) are not. The G protein of the VSV envelope spikes binds to phospholipids in coated pits, leading to endocytosis of the virion. When fusion with lysosomes lowers the pH of the resulting vesicle to less than 5, the G proteins reversibly aggregate at the ends of the virion. This aggregation may play a role in triggering fusion of the envelope with the vesicle.

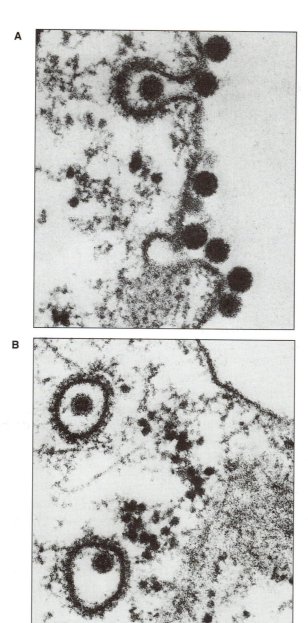

FIGURE 2.19 Penetration of the Semliki forest virus (togavirus) by endocytosis. **A,** The Semliki virus envelope tightly encircles an icosahedral nucleocapsid. Spikes on the envelope bind to plasma membrane receptors located in clathrin-coated pits (note the dark layer on the inner surface of the plasma membrane). **B,** After endocytosis the virions are contained within "coated vesicles."

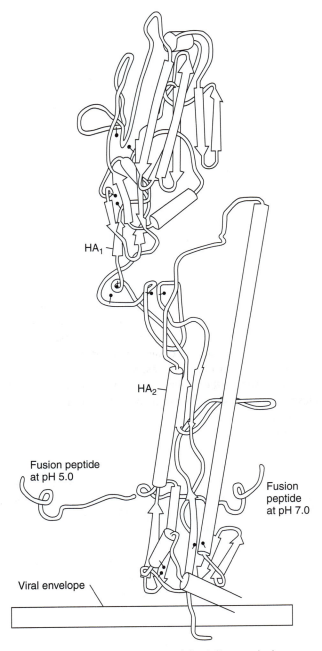

HA₁

HA₂

Fusion peptide
at pH 5.0

Fusion
peptide
at pH 7.0

Viral envelope

FIGURE 2.20 Conformation of the influenza A virus
HA fusion peptide at pH 7 and pH 5. When the virion binds
to the cell surface, the fusion peptides are located in
grooves in the HA stalk (pH 7.0). Reduction of the pH
within the endocytic vesicle to 5 causes the fusion
peptides to become extended out from the stalk and
available to interact with the vesicle membrane.

Penetration by Naked Viruses

Naked viruses must rely on endocytosis to enter their host cells. As with the enveloped viruses, the drop in pH that occurs within the resulting vesicle may play an important role in beginning the process of uncoating their genomes. In the current model proposed for picornavirus penetration and uncoating, binding of a virion to a clathrin-coated pit is followed by endocytosis into a coated vesicle. Acidification of that vesicle results in a conformation change in the capsid that releases one of the four capsomer subunits, VP4, from the capsid. This, in turn, permits the exposure of hydrophobic residues in VP1, another capsid protein, that were held within the intact capsid. These hydrophobic residues trigger fusion of the capsid with the vesicle membrane, leading to release of the viral genome into the cytoplasm.

The story is similar with adenovirus. After binding to the cell surface, the virion migrates laterally in the membrane until it encounters a clathrin-coated pit. The virion is then endocytosed into a coated vesicle. When the pH drops in the vesicle, the conformation of the capsid surface changes. The altered capsid comes into contact with the vesicle membrane, ruptures it, and releases the virion (minus the pentons and their fibers) into the cytoplasm.

Transport of the Viral Genome into the Nucleus

Most DNA viruses, as well as a few RNA viruses like the retroviruses and influenza A virus, replicate their genomes and make mRNAs in the nucleus. For these viruses, the final step in penetration is transport of the viral genome or nucleocapsid to the nuclear membrane and introduction of the viral nucleic acid into it via pores in that membrane.

While the details are sketchy for most viruses, the major features of this process can be described since they are based on the general phenomenon of directed transport of proteins from the cytoplasm into the nucleus. Proteins intended for use in the nucleus contain sequences of amino acids termed *nuclear localization signals* that are recognized by cytoplasmic receptors called importins that facilitate transport of the protein to a nuclear pore. In the case of the adenoviruses, release of the capsid (minus the fiber and pentons) from an endosome exposes nuclear localization signals on the virion hexons. The capsid is then carried along microtububules to a docking position on a pore in the nuclear envelope. Depolymerization of microtubules and inhibition of the microtubule-dependent motor protein dynein both block capsid movement. After docking, the capsid finishes disassembling and the viral DNA and its associated proteins are drawn through the pore and into the nucleoplasm.

All retroviruses make DNA copies of their RNA genomes as part of their reproductive cycles. This reaction occurs within the capsid while it is still in the cytoplasm. The new DNA molecule must then be integrated into the host cell genome. The DNA of oncogenic (tumor-producing) retroviruses cannot be transported through the nuclear pores, so these viruses can only infect cells that have just undergone mitosis and the nuclear envelope has not yet re-formed. Lentiviruses like HIV, in contrast, have nuclear localization signals on several proteins associated with the new DNA molecules that enable the nucleic acid molecules to pass through nuclear pores. HIV is therefore able to infect nondividing cells like T lymphocytes and macrophages.

HOW DO PLANT VIRUSES INFECT THEIR HOST CELLS?

Comparison of the processes of attachment and penetration in bacterial and animal viruses to those occurring in plant viruses is a study in contrasts. Plant viruses face two problems in infecting a new host organism. The first and most important of these is the mechanical barrier of the *cuticle*. The cuticle is a waxy outer covering at the surface of a plant structure like a leaf that protects the plant from desiccation. An intact cuticle appears to form an absolute barrier to the entry of viruses, in a manner analogous to the function of intact skin in protecting an animal from bacterial or viral infection. Although there are openings in the cuticle (the stomates) that might permit viral entry within the plant, each individual cell is still surrounded and protected by a cellulose wall, with layers of pectins and lignins between the walls of adjacent cells (see Figure 1.9). Plant cells do have plasmodesmata or small cytoplasmic connections to their neighbors, however, so the cell wall by itself does not prevent spread of viruses within plants after an infection has been established.

The problem of crossing the cuticle barrier has been solved in three ways by plant viruses. The first method is to avoid the problem altogether by infecting only plants that are in physical continuity with the original plant. Viruses may be passed from one generation of plants to the next through either infected seeds or infected pollen. Asexual reproduction, or vegetative propagation by runners or grafts, may also permit the spread of some viruses.

The second method is termed *mechanical inoculation*. When the cuticle is damaged by a grazing animal, for example, viruses can gain entry to the cells below the cuticle. Natural breakage of leaf hairs has also been suggested to lead to mechanical inoculation of viruses such as tobacco mosaic virus. Mechanical inoculation is the basis of the leaf local lesion assay for plant viruses described in Chapter 1.

The third and most prevalent method that plant viruses use to overcome the cuticle barrier is transmission by arthropod **vectors** (Latin for "one who carries"). Many insects like aphids and mites have long stylets that can pierce the cuticle and thread their way through the mesophyll cells and into the vascular tissues of the plant. As they feed, such insects can serve as vectors to carry viruses from plant to plant. The transmission of the viruses can be either passive or active. In passive or *noncirculative* spread, the virions are simply carried on the surface of the stylet or held within the anterior portions of the gut. The viruses are passed to a new plant when the insect feeds the next time, in a manner analogous to the carryover that occurs on the outside or within the barrel when the same pipet is used to make serial transfers. Active or *circulative* spread involves entry of the viruses into the gut or tissues of the insect vector followed by their injection from the salivary glands through the stylet into a new plant.

Some plant viruses (or are they animal viruses?) can perform the remarkable feat of multiplying in their insect vectors, especially the leafhoppers. This results in what is termed *propagative transmission* since the virus increases in numbers in the insect vector.

Penetration is strikingly different in the plant viruses compared to the bacterial and animal viruses. Plant viruses do not appear to recognize any cell surface membrane receptors. As a consequence, plant viruses are not taken into their host cells by endocytosis. Numerous studies using protoplasts, cells whose walls have been enzymatically removed, indicate that viruses can only enter their host cells through damaged plasma membranes. Although specific recognition events do not occur, the electrostatic charge of the virion does appear to play a role in the penetration process. While striking, these differences from other types of viruses are not surprising since the viruses are transmitted from plant to plant by mechanical damage or through vectors, which also breach the integrity of the plant cells, and within a plant they move from cell to cell through plasmodesmata. In neither interplant nor intraplant movement is passage through an intact plasma membrane necessary.

Since plant viruses do not have specific interactions with the cell surface membrane, host ranges for these viruses must be a result of other factors. Obviously, the feeding habits of arthropod vectors can establish a type of natural host range for viruses transmitted solely by that mechanism. At the cellular/molecular level, host range may be determined by viral genes governing cell-to-cell movement. A particular virus may be able to replicate with an individual cell when introduced into it, but may not be able to move from that cell to its neighbors, thus effectively ending the infection. In other cases, host cell functions required for viral gene expression and replication may be missing.

SUMMARY

In this chapter we have examined some of the strategies viruses use to identify and attach to their host cells and then to introduce their genomes into those cells. Recognition and attachment in bacteriophage and animal viruses are by means of electrostatic interactions between proteins/glycoproteins on the virus surface (with the capsid directly or with spikes on either the capsid itself or on an envelope) and proteins, glycoproteins, glycolipids, or phospholipids on the cell surface. The mechanism of penetration depends on the architecture of the virion and the host cell surface. Bacteriophages either use tubes associated with the virion or tubes supplied by the host (flagella or pili) to move their genomes through the layers of the cell wall and into the cytoplasm. Animal viruses rely on endocytosis by their host cells or fusion of viral and host membranes to introduce their genomes into their host's cytoplasm. Attachment and penetration may be mediated by a single viral protein binding to a cellular receptor, or the two processes may be mediated by separate binding reactions of viral proteins to an attachment receptor and a penetration coreceptor. Plant viruses, in contrast, neither directly recognize their own target cells nor bind to their surfaces, relying instead on vectors or mechanical means to enter their hosts. Viruses that replicate their genomes in the nucleus complete the process of uncoating by using cellular nuclear transport signals and systems to move their genomes into the nucleus.

SUGGESTED READINGS
General

1. "How an animal virus gets into and out of its host cell" by K. Simons, H. Garoff, and A. Helenius, *Scientific American* vol. 246, number 2:58–66, (1982), discusses the structure of Semliki Forest virus (a togavirus) and its relationship to attachment and penetration of its host cells.
2. "Plant virus-host interactions" by M. Zaitlin and R. Hull, *Ann Rev Plant Physiol* 38:291–315, (1987), reviews the full range of events that occur during a plant virus infection.

Attachment—These articles discuss the host cell receptors for various viruses:

1. "Roles of lipopolysaccharide and outer membrane protein OmpC of *Escherichia coli* K-12 in the receptor function for bacteriophage T4" by F. Yu and S. Mizushima, *J Bacteriol* 151:718–722 (1982).
2. "Specific gangliosides function as host cell receptors for sendai virus" by M.A.K. Markwell, L. Svennerholm and J.C. Paulson, *Proc Natl Acad Sci USA* 78:5406–5410 (1981).

3. "Is the acetylcholine receptor a rabies virus receptor?" by T. L. Lentz, T.G. Burrage, A.L. Smith, J. Crick and G.H. Tignor, *Science* 215:182–184 (1982).

4. "The CD4 (T4) antigen is an essential component of the receptor for the AIDS retrovirus" by A.G. Dalgleish, P.C.L. Beberley, P.R. Clapham, D.H. Crawford, M.F. Greaves, and R.A. Weiss, *Nature* 312:763–767 (1984).

5. "The T4 gene encodes the AIDS virus receptor and is expressed in the immune system and the brain" by P.J. Maddon, A.G. Dalgleish, J.S. McDougal, P.R. Clapham, R.A. Weiss, and R. Axel, *Cell* 47:333–348 (1986).

6. "Initial interaction of herpes simplex virus with cells is binding to heparan sulfate" by D. WuDunn and P.G. Spear, *J Virol* 63:52–58 (1989).

7. "Herpes simplex virus-1 entry into cells mediated by a novel member of the TNF/NGF receptor family" by R.I. Montgomery, M.S. Warner, B.J. Lum, and P.G. Spear, *Cell* 87:427–438 (1996).

Coreceptors—These papers tell the story of the discovery of the HIV, adenovirus, and herpes simplex virus coreceptors:

1. "HIV-1 entry cofactor: Functional cDNA cloning of a seven-transmembrane G protein-coupled receptor" by Y. Feng, C.C. Broder, P.E. Kennedy, and E.A. Berger, *Science* 272:872–877 (1996).

2. "Identification of a major co-receptor for primary isolates of HIV-1" by H. Deng, R. Liu, W. Ellmeier, S. Choe, D. Unutmaz, M. Burkhart, P. Di Marzio, S. Marmon, R.E. Sutton, C.M. Hill, C.G. Davis, S.C. Peiper, T.J. Schall, D.R. Littman, and N.R. Landau, *Nature* 381:661–666 (1996).

3. "HIV-1 entry into CD4⁺ cells is mediated by the chemokine receptor CC-CKR-5" by T. Dragic, V. Litwin, G.P. Allaway, S.R. Martin, Y. Huang, K.A. Nagashima, C., Cayanan, P.J. Maddon, R.A. Koup, J.P. Moore, and W.A. Paxton, *Nature* 381:667–673 (1996).

4. "CC CKR5: A RANTES, MIP-1α, MIP-1β receptor as a fusion cofactor for macrophage-tropic HIV-1" by G. Alkhatib, C. Combadiere, C.C. Broder, Y. Feng, P.E. Kennedy, P.M. Murphy, and E.A. Berger, *Science* 272:1955–1958 (1996).

5. "Evidence for cell-surface association between fusin and the CD4-gp120 complex in human cell lines" by C.K. Lapham, J. Ouyang, B. Chandrasekhar, N.Y. Nguyen, D.S. Dimitrov, and H. Golding, *Science* 274:602–605 (1996).

6. "Adenovirus interaction with distinct integrins mediates separate events in cell entry and gene delivery to hematopoietic cells" by S. Huang, T. Kamata, Y. Takada, Z.M. Ruggeri, and G.R. Nemerow, *J Virol* 70:4502–4508 (1996).

7. "Herpes simplex virus glycoprotein D can bind to poliovirus receptor-related protein 1 or herpesvirus entry mediator, two structurally unrelated mediators of virus entry" by C. Krummenacher, A.V. Nicola, J.C. Whitbeck, H. Lou, W. Hou, J.D. Lambris, R.J. Geraghty, P.G. Spear, G.H. Cohen, and R.J. Eisenberg, *J Virol* 72:7064–7774 (1998).

Penetration—These articles describe the mechanisms by which some viruses introduce their genomes into host cells:

1. "The infection of *Escherichia coli* by T2 and T4 bacteriophages as seen in the electron microscope. I. Attachment and penetration" by L.D. Simon and T.F. Anderson, *Virology* 32:279–297 (1967), and "The infection of *Escherichia coli* by T2 and T4 bacteriophages as seen in the electron microscope. II. Structure and function of the baseplate" by L.D. Simon and T.F. Anderson, *Virology* 32:298–305 (1967).

2. "The entry into host cells of sindbis virus, vesicular stomatitis virus and sendai virus" by D.P. Fan and B.M. Sefton, *Cell* 15:985–992 (1978).

3. "Mechanism of entry into the cytosol of poliovirus type 1: Requirement for low pH" by I.H. Madshus, S. Olsnes, and K. Sandvig, *J Cell Biol* 98:1194–1200 (1984).

4. "Viral and cellular membrane fusion proteins" by J.M. White, *Ann Rev Physiol* 54:675–697 (1990).

Nuclear transport—These articles describe the mechanisms by which some viruses move their capsids and genomes to the nucleus:

1. "Nuclear transport of influenza virus ribonucleoproteins: The viral matrix protein (M1) promotes export and inhibits import" by K. Martin and A. Helenius, *Cell* 67:117–130 (1991).

2. "Microtube-mediated transport of incoming herpes simplex virus 1 capsids to the nucleus" by B. Sodeik, M.W. Ebersold, and A. Helenius, *J Cell Biol* 136:1007–1021 (1997).

3. "HIV-1 nuclear import: In search of a leader" by M.I. Bukrinsky and O.K. Haffar, *Frontiers in Bioscience* 4:772–781 (1999).

4. "Nuclear import of adenovirus DNA in vitro involves the nuclear protein import pathway and hsc70" by A.C. Saphire, T. Guan, E.C. Schirmer, G.R. Nemerow, and L. Gerace, *J Biol Chem* 275(6):4298–4304 (2000).

Expression and Replication of the Viral Genome in Prokaryotic Hosts

CHAPTER OUTLINE

- What are the essential features of genome replication and expression in prokaryotes?
- What strategies do prokaryotic viruses use to facilitate replication and expression of their genomes?
- What structural features characterize the genomes of the double-stranded DNA bacteriophages?
- How do the T-odd coliphages express and replicate their genomes?
- How do the T-even coliphages express and replicate their genomes?
- How does lambda virus express and replicate its genome?
- How do single-stranded DNA bacteriophages express and replicate their genomes?
- How do RNA bacteriophages express and replicate their genomes?
- How do the RNA bacteriophages regulate the amount of each protein synthesized during their replication?
- Summary

In this and several subsequent chapters we shall consider the core events in the viral replication cycle: *replication of the viral genome* and *synthesis of viral proteins*. Much of our knowledge about the processes of the Central Dogma of Molecular Biology has been gained from investigations of these steps of viral reproduction in bacteriophages. The details of these processes are often extremely complex and are therefore beyond the scope of this book. We shall limit our examination to examples of the common themes or strategies used by bacterial viruses, especially those of *E. coli* (the *coliphages*), rather than exploring the variations; we shall view the forest in which so many have labored with zeal to describe the trees.

WHAT ARE THE ESSENTIAL FEATURES OF GENOME REPLICATION AND EXPRESSION IN PROKARYOTES?

The nature of the host plays crucial roles in the strategies viruses have evolved to reproduce themselves. We shall therefore preface consideration of the bacteriophages with a review of the features of the processes of the Central Dogma as they occur in prokaryotes.

The genome of a prokaryotic organism is in the form of a single, supercoiled, circular molecule of double-stranded DNA encoding about 6 to 10×10^3 genes. Many of these genes are grouped into clusters (**operons**) that are under the regulation of the same promoter and its associated controlling elements. In addition to a genomic DNA molecule, prokaryotes usually have one or more **plasmids,** or autonomously replicating small DNA molecules encoding functions that may be useful to the cell in particular circumstances (antibiotic resistance or the ability to utilize a scarce or unusual nutrient, for example).

A cardinal feature of the processes of the Central Dogma in prokaryotic cells is that they are not separated from each other in either space or time. Even as it is being replicated, DNA can serve as a template for transcription. The transcript contains no introns and can thus serve immediately as mRNA. Translation begins as soon as enough of an mRNA is available, even before transcription of that message is complete. None of these processes occurs in its own physically delimited compartment within the cell.

DNA Replication in Prokaryotes

DNA synthesis is a complicated process involving an array of different enzymes and DNA-binding proteins. Some of the complications result from the structure of the DNA molecule itself. Each strand of a linear double helix has a *polarity* imposed by its constituent nucleotides. One end of a linear strand has a free 5′ carbon, which usually is phosphorylated, while the other end of the strand has a free 3′ hydroxyl group. The two strands of a double helix are *antiparallel,* or arranged in opposite orientations to each other, so new daughter strands of DNA must be synthesized antiparallel to their template parent strands. A major physical task during replication is separating the two template strands over their entire lengths without creating a tangle of molecules that cannot be distributed into new daughter cells.

Genetic imperatives also govern the process of replication. Replication of the genome is the only task an organism attempts to perform perfectly each time it is done. A mistake in incorporating even a single new base in a daughter strand of DNA has the potential to kill the organism or cell that receives that DNA copy when replication is complete. For that reason *proofreading,* or checking to see that each new base is correctly base-paired, is an integral part of the process of replication and is performed by the same DNA polymerase molecule that elongates the chain. The need for proofreading is probably the reason for several of the features that distinguish DNA synthesis. First, synthesis is unidirectional (5′ ⟶ 3′), and occurs only by a condensation reaction between a nucleotide triphosphate and a 3′ hydroxyl group at the end of the growing chain. Proofreading can remove the newly added nucleotide-monophosphate and restore the original configuration of the chain when the growing point is at the 3′ end. If the growing point were at the 5′ end, in contrast, removing a newly added nucleotide would leave at most a single phosphate group rather than the original triphosphate, so synthesis would halt until two phosphates could be transferred to the growing point. A second feature of DNA synthesis that is probably a result of the requirement for proofreading is the inability of DNA polymerases to originate DNA synthesis themselves. Since DNA polymerase can only elongate preexisting nucleotide chains, which can be either DNA or RNA, it can begin proofreading with the first base inserted.

Prokaryotic DNA is replicated *semiconservatively,* with each strand of the double helix serving as the template for a new daughter strand. Replication proceeds *bidirectionally* from a single **origin of replication,** so that an intermediate structure called **theta,** from its resemblance to the Greek letter θ, is formed as the two growing points move around the circular chromosome (Figure 3.1, *A*). Since the object of replication is to copy the entire genome,

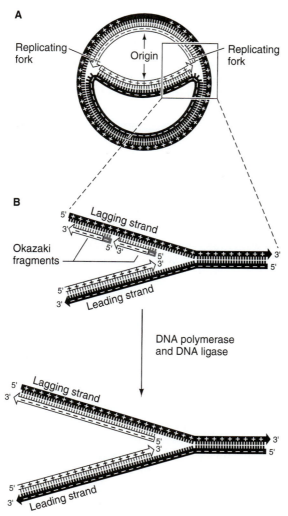

FIGURE 3.1 DNA synthesis in a prokaryotic cell. **A,** A theta (θ) intermediate is formed by bidirectional replication from a single origin. **B,** The events at a replication fork include elongation of the 3′ end of the leading strand and discontinuous synthesis of the lagging strand by means of Okazaki fragments. The conventions for labeling nucleic acid molecules used in this figure are found inside the front cover.

there are no termination signals for the process. Synthesis simply continues until the entire chromosome has served as template.

Figure 3.1, *B,* illustrates the events occurring at a replication fork during DNA synthesis. The requirements that the daughter strands be antiparallel to their template strands and that DNA synthesis be only by addition to the 3′ ends of preexisting molecules of nucleic acid creates the necessity for *discontinuous* DNA synthesis on one daughter strand (called the *lagging strand*). Synthesis of this strand begins with a short bit of *RNA* called a **primer** made by a special RNA polymerase *(primase).* **DNA polymerase III** then elongates the primer with new DNA. When it reaches the primer of the previous fragment, DNA polymerase I digests away that RNA and continues elongating until the gap left by the primer is filled. **DNA ligase** then joins the adjacent strands by creating a phosphodiester bond between the 5′ phosphate and the 3′ hydroxyl groups. Other enzymes and proteins that are also involved in replication are the *helicases,* which open the double helix to create the replication forks, the **single-strand binding proteins (SSBs),** which help to maintain the open configuration, and *topoisomerases* like **DNA gyrase,** which produce changes in supercoiling and thus prevent tangling of the DNA strands.

Transcription in Prokaryotes

The purpose of transcription is to form RNA molecules that can mediate the synthesis (translation) of specific proteins. For this reason the entire genome is not transcribed at one time, but rather discrete sequences of the DNA molecule serve as templates as needed by the cell. In prokaryotes, the **structural genes** for metabolic pathways are often physically contiguous, in a unit called an *operon,* and are under the control of a single set of controlling elements. The famous *lac* operon of *E. coli,* for example, has three controlling elements (the repressor gene, the promoter, and the operator) followed by three structural genes (the β-galactosidase, permease, and acetylase genes).

Important features of transcription (Figure 3.2) include how the cell recognizes (1) where transcription is to begin *(initiation),* (2) where transcription is to end *(termination),* and (3) when a particular set of genes is to be expressed. The first two of these are functions of the prokaryotic *RNA polymerase.* This enzyme has two significant differences from DNA polymerases: It can initiate RNA synthesis without a primer, and it does not proofread during transcription. In *E. coli* the core RNA polymerase enzyme has four subunits (two of the α peptide and one each of the β and β′ peptides). This complex has catalytic activity (adding nucleotides to the 3′ end of a chain) but requires the association of other peptides in order to recognize where to begin or end transcription. An additional peptide called **sigma (σ)** assists the polymerase in binding to the **promoter** sequence that constitutes the beginning of a transcriptional unit, while the **rho (ρ)** and nusA peptides assist in recognition of termination sequences. Timing of gene expression is usually by means of specific regulatory proteins that bind to sequences associated with the promoter regions of operons. Regulation can be either negative or positive. In negative regulation, **repressor proteins** bind to DNA to prevent RNA polymerase from moving from the promoter to the structural genes of the operon, while in positive regulation **gene activator proteins** assist RNA polymerase in attaching to promoters. An operon may have both of these forms of control; the *lac* operon of *E. coli* is an example of such an operon since lactose causes derepression, while absence of glucose permits gene activation.

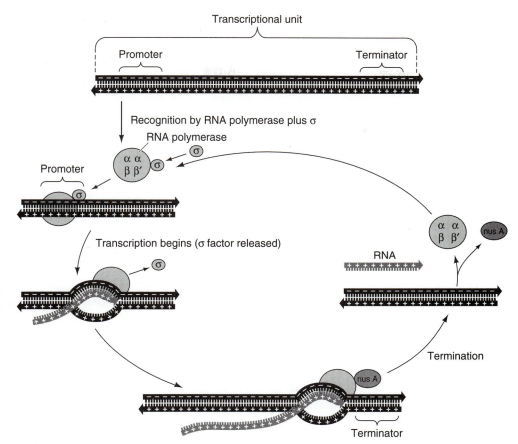

FIGURE 3.2 Transcription in a prokaryotic cell. The core RNA polymerase enzyme (ααββ′) is assisted in recognition of a promoter by a σ factor that is released after transcription begins. Two other factors, ρ and the nusA protein, assist in recognition of certain termination sequences.

Translation in Prokaryotes

The product of most transcription in prokaryotes is a *polycistronic mRNA* molecule (Figure 3.3). This molecule contains the information coding for several proteins and does not have the 5′ "cap" or 3′ "tail" structures characteristic of eukaryotic mRNAs. As with transcription, translation requires identification of particular starting and stopping signals within the mRNA. An **initiation codon** (AUG), in association with other sequences for ribosome binding, marks the beginning of a protein-coding section of the mRNA, while any of three *termination* or **"nonsense" codons** (the *amber* triplet UAG, the *ochre* triplet UAA, and the *opal* triplet UGA) stop translation. Each initiation codon determines the **reading frame,** or groups of three bases to be translated, for that particular gene within the polycistronic mRNA.

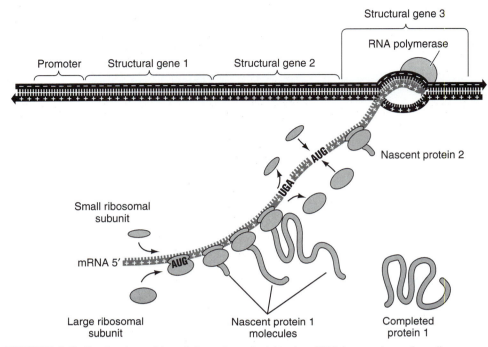

FIGURE 3.3 Synthesis and translation of a polycistronic mRNA in a prokaryotic cell. Transcription and translation are coupled so that translation begins as soon as the first AUG initiation codon becomes available. Translation of each subsequent protein coding region begins when each initiation codon has been transcribed.

WHAT STRATEGIES DO PROKARYOTIC VIRUSES USE TO FACILITATE REPLICATION AND EXPRESSION OF THEIR GENOMES?

The events during this stage of the virus replication cycle have evolved to facilitate rapid and efficient synthesis of new viral nucleic acids and proteins. As often occurs in nature when a small event leads to very large consequences, several stages of *amplification* characterize this part of the replication cycle. Often only a single virus infects a host cell. That virus produces numerous progeny viruses by a series of steps in which a given molecule mediates the synthesis of multiple new molecules. The first viral transcription and translation events to occur after penetration and uncoating usually involve preparation for synthesis of more copies of the viral genome. In DNA viruses (Figure 3.4), the new copies of the genome can then serve as templates for transcription of many copies of each viral mRNA. These, in turn, are translated many times to synthesize numerous copies of the viral structural proteins. The copies of the genome are then already available for incorporation into new virions.

RNA viruses use a similar series of steps, but the nature of their genome creates the need for an additional round of RNA synthesis. Nearly all the RNA viruses of prokaryotes have *positive-strand* genomes capable of serving directly as mRNA, so the first process after

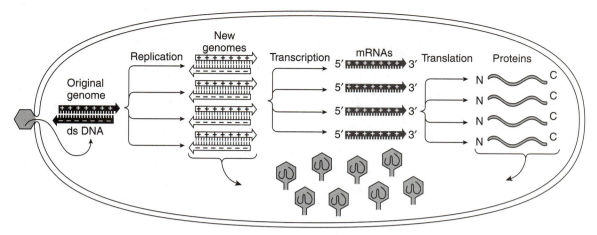

FIGURE 3.4 Amplification steps in DNA virus replication. The original double-stranded *(ds)* DNA genome is the template for synthesis of numerous copies, each of which can be used for the synthesis of multiple mRNAs. Each of these is then translated numerous times.

uncoating is translation rather than synthesis of new nucleic acids (Figure 3.5, *A*). Prokaryotic cells have no **RNA-dependent RNA-polymerase** that would permit the synthesis of an RNA molecule using another RNA molecule as template. Every RNA virus, regardless of whether it is positive- or negative-strand RNA, *must* carry a gene for an **RNA replicase** in order to reproduce its own RNA genome. Synthesis of the replicase permits a two-stage amplification of the RNA genome and thus copious protein synthesis. The genomes and structural proteins can then be assembled into new virions.

Negative-strand RNA viruses bring in viral-encoded replicase molecules along with their genomes when they infect their host cells since that enzyme cannot be translated directly from the genome. The replicase makes positive-strand antigenomes that serve both as a template for the synthesis of new negative-strand genomes and as mRNAs for the synthesis of viral proteins (Figure 3.5, *B*).

As the examples discussed later in this chapter will illustrate, bacterial viruses may encode a variety of different enzymes or regulatory proteins to facilitate the replication and expression of the genomes. Table 3.1 presents just a few examples of types of enzyme activities that a virus can alter in a cell, either by encoding a new enzyme or by altering the activity of a host cell enzyme. As mentioned earlier, RNA viruses must encode a novel enzyme, the RNA replicase, but other viruses may produce new polymerases or modify the recognition abilities of host enzymes to convert them from host cell-directed synthesis to viral-directed synthesis. To increase the efficiency of synthesis of viral gene products, many viruses subvert or destroy host cell activities that would compete with the same viral activities. Nucleases that cleave host DNA and enzymes that convert biosynthetic pathways to the formation of virus-specific precursors are two types of enzymes that may be involved in such activities. Other new enzymes are involved solely in catalyzing reactions particular to the synthesis and assembly of new virions.

A Positive-strand RNA viruses

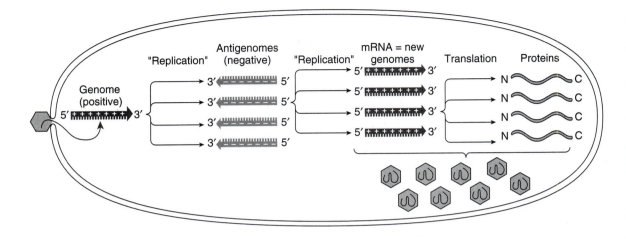

B Negative-strand RNA viruses

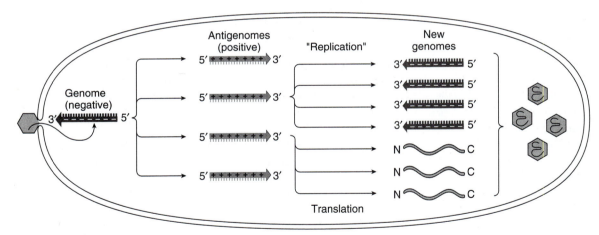

FIGURE 3.5 Amplification steps in RNA virus replication. **A,** A positive-strand RNA genome is amplified in two steps as the positive-strand genome serves as the template for synthesis of numerous negative-strand "antigenomes," which in turn are the templates for synthesis of many positive-strand mRNAs for translation of viral proteins. **B,** The positive-strand antigenomes made from a negative-strand viral genome serve both as the template for synthesis of new genomes and as mRNA molecules for the synthesis of viral proteins.

Table 3.1 | Novel or Modified Enzyme Activities in Bacteriophage-infected Cells

Polymerases
- DNA polymerase
  - New enzymes that recognize the virus DNA
  - Altered or new proteins associated with DNA synthesis (single-stranded binding proteins, primases, helicases) that shift the process from host synthesis to viral synthesis
- DNA-dependent RNA polymerase [transcriptase]
  - New enzymes
  - Altered recognition of regulatory sequences to shift from transcription of host genes to viral genes
- RNA-dependent RNA polymerase [replicase]

Nucleases
- DNases and RNases to degrade host nucleic acids
- Endonucleases to make specific cleavages in viral nucleic acids during maturation

Biosynthetic enzymes
- Make phage-specific precursors
- Convert nucleotides-triphosphates into nucleotide monophosphates
- Covalently modify nucleic acids after polymerization

WHAT STRUCTURAL FEATURES CHARACTERIZE THE GENOMES OF THE DOUBLE-STRANDED DNA BACTERIOPHAGES?

Only two forms of DNA genomes have been found in the viruses of prokaryotic cells: linear double-stranded DNA and circular single-stranded DNA. The circular single-stranded molecules are much smaller (in the range of 1.7 to 2.7×10^6 daltons) than the linear double-stranded genomes (in the range of 25 to 120×10^6 daltons). As might be expected, the complexity of the viral capsids parallels the size of their genomes, with the circular single-stranded DNA viruses having simple icosahedral or helical capsids, while the linear double-stranded DNA viruses enjoy more complicated structures. Complexity is not limited to the morphological level. Linear double-stranded DNA molecules exhibit a variety of differences in their ends that reflect their different solutions to the problems inherent in replication of linear DNA molecules.

The T-Series Coliphages

The most thoroughly studied bacteriophages over the past half century have been the *T-series* viruses of *E. coli*. These coliphages, despite their simple designations of T1 through T7 based on their order of isolation, actually include members of several different groups (Table 3.2). Based on their morphologies, immunological properties, and genomes, there are four groups: T1; T2, T4, and T6 (often called the *T-even* phages); T5; and T3 and T7. The *T-odd* phages have icosahedral heads with tails of different lengths, none of which is contractile. The T-even phages have very elaborate capsids with contractile tails. The size of the genomes of the T-series viruses varies by more than a factor of four, encoding from a little more than two dozen proteins in the small T3 and T5 phages to more than 135 in the T-even viruses. All of the T-series phages have **terminal redundancies** of their DNA molecules; that is, some

Table 3.2 | Features of the T-Series Viruses of *E. coli*

Phage	Morphology		DNA Length		Terminal Redundancy
	Head	**Tail**	**Length**	**Weight**	
1	70	10 × 150	16	25	2800 ± 530 (6.5 ± 1.2%)
5	65	10 × 180	38	75	10,000 (9%)
3,7	50	10 × 15	12	24	260–300 (0.7%)
2,4,6	85 × 110	25 × 110	52	120	3000–6000 (2%–4%)

Head size, tail dimensions (diameter and length), and DNA length are in nm, DNA molecular weight in millions of daltons, and terminal redundancies in numbers of base pairs and their percent of the total genome.

number of base pairs are repeated at both ends of the linear molecule. In the T-odd viruses, the terminal redundancy of a particular type of virus is always the same. In contrast, when many individual virions of a T-even phage are examined, each has a different set of bases repeated at the ends of its genome (Figure 3.6). In genetic crosses, this structural feature produces genetic maps for these viruses that are *circular,* even though their physical genomes are *linear* molecules! The genomes are therefore termed **circularly permuted.**

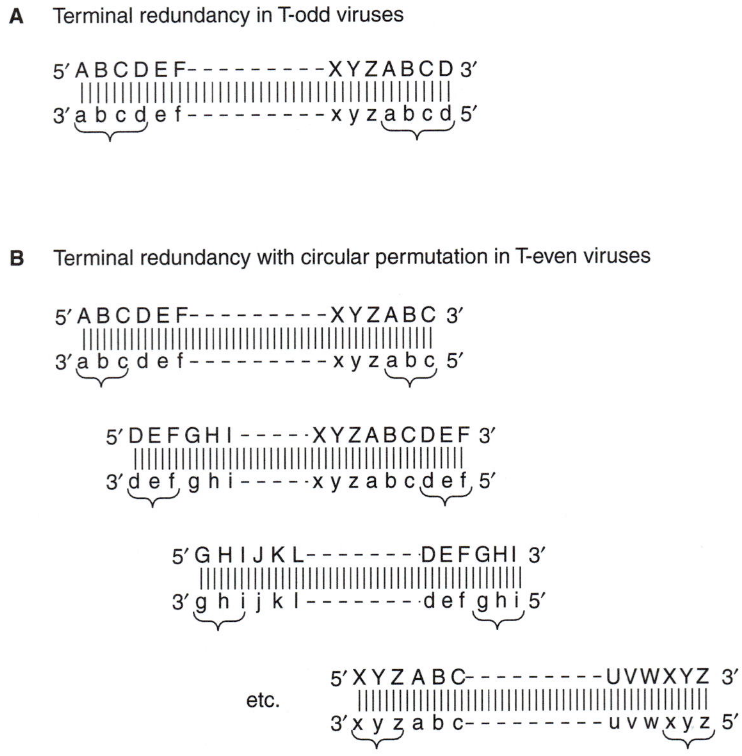

A Terminal redundancy in T-odd viruses

B Terminal redundancy with circular permutation in T-even viruses

etc.

FIGURE 3.6 Genetic structures at the ends of linear double-stranded DNA viral genomes. In a unique terminal redundancy **(A),** the same sequence occurs at both ends of all genomic molecules. Terminal redundancy with circular permutation occurs when each terminally redundant genomic molecule begins at a different location **(B).**

Since bacterial cells have all the necessary machinery to replicate double-stranded DNA, viruses that have this form of genome theoretically need only utilize their host's synthetic systems to generate new genomic DNA molecules. Immediately after penetration, however, even the smallest of the T-series viruses begins both to modify host processes to their own ends and to synthesize their own enzymes for replication and transcription of the viral genome. In most viruses the genes expressed before or during replication of the genome are termed **early genes,** while those expressed after replication (generally structural proteins of the virion and other proteins involved in maturation and release) are called **late genes.** Further subdivisions of these categories are frequently made. For example, early genes may be grouped into "immediate early" genes that are expressed before replication and "delayed early" genes that are expressed concurrently with replication.

HOW DO THE T-ODD COLIPHAGES EXPRESS AND REPLICATE THEIR GENOMES?

The Bacteriophage T7

The simplest of the T-odd viruses is T7. A map of the T7 genome is presented in Figure 3.7. During penetration one end of the viral genome always enters the host cell first, so the map is calibrated (in percent of the total length) starting from that end. As soon as enough of the genome enters the cytoplasm, the normal *E. coli* DNA-dependent RNA polymerase identifies three immediate early gene promoters located between positions 1% and 2% on the 3′ end of the coding strand of the genome. In all three cases, transcription ends at a termination sequence between 19% and 20%. The early gene products are all involved in establishing control of the host cell and preparing for replication and expression of the viral genome. For example, the probable role of the protein kinase (gp [for "gene product"] 0.7) is shutting down transcription of *E. coli* genes so that all further transcription and translation will be directed at viral synthesis. The viral RNA polymerase (gp 1.0) differs from the host cell enzyme with the same function in that it is composed of only a single peptide rather than having several subunits. Mutations inactivating this polymerase are lethal, since the virus can produce only early gene products and can neither replicate its genome nor make structural proteins. This indicates that the viral RNA polymerase rather than the host enzyme is responsible for transcription from the viral promoters of both delayed early genes and late genes. Figure 3.8 illustrates that different T7 proteins are synthesized at different times. When the proteins of infected cells are pulse-labeled and then separated on SDS-polyacrylamide gels, three classes are evident. Class I proteins are made immediately and cease being made after about 6 minutes. Class II proteins are synthesized from about 6 to 12 minutes after infection. These proteins are involved primarily in DNA synthesis but also include several nucleases that degrade the host cell DNA and thus make all the synthetic resources of the host available for viral processes. Class III protein synthesis makes viral structural proteins beginning at about 8 minutes and proceeding until the cell is lysed at about 20 minutes. Note also that cellular protein synthesis ceases very rapidly after infection.

Class I genes are transcribed by *E. coli* RNA polymerase, while the genes for both Class II and Class III proteins are transcribed by T7 RNA polymerase. How are the shifts between the Class I and Class II and then between the Class II and Class III genes made? Sequencing the T7 promoters has provided the answer. Class I promoters have sequences similar enough to those of its host to be recognized readily by the *E. coli* RNA polymerase, while

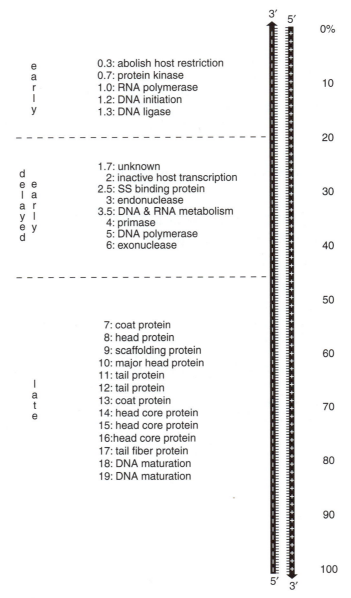

FIGURE 3.7 Genetic map of the bacteriophage T7. The genome is calibrated in percent of its length beginning at the end of the molecule that always enters the host cell first. Genes in T7 (and many other viruses) were numbered according to their position on the genome (e.g., gp 1, gp 2 ...). As additional genes between these regions were discovered, they were numbered on that basis (e.g., gp 2.5 is midway between gp 2 and gp 3).

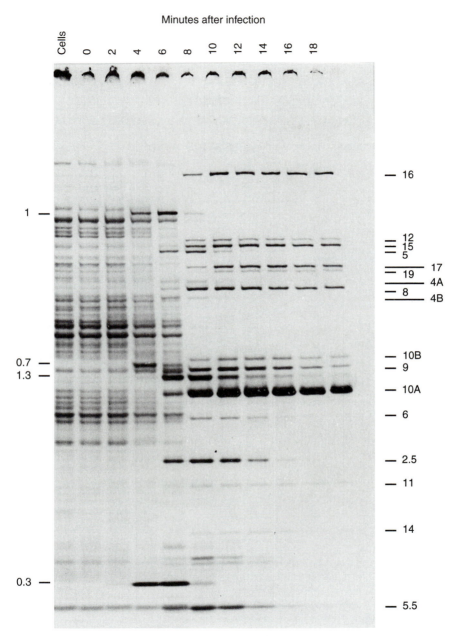

FIGURE 3.8 The temporal expression of T7 genes. *E. coli* cells growing at 30° were infected with T7. Proteins were pulse labeled with ^{35}S-methionine at 2-minute intervals. The cells were lysed, their proteins separated by SDS-polyacrylamide gel electrophoresis, and the proteins visualized by autoradiography. The proteins labeled by a 2-minute pulse of uninfected cells are shown on the left side of the figure. Labels beside the autoradiogram identify the viral proteins in specific bands. Host-specific protein synthesis ceases within the first few minutes following infection as the cell begins to synthesize viral-specific proteins. Class I genes (0.3 to 1.3) are expressed up to about 8 minutes post infection, Class II genes (1.7 to 6) from 6 to 12 minutes, and Class III genes (7 to 19) from 8 to 18 minutes.

Class I promoters: Recognized by *E. coli* RNA polymerase

Position	−40	−30	−20	−10	+1	+10

Consensus <u>A</u>   TTGAC   TATAAT

A1   <u>AAAAGAGTA</u><u>TTGAC</u>TTAAAGT CTAACCTATAG<u>gATAc</u>TTACAGCC<u>AT</u>ACGAGAGG

A2   <u>AAACAGGTA</u><u>TTGAC</u>AACA<u>R</u>GAAGTAACATGGAGT<u>Aag</u>ATACAAATC<u>GC</u>TAGGTAAC

A3   <u>ACAAAACGG</u><u>TTGAC</u>AACATGA AGTAAACACGG<u>TACGAT</u>GTACCAC<u>AT</u>GAAACGAC

Class II promoters: Recognized by T7 RNA polymerase

Position	−40	−30	−20	−10	+1	+10

Consensus   TAATAC−

Φ1.1A   AACGCCAAAT <u>cAATACGACTCACTATAGaGgGACA</u>

Φ1.1B   TTCTTCCGGT <u>TAATACGACTCACTATAGGagGACC</u>

Φ1.3   GGACTG−   <u>TAATACGACTCAgTATAGGGAGAAT</u>

Φ1.5   GAAG   <u>TAATACGACTCACTAaAGGagGtAC</u>

Φ1.6   AGTTAACTGG <u>TAATACGACTCACTAaAGGagacAC</u>

Φ2.5   TGGTCACGCT <u>TAATACGACTCACTATtaGGgaAGA</u>

Φ3.8   AGCACC−   <u>TAATtgaACTCACTAaAGGGAGACC</u>

Φ4c   GAAG   <u>cAATcCGACTCACTAaAGaGAGAGA</u>

Φ4.3   CGTGGATAAT <u>TAATACGACTCACTAaAGGagacAC</u>

Φ4.7   CCGACTGAGS <u>cTATtCGACTCACTATAGGagatAT</u>

Class III promoters: Recognized by T7 RNA polymerase

Position	−40	−30	−20	−10	+1	+10

Φ6.5   GTCCCTAAAT <u>TAATACGACTCACTATAGGGAGATA</u>

Φ9   GCCGGGAATT <u>TAATACGACTCACTATAGGGAGACC</u>

Φ10   ACTTCGAAAT <u>TAATACGACTCACTATAGGGAGACC</u>

Φ13   GGCTCGAAAT <u>TAATACGACTCACTATAGGGAGAAC</u>

Φ17   GCGTAGGAAA <u>TAATACGACTCACTATAGGGAGAGG</u>

FIGURE 3.9 The promoter sequences of T7 bacteriophage genes. Consensus sequences for Class I and Class II/III promoters are underlined. Bases that differ from these consensus sequences are written in lowercase letters. Class I promoters contain consensus sequences in the −10 and −35 regions usually recognized by host cell RNA polymerase. The Class II/III promoters are missing those sequences but exhibit a 23-base-long consensus sequence between −15 and +8, which is recognized by the viral RNA polymerase.

Class II and III promoter sequences (Figure 3.9) are so different from those of *E. coli* that the host enzyme fails to recognize them. T7 polymerase, however, can recognize only those sequences. The shift from Class II to Class III genes involves only the viral polymerase. As noted earlier, the T7 DNA molecule is always injected into the host cell from the same end. Since the promoters for the Class II genes are less like the T7 consensus sequence than those for Class III genes, which are perfect matches, they are probably somewhat weaker.

It is likely, therefore, that synthesis begins at the Class II promoters when they enter but shifts to the stronger Class III promoters when they become available.

Replication of the T7 genome is carried out by a viral DNA polymerase (gp 5). In addition to this enzyme, a multifunctional viral protein (gp 4) with primase, nuclease, and helicase activities, single-strand DNA binding proteins (either host or gp 2.5), and the bacterial protein thioredoxin are also involved. DNA synthesis requires RNA priming activity by a phage enzyme and proceeds bidirectionally from the origin located between genes 2 and 3 (about map position 15%). Replication of any linear DNA molecule leaves a stretch of uncopied bases at the 3′ ends of the template molecules. An RNA primer complementary to the 3′ end of the template strand cannot be removed and replaced by DNA, so the usual round of priming, elongation, and replacement of the primer does not occur at that position on the lagging strand. The result is the production of two double-stranded molecules with single-strand tails (Figure 3.10).

Note that in molecules with terminal redundancies like the T7 genome, the unpaired bases of one new molecule are complementary to those of the other new molecule. The two molecules are therefore able to recombine with one another, or to form **concatamers** *(con* is Latin for "together" + *catena* is Latin for "chain" + *mer* is Greek for "part"), to form a two-genome-long molecule held together by hydrogen-bonded base pairs. The nicks in the phosphodiester backbones of the two chains in the concatamer are closed by DNA ligase to complete the process. This series of steps can be repeated to form concatamers four genomes long, and so forth. During the process of maturation (to be described in Chapter 6), the concatamers will be processed into individual genome-length molecules.

The Bacteriophage T5

The coliphage T5 is in a class by itself in several respects. It is larger than T3 and T7, with about three times more DNA. The arrangement of this DNA is unusual: Denaturation of the genome results in the formation of one full-length strand and six short strands, so it appears that one strand is nicked in five places and is held to the other unnicked strand only by base-pair hydrogen bonds. If the nicks are closed by DNA ligase, however, there is no change in the activity of the virus, so the nicks appear to be unusual but not important. Although the details of T5 replication have not been described, the genome has a large terminal redundancy (9%), which suggests that concatamerization may play a role in the process.

A second unusual feature of the replication cycle of T5 is that penetration of the genome is a two-part process. Penetration begins by entry of about 8% of the length of the genome, including three viral early genes that must be transcribed and translated before the rest of the genome is drawn into the cell.

HOW DO THE T-EVEN COLIPHAGES EXPRESS AND REPLICATE THEIR GENOMES?

The T-even coliphages, particularly T4, are among the best understood of all viruses since they have been extensively studied for more than 50 years. The T-even phages have enormous genomes encoding more than 135 gene products and a corresponding complexity in the architectures in their virions and in the replication and expression of their genomes. Figure 3.11 presents an abbreviated genetic map of T4; the circular permutation of the genome results

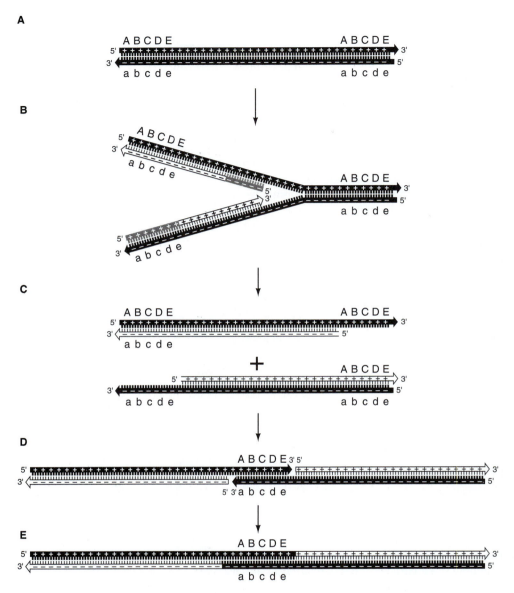

FIGURE 3.10 Formation of concatamers during replication of T7 bacteriophage. **A,** The terminal redundancy of the genome is represented with uppercase and lowercase letters A through E, indicating bases complementary to each other. **B,** Bidirectional DNA synthesis begins at an origin near the left side of the genome. **C,** Synthesis of the 5′ ends of both daughter strands cannot be completed since an Okazaki fragment at that location cannot be processed to replace its RNA primer with DNA. **D,** The complementary bases at the 3′ ends of the template strands base pair to form a double-length molecule with a nick in each strand. **E,** DNA ligase closes the nicks.

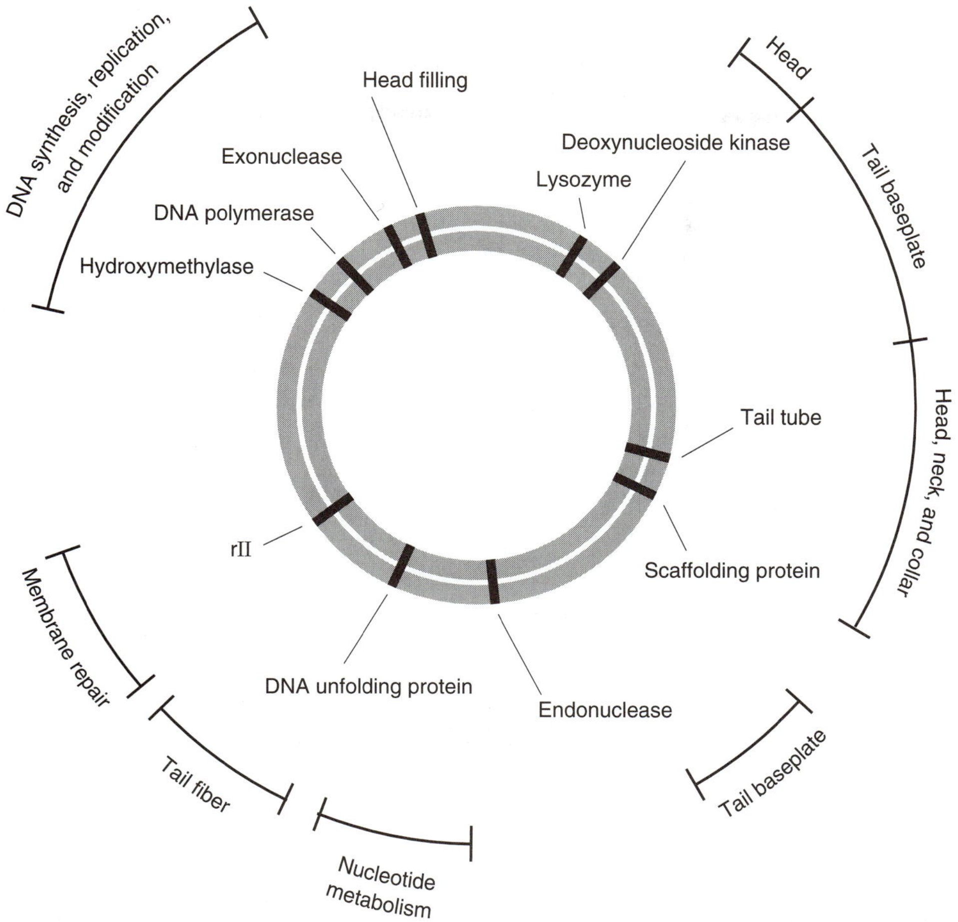

FIGURE 3.11 An abbreviated genetic map of T4 bacteriophage showing the functions of clusters of genes.

in a circular map rather than the linear one expected from the physical structure of T4's DNA. Nearly half the genes described thus far in the T4 genome are nonstructural, coding for degradative enzymes, components of biosynthetic pathways and regulatory proteins, and even tRNA species. Many of these genes are located near one another on one side of the map, while most of the genes for the viral capsid are grouped on the other side of the map. Although not evident at this level of map detail, these viruses appear to be large enough that they can expend their coding power in duplicating numerous host cell enzymatic functions, particularly nucleases, since fully 44% of the genes in T4 code for nonessential metabolic functions.

A striking biochemical feature of the T-even phage genomes is the quantitative replacement of the typical DNA base cytosine with a modified base, 5-hydroxymethylcytosine (HMC)

HMC forms normal base pairs with guanine, so it does not alter the template functions of the viral genome, but its presence in viral DNA serves as a convenient way for viral enzymes to distinguish between host and viral nucleic acids. Proteins from dozens of phage genes are involved in converting the host cell's cytosine biosynthetic pathway to the production of HMC, as well as in degrading existing pools of dCTP and dCDP to dCMP. Other enzymes digest host DNA containing unmodified cytosines into mononucleotides. All of these processes effectively prevent any further host DNA synthesis or the transcription and translation of host genes, leaving all the host cell's biosynthetic machinery available for viral synthesis. After HMC is incorporated into phage DNA, it is further modified by addition of two glucose molecules via the hydroxyl group on the 5-methyl substituent. This glycosylation prevents degradation of T4 DNA by an *E. coli* endonuclease that attacks certain sequences containing HMC.

The reproductive cycle of T4 takes about 20 to 25 minutes at 30°, with DNA synthesis beginning about 5 minutes after infection. Viral genes appear to be expressed in four groups: immediate-early (approximately 0 to 2 minutes post infection), delayed-early (2 to 5 minutes), quasi-late (5 to 12 minutes), and late (12 minutes and after). The mechanism T4 uses to shift from one set to the next is a complicated cascade of alterations to *E. coli's* RNA polymerase (Figure 3.12).

Immediate-early genes are transcribed by native RNA polymerase in its $\alpha\alpha\beta\beta'\sigma$ configuration. Immediately upon entry of the T4 genome, however, a protein (the *alt* gene product) carried into the host with the viral DNA molecule begins the sequential modification of RNA polymerase. The first step, called *alteration,* is the transfer of an ADP-ribosyl group from NAD to the arginyl residue 265 of one of the two α subunits of the *E. coli* RNA polymerase. The effect of ADP-ribosylation of the α subunit is to reduce the affinity of the σ factor for binding to the core enzyme. This probably plays a role both in shutting off host cell transcription, which requires recognition of promoter sequences by σ, and in the efficiency of recognition of early viral promoters. About a minute after infection the second α peptide is ADP-ribosylated by the T4 gene *mod* protein in a process termed *modification.* Modification further reduces the affinity of σ for the core enzyme, and results in the selective shutoff of some viral immediate-early genes that require the presence of σ for their recognition.

The remaining alterations to RNA polymerase involve noncovalent binding of T4 proteins to core enzyme rather than covalent modifications of its subunit peptides. By 5 minutes into the infective cycle, two peptides, called the 10K and 15K proteins for their molecular weights, become associated with the modified core enzyme. The amount of synthesis from the delayed-early promoters decreases, while that from the quasi-late promoters increases. At the same time (about 5 minutes post infection) DNA synthesis begins. Late gene transcription is

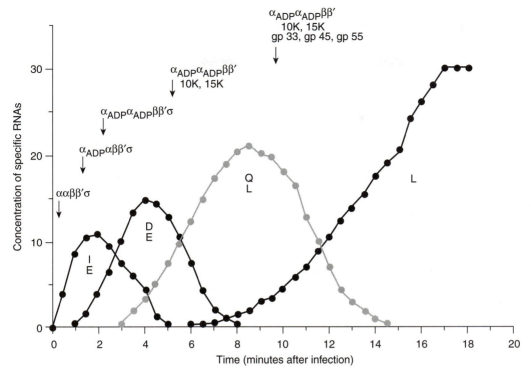

FIGURE 3.12 Sequential modification of *E. coli* RNA polymerase by T4 bacteriophage gene products alters transcription of viral genes. The original polymerase ($\alpha\alpha\beta\beta'\sigma$) is first modified by addition of ADP to the α peptides and then by the association of various T4 proteins (10*K*, 15*K*, etc.) to shift the expression of viral genes from immediate-early genes $\longrightarrow$ delayed-early genes $\longrightarrow$ quasi-late genes $\longrightarrow$ late genes.

absolutely dependent on the presence of newly synthesized DNA, which is probably in a looser configuration that is amenable to transcription. Three peptides, the products of T4 genes 33, 45, and 55, are associated with RNA polymerase during transcription of the late genes. The exact role of these peptides is unclear, but they seem to function as positive regulators (gene activators) permitting the recognition of late promoters. The late gene promoters are totally unlike the usual prokaryotic promoter, having long AT-rich stretches and none of the usual consensus sequences characteristic of promoters. In addition to their roles as positive regulators, gp 33 and gp 55 also seem to be negative regulators of some early transcriptional events and appear to be involved in some aspects of DNA replication as well.

The details of sequential modification of host cell RNA polymerase to regulate the expression of a viral genome are especially complex in T4, but many bacteriophages use simpler variations on the same theme. The phage SPO1 of the gram-positive organism *Bacillus subtilis,* for example, has three sets of promoters. The first are recognized by the native *B. subtilis* RNA polymerase. One of the early gene products of SPO1 is a recognition factor peptide called gp 28 that then replaces the normal σ factor associated with the core RNA polymerase, leading to transcription of the second set of promoters. The products of genes 33 and 34 from this transcriptional event in turn replace gp 28 on the core polymerase and permit transcription of the final set of phage genes.

The replication of the T4 genome is also an extremely complex phenomenon. While numerous phage gene products appear to be involved in the process, many of them are not essential since they largely duplicate functions that can be provided by the host cell. The proteins encoded by seven phage genes (32, 41, 43, 44, 45, 61, and 62) have been demonstrated in *in vitro* replication experiments to be sufficient to reproduce all the events that occur *in vivo*. These seven proteins, whose functions are presented in Table 3.3, form a multimeric complex with unwinding, priming, and polymerizing activities. The phage DNA polymerase is a very large protein with the same 3′ and 5′ exonuclease activities, and therefore proofreading and nick-translating functions, found in the *E. coli* DNA polymerases.

Table 3.3 | Proteins Forming the T4 DNA Replication Complex

Gene	Protein	Function in Replication
32	DNA-binding protein	Unwinding DNA
41	Primase	RNA primer synthesis
43	DNA polymerase	DNA synthesis
44		Polymerase accessory protein
45		Polymerase accessory protein
61	DNA helicase	Open helix for RNA primer synthesis
62		Polymerase accessory protein

Replication of the linear T4 DNA molecules with their terminal redundancies leads to the formation of concatamers by recombinational events between the redundant ends of daughter molecules, as described previously for the phage T7. In addition to recombination at the ends of daughter molecules, however, internal recombinational events are common, so it is possible to isolate enormous interconnected molecules of DNA from cells actively replicating viral genomes. The cleavage of these molecules to form sets of single DNA molecules whose different terminal redundancies produce the circular permutation of the T4 genome will be presented in Chapter 6.

HOW DOES LAMBDA VIRUS EXPRESS AND REPLICATE ITS GENOME?

The coliphage λ has been as intensively studied as T4, with these studies yielding a veritable "who's who" list of significant discoveries in molecular biology. A number of features of DNA synthesis, including bidirectional replication and the activities of DNA gyrase and DNA ligase, as well as the nature of genetic recombination, were first described in experiments with λ virus. Study of λ has also been an important vehicle for understanding the roles of DNA binding proteins in the regulation of gene activities.

λ is about the size of T7, both in its virion (an icosahedral head about 55 nm in diameter and a noncontractile tail about 15×135 nm long), and in its genome of linear double-stranded DNA (weighing 31×10^6 daltons). Rather than terminal redundancies, the 5' ends of each strand consist of 12 unpaired bases that are called **cohesive ends** (or **sticky ends**) since the sequences on opposite strands are complementary and can therefore base pair with one another (Figure 3.13). Such pairing occurs within 2 minutes after penetration to convert a linear molecule into a circular one with two nicks. The single-strand nicks are closed

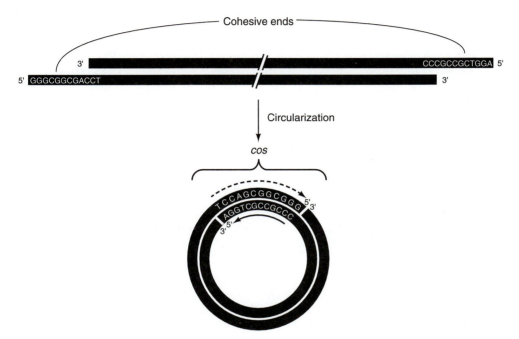

FIGURE 3.13 Twelve complementary unpaired bases at the 5' ends of the λ bacteriophage genome permit conversion of the linear genome into a circular molecule. Following circularization, DNA ligase closes the nicks at the ends of the *cos* site.

by DNA ligase and the entire molecule supercoiled by DNA gyrase within 3 to 6 minutes. The biologically active form of λ is thus *circular* rather than linear. For that reason the genetic map for λ (Figure 3.14) is usually presented as circular, with the base-paired cohesive ends labelled as the *cos* site. Forty-six genes have been identified in λ, and since the entire DNA molecule has been sequenced, it is unlikely that many more will be found. The clustering of genes by function is clearly evident in the map.

The bacteriophages discussed thus far all enter host cells, replicate themselves and then disrupt or lyse their hosts. This type of replication cycle is therefore termed a **lytic cycle,** and the viruses are identified as lytic viruses. Viruses like λ have a lytic cycle to produce more progeny viruses, but in addition they have the ability to recombine their genomes with those of their host cells in a way that leads to insertion of the phage genome into the bacterial

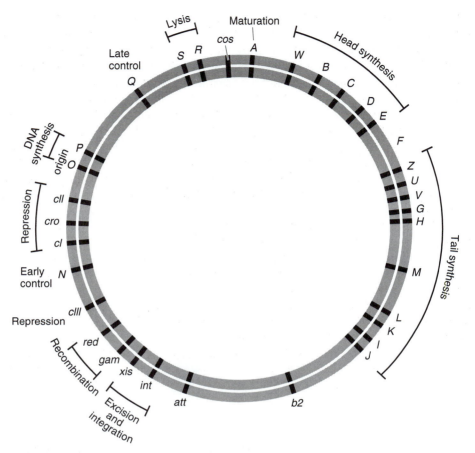

FIGURE 3.14 An abbreviated genetic map of λ bacteriophage showing the functions of various clusters of genes.

chromosome. In this integrated state, the virus is replicated along with the chromosome each time the cell divides and is thus passed from generation to generation. Only the few viral genes necessary to maintain the integrated state are expressed as long as the host cell is healthy. When conditions such as damage to the host cell by UV irradiation warrant it, however, the viral genome shifts from maintaining its integrated state to a lytic cycle, leading to the production of progeny viruses and eventual lysis of the host cell. The observation that apparently uninfected cells could be induced to lyse, suddenly producing a virus, led to those viruses being termed **lysogenic** (*lysis* is Greek for "dissolution" + *gennan,* Greek for "to produce"). Another term applied to the same viruses that emphasizes the passive, nonharmful aspect of the inserted state is **temperate** (*temperatus,* which is Latin for "controlled").

λ virus makes the "choice" between the lytic cycle or lysogeny immediately after infection begins. Only the lytic cycle will be described here. The details of how the choice between the lytic cycle and lysogeny is made, and how lysogeny is maintained, will be discussed in Chapter 8, which considers the variety of ways in which the viruses of both prokaryotes and eukaryotes can establish nonlytic relationships with their host cells.

As with all viruses, an important feature of the λ lytic cycle is regulation of the time at which groups of genes (early, delayed-early, and late) are transcribed. T7 did this by synthesizing a novel RNA polymerase, which recognized special viral promoters, while T4 sequentially modified the host cell RNA polymerase to change its recognition functions. In both cases, *initiation* of transcription was the regulatory target. Like T4, λ phage also modifies the activity of its host's RNA polymerase, but it uses the opposite strategy of altering the other end of the process of transcription, *termination*. In addition, λ also employs repressor proteins to block transcription, a very common regulatory device in its prokaryotic hosts.

Figure 3.15 presents a simplified version of the events that occur during the first 10 minutes of the 45-minute λ lytic cycle (genes and gene products solely involved in establishing lysogeny are omitted). In the first stage of the lytic cycle, three promoters are recognized by *E. coli*'s RNA polymerase: P_{L1}, P_R, and $P_{R'}$. Transcription from P_{L1} ends at T_{L1} and yields only a single protein called N. Transcription from P_R ends at a weak termination sequence, T_{R1}, most of the time, again giving rise to only one protein called Cro (for *c*ontrol of *r*epressor and *o*ther things). Other times transcription continues on to the strong terminator, T_{R2}. In that case, small amounts of two proteins, O and P, required for DNA synthesis are made in addition to the Cro protein. There is also synthesis of a tiny fragment of nonfunctional RNA from a third promoter, $P_{R'}$, to a nearby terminator, T_{R4}.

The immediate-early proteins Cro and N work in opposition to each other. Cro operates as a repressor protein binding to P_{L1} and P_R to stop further transcription from those locations. The N protein, in contrast, is an **antiterminator** protein that acts in concert with an *E. coli* regulatory protein called nusA to modify the activity of RNA polymerase. The modified enzyme is able to read through the termination sequences of the early genes and thus begin transcription of the delayed early genes.

With the N protein facilitating antitermination at T_{L1}, transcription continues on to a set of delayed-early genes that genetic studies with deletion mutants indicate are not essential for λ reproduction. On the P_R side, N permits transcription through the T_{R1} and T_{R2} termination sequences and on to T_{R3}, a termination sequence that is not sensitive to N. Transcription to T_{R3} allows synthesis of a second antiterminator protein called Q. In addition, sufficient quantities of O and P proteins are produced to permit the initiation of active DNA replication. While these delayed-early gene products are being made, the amount of the Cro protein gradually becomes high enough to block further transcription from P_L, and slightly later, at P_R. Synthesis of all immediate-early and delayed-early gene products is therefore inhibited.

The target of the second antiterminator protein Q is the T_{R4} sequence. In the presence of Q, transcription can continue from $P_{R''}$ through T_{R4} and on to all the late genes whose protein products are required for production of new virions and their release from the cell. This transcription is delayed somewhat by the action of a delayed-early protein CII, so that synthesis of structural gene products lags behind DNA replication.

DNA replication in λ occurs in two stages, using two different mechanisms. The initial replication of the circularized genome is during the delayed-early stage of transcription and requires the O and P gene products. Bidirectional DNA synthesis from a single origin produces the theta structures typical of replication of a circular DNA molecule. About 15 minutes after infection, DNA synthesis shifts to a form of *asymmetrical* replication called **rolling circle replication.** Figure 3.16 illustrates that rolling circle replication produces a **sigma intermediate,** so called from its resemblance to the Greek letter σ. A nick in one strand of the double-stranded circular molecule, probably at or near the origin, permits the free $5'$ end of the nicked strand to be peeled away from its complementary strand. That leaves the

A: T_0 - T_3 min: Immediate-early gene expression

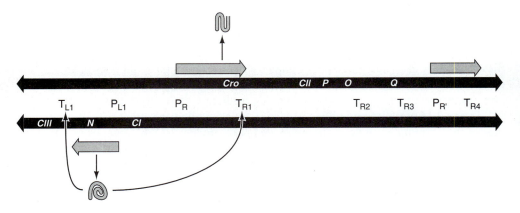

B: T_3 - T_9 min: Delayed-early gene expression

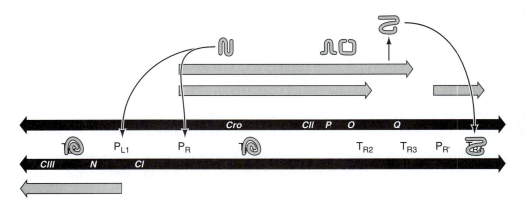

C: T_9 - T_{45} min: Late gene expression

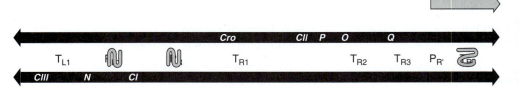

FIGURE 3.15 Regulation of gene expression in the lytic cycle of λ bacteriophage. **A,** Immediate-early gene expression allows the synthesis of two proteins: the antiterminator N and the repressor protein Cro. **B,** The N protein permits transcription to continue past the T_{R1} and T_{L1} sites and into the delayed-early genes, two of which (O and P) are involved in DNA replication. A second antiterminator, Q, is also synthesized. **C,** The binding of the Q protein to T_{R4} allows transcription of the viral late genes. The Cro protein blocks P_{R1} and P_{L1} so that both immediate-early and delayed-early gene expression is blocked.

bases of the unnicked strand open to serve as template for continuous elongation of the 3′ end of the nicked strand as the 5′ end is further peeled away. The complement of the peeled away DNA strand is synthesized by the usual formation of Okazaki fragments. Rolling circle replication, therefore, leads to the creation of concatamers by continuous DNA synthesis, rather than by recombination events between terminal redundancies as seen in the T-series viruses. Site-specific cleavage of the sigma intermediate during maturation creates the linear genomes with their cohesive ends.

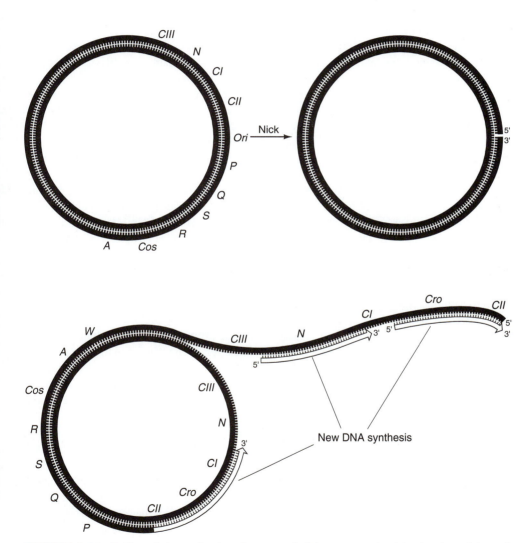

FIGURE 3.16 Rolling circle replication. One strand of the genome is nicked at the origin and its 5′ end peeled away from its unnicked complement. The 3′ end can then be elongated continuously while Okasaki fragments are synthesized using the free 5′ strand as a template. The replicating structure resembles the Greek letter sigma (σ).

HOW DO SINGLE-STRANDED DNA BACTERIOPHAGES EXPRESS AND REPLICATE THEIR GENOMES?

With this group of bacteriophages we begin consideration of viruses with genomic architectures not found in either prokaryotic or eukaryotic cells. As the viruses are apt to do, the coliphage ϕX174 (ϕ = phi) has provided a number of surprises since its first description in the inaugural issue of the *Journal of Molecular Biology* in 1959. First of all, the DNA genome was found to have nucleotides in the ratio of 1A: 1.33T: 0.98G: 0.75C. Robert Sinsheimer, the discoverer of these data, concluded that the genome of ϕX174 had to be *single-stranded* DNA rather than the double-stranded molecules that had become accepted as the rule. He said this was like finding "a unicorn in the ruminant section of the zoo!" The next surprise was that the single-stranded DNA molecule was *circular* rather than linear like the other bacteriophages known at that time. Further wonders were to come.

ϕX174 is a very small virus. The virion is composed of 60 copies of a single protein making up the faces of an icosahedron, with two other proteins forming short spikes at the vertices. Its genome, only 5387 bases long (1.7×10^6 daltons), was the first DNA molecule to be completely sequenced (in 1977), earning Frederick Sanger his second Nobel Prize (his first was for developing a sequencing method for proteins and determining the sequence of amino acids in insulin).

This sequence helped explain several paradoxes concerning the size of the ϕX174 genome versus the number of proteins it appeared to encode. Genetic studies indicated that at least eight different proteins were produced from the ϕX174 genome. That number is now known to be even higher, with a total of 11 proteins identified. Even if every single one of its 5387 nucleotides were used, however, only 1795 triplets, and hence amino acids, could be encoded. This is equivalent to only four or five average-sized (400 amino acids) proteins, about half the number required. Genetic mapping studies raised additional questions since the known genes did not seem to form the end-to-end string expected from an unbranched DNA molecule.

The DNA sequence revealed how ϕX174 did the seemingly impossible. The sequence indicated that the genome was transcribed from three promoters, each of which yielded a polycistronic mRNA for a set of functionally related genes, the usual situation in prokaryotic viruses. The surprises came from the two different strategies by which those mRNA were then translated to produce more than one protein from the *same* sequence of bases. In the first case, two proteins called A (59 K daltons) and A* (37 K daltons) are read from the same section of mRNA by beginning translation at *two different initiation sequences*. Since both A and A* are read in the same reading frame and end at the same termination codon, A* is a truncated version of A made at the level of translation rather than by the common posttranslational method of cleaving a larger protein to form a smaller one.

More striking than this was the synthesis of the B, K, and E proteins. The information for the B protein is entirely contained within the sequence of bases that also encoded both the A and A* proteins—a viral three-for-one sale! The B protein is not merely a shorter version of A and/or A*, however. Instead, the initiation codon for the B gene allows the common set of bases to be translated in a *different reading frame* from the A/A* proteins. Similarly, the E protein is translated from the same sequence as the D protein, but using a shifted reading frame. The K protein gene likewise overlaps the end of the code for the A and the beginning of the C gene, and is translated in a different reading frame than either of them. Figure 3.17 presents the sequence for a part of the ϕX174 genome encoding portions of the A, B, K, and C proteins to illustrate how **overlapping genes** encode their information so compactly.

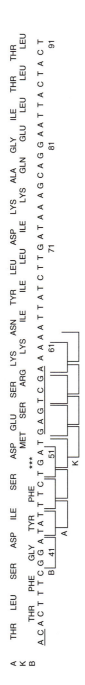

FIGURE 3.17 The DNA sequence of the bacteriophage φX174 genes B and K with the portions of the A/A* and C genes that they overlap. The reading frames at the beginnings and ends of genes have been indicated by brackets. Note especially the economy of base use around residues 45–55 and 130–135.

Figure 3.18 illustrates the genetic map of φX174 and gives the functions for each gene. Note that nonoverlapping genes depicted on the left side of the map encode structural components of the virion that are required in large amounts during maturation. Although the exact functions of some of the overlapping genes are not yet known, it is clear that they are either enzymatic or regulatory proteins that are needed in limited amounts during the viral replication cycle. While overlapping genes allow information to be stored with maximum efficiency, translation of the mRNAs from these genes is probably inefficient since ribosomes in different reading frames would have to compete with one another for the same sequence. Overlapping genes also pose an obvious genetic risk since a single mutation has the potential to alter two different proteins, but the redundancy of the genetic code probably reduces that risk somewhat. They also place considerable restraints on evolution, since a single base change would have to create beneficial changes in two proteins concurrently (or at least a large enough positive effect in one to offset a negative effect in the other). Despite these risks, overlapping genes have been found not only in bacteriophages but in some small animal viruses as well.

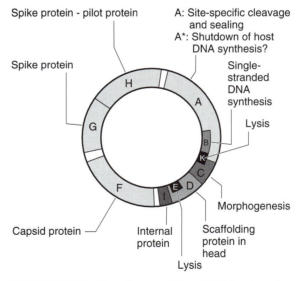

FIGURE 3.18 Genetic map of the φX174 bacteriophage. Overlapping reading frames are shown by smaller wedges within the larger wedges. Note that the viral structural proteins (left side of map) are not encoded by overlapping reading frames.

DNA replication is required before RNA transcription can occur in φX174 because the single-strand genome is a positive strand; that is, its sequence is the same as that of the viral mRNA. Figure 3.19 illustrates the φX174 replication cycle. The first event following penetration is binding of *E. coli* single-strand binding proteins to the genomic DNA molecule, followed by synthesis of a complementary negative strand of DNA. This synthesis involves first a priming reaction by host primase, then elongation by DNA polymerase III, and finally removal and replacement of the primer by DNA polymerase I when synthesis has encircled the tem-

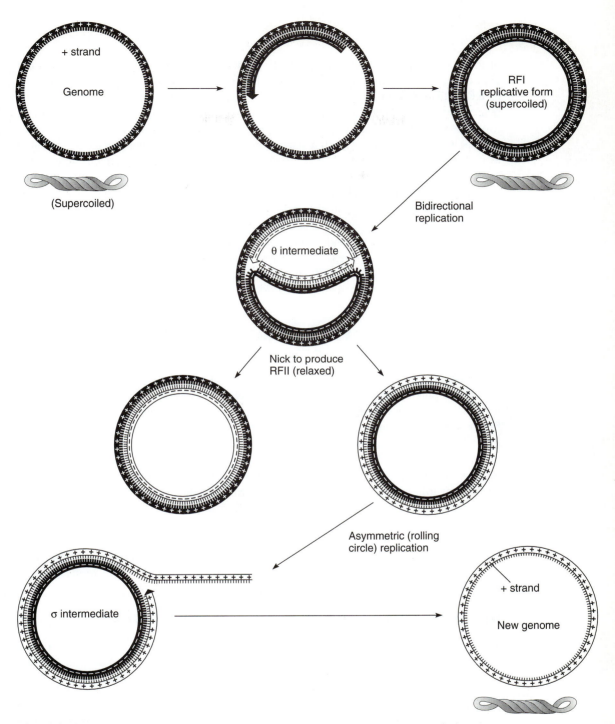

FIGURE 3.19 Replication of φX174 occurs in three stages. First, the single-stranded genome is converted into a supercoiled double-stranded replicative form (*RF*), which is then replicated bidirectionally. When a sufficient number of RFIs accumulate, they are nicked in one strand to convert them into relaxed RFII molecules, which are then replicated asymmetrically by the rolling circle mechanism.

plate molecule. DNA ligase then closes the nick between the ends of the negative strand to generate a complete double-stranded DNA molecule. This structure is called a **replicative form,** or **RF,** to distinguish it from the single-stranded genomic DNA molecule. The RF may be relaxed; that is, without any supercoiling, or topoisomerase II (DNA gyrase) may introduce positive supercoils. The relaxed form is sometimes called RFII and the supercoiled form, RFI.

As was seen previously with λ phage, φX174 uses two different modes of replication of its RFI. Early replication is bidirectional, producing the theta structure characteristic of that process in a circular molecule. After a number of RFIs accumulate to provide sufficient negative-strand templates, replication shifts to the rolling circle mode. Gene product A produces a site-specific nick in the positive strand that converts the RFI into a relaxed RFII. The A protein remains associated with the 5′ end of the nicked strand as it is peeled away from the negative-strand template. During maturation the A protein is responsible for nicking the positive strand to generate a separate genome from the rolling circle concatamer, and then for ligating that linear molecule into a circular one. A phage protein is required for that ligation since host cell enzymes work only on double-stranded molecules.

HOW DO RNA BACTERIOPHAGES REPLICATE AND EXPRESS THEIR GENOMES?

The RNA bacteriophages of *E. coli* are the smallest and simplest autonomous viruses known, yet like the single-stranded DNA phage, they have proven capable of surprising molecular ingenuity in solving the problems entailed in replicating and expressing their genomes. The first problem is inherent in all viruses having RNA as their genetic material: No host cell normally has the enzymes necessary to make RNA from an RNA template. All RNA viruses, therefore, *must* carry the genetic information for the synthesis of an *RNA-dependent RNA polymerase* (usually called simply RNA replicase). The other problems are common to all viruses, regardless of their type of genetic material—producing sufficient amounts of each protein at the proper time during the viral reproductive cycle.

All but one of the RNA bacteriophages have positive-strand RNA as their genome, so the genomes serve directly as mRNA for protein synthesis as well as template for new genome synthesis. All RNA bacteriophages also have linear genomes. These commonalities allow several generalizations to be made regarding the processes of replication and expression of the genomes.

General Features of RNA Bacteriophage Replication

When viral RNA molecules are purified from RNA-phage infected cells and then separated into size classes by centrifugation or electrophoresis, three groups of molecules are seen (Figure 3.20). Two of the peaks seem to represent populations of homogeneous size, one equivalent in size to the genome itself and the other equivalent in size to the genome and its complementary strand base-paired to each other. As with the DNA viruses, this double-stranded molecule is called a *replicative form,* or RF. The remaining group of molecules appears to be larger in aggregate molecular weight than either the genome or the RF, and to be more heterogeneous as well. A gentle treatment of this group of molecules with an RNase that will degrade only single-stranded RNA reduces the size of the molecules to that

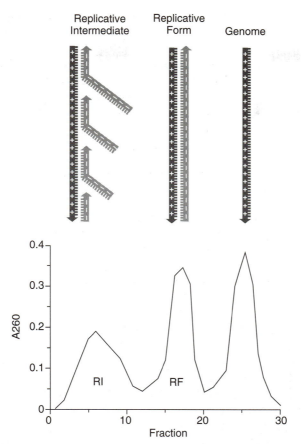

FIGURE 3.20 Types of viral RNAs isolated from cells infected with RNA bacteriophage. The graph illustrates the type of result obtained if RNAs are separated by density-gradient centrifugation. The structures corresponding to each class are drawn above the peaks. RF = replicative form (a genome and one complete antigenome); RI = replicative intermediate (a genome with several antigenomes being synthesized concurrently).

of the RF, while a more potent treatment with RNase completely degrades the material. These data suggest that the molecules in the third peak are a collection of template molecules with numerous daughter molecules in the process of synthesis. This arrangement is termed a **replicative intermediate,** or **RI.** The daughter molecules are base paired near their sites of synthesis but have "tails" free in the environment. Gentle RNase treatment digests the tails and produces essentially the double-stranded RF form, while more stringent treatment digests the entire molecule with its many nicks and gaps.

These data suggest that RNA replication occurs in two stages. The positive-strand genome is first replicated to generate negative RNA strands (Figure 3.21, *A*). During this

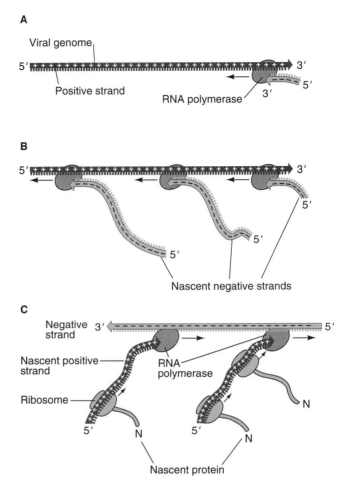

A

Viral genome

5′ ———————————————— 3′

Positive strand

RNA polymerase 3′
5′

B

5′ ———————————————— 3′

5′

5′

5′

Nascent negative strands

C

Negative 3′ ←———————————— 5′
strand

Nascent positive
strand

RNA
polymerase

Ribosome

5′ N 5′ N N

Nascent protein

FIGURE 3.21 Stages in the replication of RNA bacteriophage genomes. The positive-strand genome **A,** is transcribed numerous times in the RI **B.** Each new negative-strand antigenome is then transcribed to form new genomes (mRNAs), which are immediately translated **C.**

process, an RI form would occur as several replicases move down the genome, each displacing the base-paired RNA ahead of itself. It is likely that the RF form does not occur naturally during this process, and may, in fact, be an artifact produced by the deproteination steps used in nucleic acid purification schemes. The second stage of replication involves numerous RIs in which each new negative-strand RNA molecule serves as template for synthesis of many positive-strand RNAs (Figure 3.21, *B*).

A significant feature of the synthesis of new positive strands is that they are also new mRNAs. Replication is therefore tightly coupled to the process of translation since there is no physical compartmentalization of these processes in prokaryotic cells. As soon as the first initiation codon on each new genome is available, ribosomes bind to it and begin the synthesis of viral proteins (Figure 3.21, *C*).

The Genomes of the RNA Bacteriophages Qβ and MS2

The RNA bacteriophages of *E. coli* all have small (25 nm in diameter) icosahedral capsids composed of 180 copies of a single species of peptide called in each case the **coat protein.** Although the coat proteins of Qβ and MS2 are of very similar lengths (131 and 129 amino acids, respectively), their amino acid sequences show that they are only somewhat related, having about a quarter of their residues the same. In addition to the coat protein, the virion also contains a single copy of a peptide called the **A protein** associated with the RNA genome. The A protein is named because of its probable involvement in the attachment of the virion and then the penetration of the genome into a host cell.

The genomes of both Qβ (4700 bases) and MS2 (3569 bases) have been sequenced. Since these genomes can direct the synthesis of proteins in *in vitro* translation assays, they are unquestionably positive-strand or messenger RNAs. Analysis of their sequences both confirms this and allows explanations of several genetic puzzles as well.

The major puzzle was like that seen in the single-stranded DNA phages: How can so many peptides be synthesized from such a small molecule? In addition to the two proteins that are found in the capsid, the genome *must* encode the information for an RNA replicase. In MS2 the A protein is 393 amino acids long, the coat protein is 129 amino acids long, and the replicase, as might be expected of an enzyme involved in nucleic acid synthesis, is 544 amino acids long. Thus, 1066 amino acids × 3 bases/codon = 3198 bases of the 3569 available are needed for these three proteins alone. This means 90% of the genome would be used directly for coding.

Genetic recombination occurs only very rarely in RNA viruses, but it is nevertheless possible to do many kinds of genetic analysis. Genetic evidence from mutants indicated that MS2 had not only genes for the A protein, the coat protein, and the replicase, but also a *fourth* gene whose product was involved in the lysis of the host cell to release progeny virions. Where was the *lysis* protein encoded? Sequence data allowed the generation of genetic maps of MS2 and Qβ, which are presented in Figure 3.22. The coding sequence for the lysis protein of MS2 is formed from the 3′ end of the coat protein gene, the spacer between that gene and the next, and then the 5′ end of the replicase gene. Qβ also has two overlapping genes according to its sequence. Rather than a separate lysis protein, however, that function is one of the A (sometimes also called the "maturation") protein's activities. The overlapping genes in Qβ are those for a minor virion constituent protein called A1 and for the coat protein. The coat protein gene is encoded entirely within the A1 gene.

The phage φX174 had two strategies to encode more than one protein from the same sequence: (1) overlapping genes translated from the same bases using different reading frames, and (2) initiation of translation at two different locations within the same sequence. MS2 uses the first of the strategies to encode its lysis protein. The reading frame for that protein is shifted one base toward the 3′ direction compared to that for the coat protein. Neither overlapping genes nor different initiation sites are the solution that Qβ uses to encode its coat protein, however. Instead, the Qβ coat and A1 proteins are exactly the same for their first 132 amino acids. At that location there is a relatively weak termination sequence (UGA) that permits occasional readthrough to add another 200 amino acids to the growing protein chain. This is reminiscent of the antitermination by proteins N and Q seen in λ phage, but Qβ readthrough is not mediated by any novel viral proteins as was the case with λ.

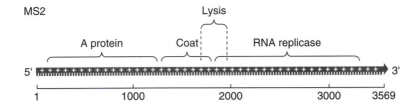

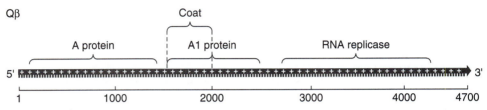

FIGURE 3.22 Genetic maps of the RNA bacteriophages MS2 and Qβ. Each virus encodes an "extra" protein: the MS2 lysis protein gene overlaps both the coat and replicase genes, while the Qβ coat gene is contained entirely within the A1 gene. MS2 uses different reading frames to translate overlapping genes, while Qβ uses readthrough of a weak terminator during translation to make a longer protein.

HOW DO THE RNA BACTERIOPHAGES REGULATE THE AMOUNT OF EACH PROTEIN SYNTHESIZED DURING THEIR REPLICATION?

In all viruses many individual capsid proteins surround a single genome during the final maturation process to form a virion. The RNA phages thus need far more translation from their capsid protein genes (the coat protein) than from their replicase or maturation protein genes. How do these phages achieve the proper ratio of proteins?

Even before sequences became available it was evident that the three-dimensional configuration of the viral RNA genome played a crucial role in regulating the translation of its genes. When MS2 genomic RNA was tested in an *in vitro* translation system, only the coat protein gene was translated. If the genome was broken into fragments, however, synthesis of the replicase and maturation proteins was also initiated. These experiments suggested that only the initiation site for the coat protein was accessible to ribosomes in native genomic RNA, while the other initiation sites were protected by folding or base pairing of the single-stranded RNA molecules with itself.

Computer models that generate possible secondary structures of the MS2 genome support this hypothesis. One set of configurations for the coat and A protein genes, as well as the initiation sites for the replicase and lysis genes, are shown in Figure 3.23. The initiation codon for the coat protein is prominently displayed on the end of a hairpin loop, so that ribosomes can readily find it. The initiation codons for the other three genes are in less accessible locations where they appear either to be involved in pairing with other bases in the genome or possibly buried at the bottom of hairpin loops. Note especially that the initiation codon for the A protein is not only located in an area likely to be difficult for a ribosome to enter, but also that it is GUG rather than the standard AUG.

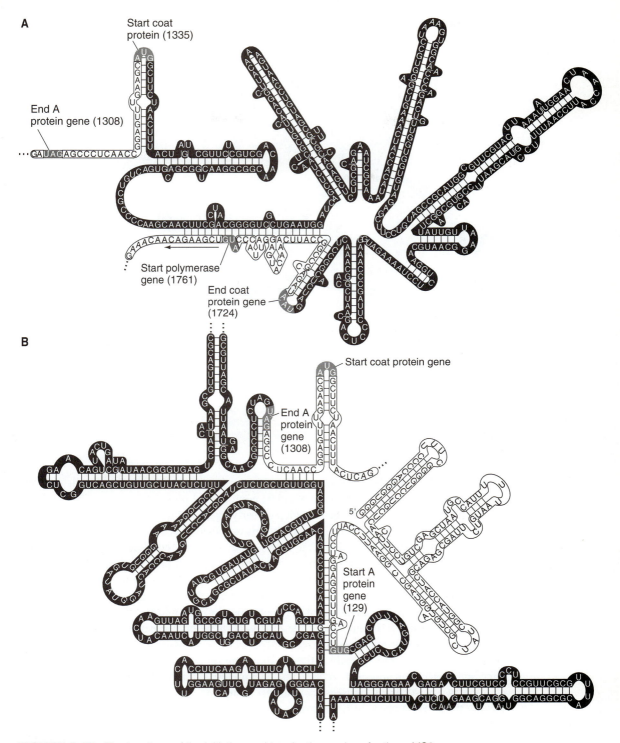

FIGURE 3.23 The locations of the initiation and termination codons for three MS2 genes. **A,** The initiation codon for the coat protein gene is at the end of a hairpin loop, and thus is readily accessible to ribosomes. The replicase (polymerase) protein gene initiation codon is more hidden in a region of paired bases. **B,** The GUG initiation codon for the A protein is at the base of a hairpin loop. (Continued.)

C

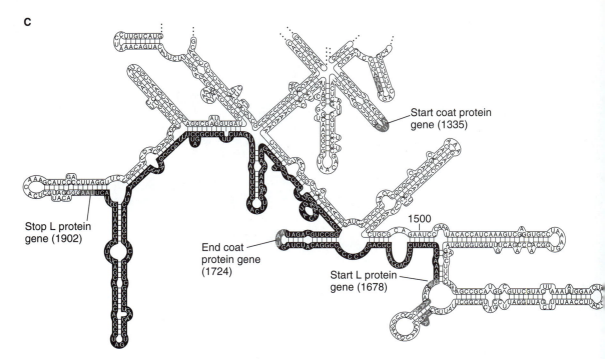

Start coat protein gene (1335)

Stop L protein gene (1902)

End coat protein gene (1724)

Start L protein gene (1678)

1500

FIGURE 3.23, CONT'D. C, The initiation codon for the lysis gene is immediately down-stream from two out-of-frame termination codons in the coat protein gene. The initiation codon for the replicase protein gene is at position 1761, about a third of the way down the lysis gene.

Figure 3.24 illustrates the various types of regulation of translation that occur in MS2 and the role that replication of the genome and its complement play in that regulation. The coat protein and the replicase could be considered "early" genes of MS2 since they are translated before replication of the genome occurs, but this designation is blurred by the close linkage between translation and replication. Translation of the replicase gene requires that the coat protein first be translated. Since the termination codon for the coat protein is only 33 bases from the replicase gene initiation codon, it appears that the process of translation alters the secondary structure of the genome to make the replication initiation codon accessible to ribosomal binding.

The other effect of translation of the coat gene is to prevent the translation of the lysis gene that overlaps its 3′ end. While the coat gene is being correctly translated, the lysis gene initiation codon is not recognized, since incoming ribosomal subunits would have to compete with ribosomes already active on the same set of bases. Synthesis from that lysis gene initiation codon appears to require that coat protein synthesis be prematurely terminated at one of two out-of-frame nonsense codons just upstream from the lysis gene initiation codon, leaving those bases open for ribosomal binding.

Translation of the A protein gene of MS2 is absolutely dependent on replication of the genome. MS2 replicase synthesizes numerous negative-strand complements to the MS2 genome, each of which in turn serves as a template in an RI to make numerous positive-strand

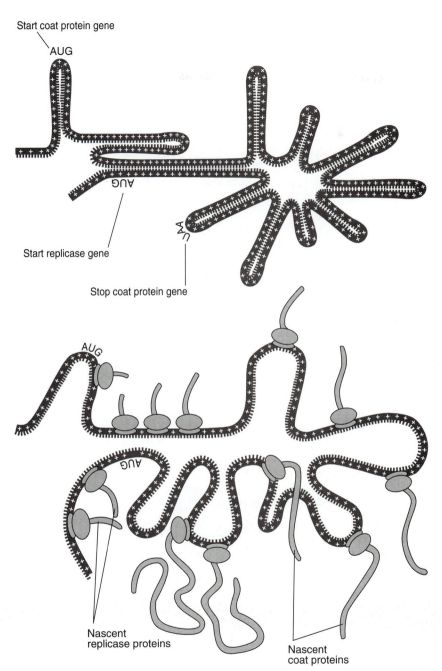

FIGURE 3.24 Initiation of translation of the MS2 replicase gene requires translation of the coat gene first. The presence of ribosomes on the coat gene releases the replicase gene initiation codon from its base-paired configuration, making it available to bind ribosomal subunits.

genomes. It is at this point that the order of the MS2 genes along the genome appears to be crucial. The first gene to be replicated on each new genome is the A protein gene (Figure 3.25). Since A protein synthesis occurs only as new positive strands are also being synthesized, it appears that the A protein initiation codon is transiently available for ribosomal binding as soon as it is synthesized and before sufficient additional bases have been synthesized to allow for that end of the molecule to assume its usual occluded configuration. Thus, it

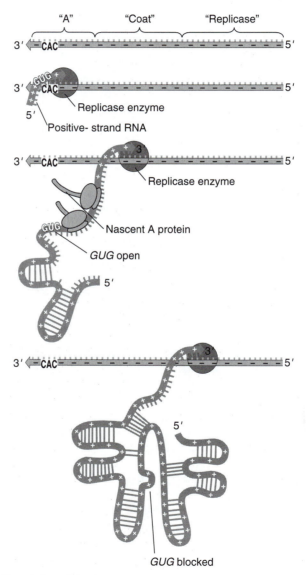

FIGURE 3.25 Translation of the A gene occurs only on nascent positive strands as they are synthesized in the RI. The nonstandard GUG initiation codon is probably available for ribosome binding only until the new strand has been elongated to contain the bases that permit the A protein gene to assume its usual base-paired configuration.

appears that the A gene of each new genome will be translated only a limited number of times (perhaps only once since only one copy of the A protein is needed to bind to the 5′ end of each genomic molecule), while the coat and replicase genes are translated more often.

MS2 has one final regulatory trick up its sleeve. Since two rounds of amplification occur during replication (positive strands to negative strands, and then negative strands to positive strands), synthesis of a replicase peptide each time a coat protein is made is unnecessary and wasteful of the host cell's synthetic capacity. The coat protein, therefore, functions in a feedback loop as *translation repressor* by binding to a 20-base sequence at the replicase initiation site and preventing further translation from that location.

SUMMARY

Bacterial viruses must replicate and express their genomes within the constraints created by the processes of the Central Dogma as performed in their prokaryotic hosts. Within these constraints, however, the bacteriophages have evolved a number of intriguing strategies to carry out those processes. The strategies discussed in this chapter are summarized in Table 3.4. Different viruses may have different types of nucleic acid (DNA or RNA) as their genomes, and that nucleic acid may be arranged in different ways (circular or linear, double-stranded, or single-stranded). Several types of replication occur in the viruses that are not seen in their host cells. The need for the virus to be able to distinguish its own transcription and translation from that of its host cell is a major requirement of this stage of the viral replication cycle. This is often coupled with the need to express different viral genes in the proper temporal sequence. Different viruses have evolved a variety of strategies to meet those needs, some of which occur in the prokaryotes as well. For example, modification of RNA polymerase so that a new set of promoters is recognized, seen in the T-even bacteriophages, is used by sporeforming bacteria to change from vegetative growth to spore production and back again. It is in the efficient use of

Table 3.4 | Summary of Strategies for Replication, Transcription, and Translation of Bacteriophage Genomes

DNA Synthesis
- Bidirectional replication of circular genomes: the theta intermediate
- Concatamerization via terminal redundancies
- Concatamerization via asymmetric replication of circular genomes: the sigma intermediate

Transcription
- Novel viral RNA polymerases that recognize only viral promoters
- Sequential modification of host transcriptase to recognize different viral promoters
- Antitermination proteins
- Repressor proteins

Translation
- Overlapping genes translated in one reading frame starting at different initiation codons and ending at same termination codons
- Overlapping genes translated in one reading frame starting at the same initiation codon and ending at different termination codons
- Overlapping genes translated in different reading frames
- Internal base pairing to prevent initiation
- "Repressor proteins" to prevent initiation

limited coding power that the viruses display their splendid "creativity" most, utilizing different initiation points, termination points, and reading frames. As will be presented in the next chapters, many of these strategies are not unique to the bacteriophages.

SUGGESTED READINGS
Double-Stranded DNA Bacteriophages

1. "Bacteriophage T3 and bacteriophage T7 virus-host cell interactions" by D.H. Kruger and C. Schroeder, *Micro Rev* 45:9–51 (1981), presents the details of all stages of the replication cycles of these T-odd viruses.

2. "Complete nucleotide sequence of bacteriophage T7 DNA and the locations of T7 genetic elements" by J.J. Dunn and F.W. Studier, *J Molecular Biol* 166:477–535 (1983), contains a wealth of interesting studies on gene expression in T7 in addition to the sequence data.

3. "Gene expression: A paradigm of integrated circuits" by B. Mosig and D.H. Hall, pages 127–131 in *Molecular Biology of Bacteriophage T4,* edited by J.D. Karam, American Society for Microbiology, Washington, D.C. (1994), summarizes the sequential modifications made to RNA polymerase leading to the expression of different sets of phage genes. The next three chapters in this splendid book describe the details of these processes.

4. "Lytic mode of lambda development" by D. Freidman and M. Gottesman, pages 21–52 in *Lambda II,* edited by R.W. Hendrix, J.W. Roberts, F.W. Stahl, and R.A. Weisberg, Cold Spring Harbor Laboratory, Cold Spring Harbor, N.Y. (1983), emphasizes the role of the N anti-terminator protein, but also describes contributions of other viral proteins in beginning the lambda lytic cycle.

Single-Stranded DNA Bacteriophages

1. "Single-stranded DNA" by R.L. Sinsheimer, *Scientific American* 207 (July):109–116 (1962), contrasts the bacteriophage φX174 with the DNA model of Watson and Crick, then only 9 years old.

2. "Electron microscopy of single-stranded DNA: Circularity of DNA of bacteriophage φX174" by D. Freifelder, A.K. Kleinschmidt, and R.L. Sinsheimer, *Science* 146:254–255 (1964), continues the story of φX174's surprising genome.

3. "Overlapping gene in bacteriophage φX174" by B.G. Barrell, G.M. Air, and C.A. Hutchison III, *Nature* 264:34–41 (1976), presents evidence that genes D and E are translated from the same DNA sequence in different reading frames.

4. "Nucleotide sequence of bacteriophage φX174 DNA" by F. Sanger, et al., *Nature* 265:687–695 (1977), presents the entire genome sequence and describes the coding features for nine φX174 genes.

5. "The rolling circle-capsid complex as an intermediate in φX DNA replication and viral assembly" by K. Koths and D. Dressler, *J Biol Chem* 255:4328–4338 (1980), describes, with some excellent electron micrographs, asymmetric replication of the φX174 genome.

RNA Bacteriophages

1. "Nucleotide sequence of the gene coding for the bacteriophage MS2 coat protein" by W. Min Jou, G. Haegeman, M. Ysebaert, and W. Fiers, *Nature* 237:82–88 (1972).

2. "Complete nucleotide sequence of bacteriophage MS2 RNA: primary and secondary structure of the replicase gene" by W. Fiers, et al., *Nature* 260:500–507 (1976), presents folding models of the genome that suggest how translation of the coat and replicase genes are regulated.

3. "Lysis gene expression of RNA phage MS2 depends on a frameshift during translation of the overlapping coat protein gene" by R.A. Kastelein, E. Remaut, W. Fiers, and J. van Duin, *Nature* 295:35–41 (1982), describes the relationship between coat and lysis gene translation in MS2.

Expression and Replication of the Viral Genome in Eukaryotic Hosts: The RNA Viruses

In this chapter we shall consider how various viruses with RNA genomes have solved the problems of replicating their genomes and synthesizing virus-specific proteins in eukaryotic host cells. We shall begin with the *positive-strand RNA viruses* whose genomes can serve directly as mRNAs for the synthesis of viral proteins. Such viruses must solve two problems: (1) making individual proteins from a single genomic RNA molecule and (2) making new RNAs using an RNA template. The *negative-strand RNA viruses,* whose genomes are the complement of the mRNAs necessary for protein synthesis, must solve the same problems but do it in the reverse order. They must make mRNA from their genomes before their genes can be expressed. The third set of viruses we shall consider have *double-stranded RNA* genomes. They express and replicate their genomes in a novel fashion compared to both positive- and negative-strand viruses. The last group of viruses we shall discuss, the *retroviruses,* have the most striking "lifestyle" of all the RNA viruses. They create a double-stranded DNA copy of their positive-strand RNA genomes and then insert that DNA copy into their host cell's genome.

WHAT FEATURES OF A EUKARYOTIC CELL'S GENETIC SYSTEM INFLUENCE VIRAL REPLICATION AND GENE EXPRESSION?

The basic biochemical steps of the processes of the Central Dogma, reviewed in Chapter 3, are the same in prokaryotic and eukaryotic cells. Very significant differences in the details of the processes, as well as in subcellular architecture, create markedly different environments for viral replication and gene expression in prokaryotic and eukaryotic cells.

DNA replication is the process that is most similar in prokaryotes and eukaryotes. The eukaryotic genome, contained within a membrane-delimited nucleus, is a set of long, linear DNA molecules that are replicated semiconservatively from multiple origins during the S phase of the cell cycle. A major difficulty in this process, as with any linear double-stranded DNA molecule, comes in replicating the ends (telomers) of the linear chromosomes, since the usual round of priming, elongation, and nick translation cannot occur on the lagging strand.

The steps leading to the synthesis of a protein in a eukaryotic cell differ in many ways from the same steps in a prokaryotic cell, as illustrated in Figure 4.1. First, unlike the *polycistronic* mRNAs of prokaryotes that carry the information from several genes, each individual protein-coding gene in the nuclear DNA genome of a eukaryote is transcribed into a high-molecular weight RNA (hnRNA, for *heterogeneous nuclear* RNA) containing only the information for that particular gene (that is, it is *monocistronic*). In the next posttranscriptional modification, this transcript is processed by addition of a polyadenosine (poly-A) "tail" to the 3′ end and a "cap" of 7-methyl-guanosine (7mG) attached 5′ to 5′ at the other end. Then noncoding sequences *(introns)* are *spliced* out and the coding sequences *(exons)* are brought together to form the final mRNA product. Finally, this mRNA must be transported from the nucleus into the cytoplasm where it is translated into a single protein species. The newly synthesized protein may then be processed into an active form by cleavages, phosphorylations or glycosylations, or by formation of disulfide linkages.

The processes of the Central Dogma create the environment within which all viruses must produce the proteins encoded in their genomes, as well as make copies of their genomes for inclusion in new virions. Only those viruses that (1) enter the nucleus and (2) have dou-

A   Prokaryotic cell

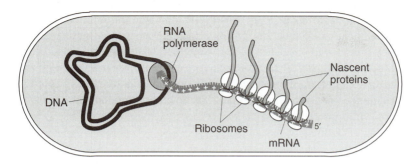

B   Eukaryotic cell

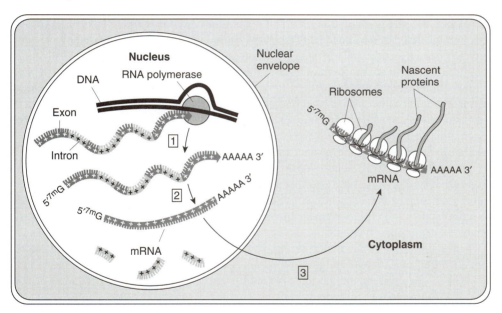

FIGURE 4.1 Comparison of transcription in prokaryotic **A** and eukaryotic **B** cells. In eukaryotes, posttranscriptional modification (addition of a 5′ cap and 3′ poly-A tail) (1), and splicing out of introns (2) and transport out of the nucleus (3) occur between transcription and translation.

ble-stranded DNA genomes containing the appropriate types of controlling sequences are able to reproduce themselves using all the normal cellular machinery in the normal fashion. Viruses with other genomic organizations must modify the usual cellular machinery or processes in order to reproduce.

   This is particularly true of viruses that have genomes of RNA rather than DNA. RNA viruses face two significant problems with replication in eukaryotic cells. First, eukaryotic cells synthesize RNA using information contained in a DNA template molecule. They contain no enzymes for synthesizing RNA molecules using an *RNA* template (although some plant cells may be a possible exception). RNA viruses must, therefore, provide the RNA-dependent–RNA-polymerase necessary to copy their own genomes, and, in some cases, to synthesize mRNA molecules. Second, eukaryotic cells can initiate translation only from the 5′ end of an mRNA molecule and cannot recognize internal initiation sequences. This means that they translate all mRNAs as if they were monocistronic messages containing only the information for a single protein. An RNA virus with a polycistronic genome must therefore either have (1) a means of creating functionally monocistronic mRNAs for each viral protein in its genome, or (2) an ability to process the protein products of its genome to generate individual viral proteins.

WHAT STRUCTURAL FEATURES DISTINGUISH THE POSITIVE-STRAND RNA VIRUSES OF ANIMALS?

Positive-strand RNA viruses are those whose genomes can serve directly as the source of information for protein synthesis; that is, the genomes are messenger RNAs. This means, by definition, that the RNA genome of a positive-strand virus, stripped of all other molecules, must be *infectious* or capable of being translated when it is introduced into a host cell, or when it is placed in an appropriate *in vitro* translation system. The most rigorous demonstration of this usually begins with mixing an aqueous suspension of virions with liquid phenol. When the aqueous and phenolic phases again separate, the RNA genome remains in the aqueous phase while the proteins are extracted into the phenol layer. This purified RNA is then either transfected into host cells and the synthesis of new virions observed, or it is added to an *in vitro* translation system, and the synthesis of specific viral proteins is detected.

While the ability of the pure RNA genome to be translated is the ultimate indication of a positive-strand virus, there are other structural features that also suggest that a virus may belong to this category. These structural features are the same ones that identify a cellular RNA molecule as being a eukaryotic messenger RNA: the presence of a 7-methyl-guanosine cap on the 5′ end of the molecule and/or a poly-A tail (a chain of adenosines) attached to the 3′ end of the molecule. As we shall see, however, neither of these structures is an absolute requirement in a positive-strand virus. Exceptions like poliovirus, in fact, have helped us to understand the roles played by these structures in the translation of normal mRNA molecules in eukaryotic cells.

We shall consider in some detail three viruses (poliovirus, Sindbis virus, and murine hepatitis virus) as representatives of the major groups of positive-strand RNA viruses. These groups are identified by their similar genomic organizations and strategies for replication and expression of those genomes, as well as the amino acid sequences of certain viral proteins. An interesting feature of these groups is that each contains both animal and plant cell viruses, which suggests that a common evolutionary ancestor diverged into individual viral strains with different hosts. This fact emphasizes that the essential, significant feature of both plant and animal cells, to the viruses at least, is that they are *eukaryotic,* and that the differences in the "lifestyles" of these organisms that obsess botanists and zoologists are merely superficial.

Poliovirus represents the *Picornaviridae* (*pico* is Italian for "small" + RNA), a group whose name clearly describes their nature. The two major subdivisions of the picornavirus group are (1) the *enteroviruses* (*entero* is Greek for "intestine" which include poliovirus, among the most widely studied of all viruses, and the hepatitis A virus that causes infectious hepatitis; and (2) the *rhinoviruses* (*rhino* is Greek for "nose") that cause many of the common colds that afflict humankind. Several groups of plant viruses (the *comoviruses, nepoviruses,* and *potyviruses*) are similar to the picornaviruses in many aspects of the reproduction and expression of their genomes.

The picornaviruses are small and simple in their architecture, as illustrated in Figure 4.2, *A*. The virions typically have capsids about 27 to 30 nm in diameter, which are formed from 240 molecules (60 molecules each of four different peptides) arranged in icosahedral symmetry. There is no envelope. The picornavirus capsid encloses a molecule of single-stranded RNA about 2.4×10^6 daltons in size. The genomes of several picornaviruses have

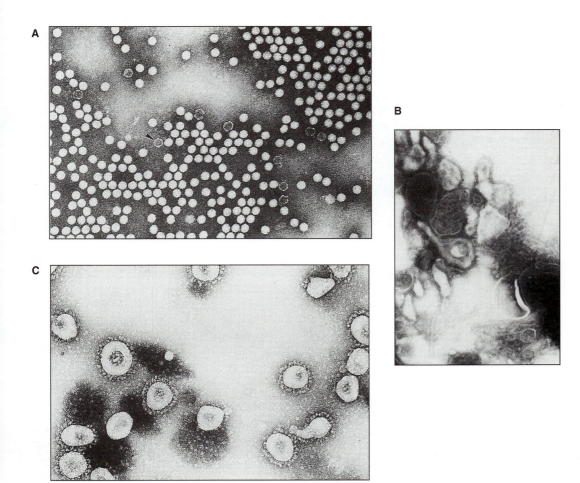

FIGURE 4.2 Virions of three positive-strand RNA viruses. **A,** Poliovirus *(Picornaviridae).* **B,** Sindbis virus *(Togaviridae).* **C,** Murine hepatitis virus *(Coronaviridae).*

been sequenced and range from 7209 to 8450 bases long. All of the picornaviruses have the poly-A tail characteristic of a eukaryotic mRNA attached to the 3′ end of their genome, although its length varies considerably. The genomes do not have 7-methyl-guanosine caps on their 5′ ends, however. Instead there is a viral protein about 20 to 24 amino acids long called VPg (*v*irus *p*rotein, *g*enome) covalently attached to the 5′ end via a phosphodiester bond between a tyrosine in the VPg and the terminal uridine base in the genome. The possible role of this protein will be discussed in a later section.

Sindbis virus represents the second group, the *Togaviridae* (a *toga* is the loose outer garment worn by ancient Roman citizens), named for the membranous envelope that surrounds the nucleocapsid in these viruses. These small viruses include the *alphaviruses,* which are responsible for a variety of types of encephalitis in humans; the *rubiviruses* (*rubi* is Latin for "red"), whose sole member, rubella virus, causes German measles; and the *pestiviruses,* which include economically important viruses such as those that cause hog cholera and other enteric diseases in animals. The *flaviviruses* (*flavo* is Latin for "yellow") were formerly included in the *Togaviridae* but are now a family of their own *(Flaviviridae).* Members of this family that cause human disease include yellow fever virus, Dengue virus, and hepatitis C virus.

The togaviruses, like the picornaviruses, have icosahedral nucleocapsids. In the alphavirus, Sindbis, the nucleocapsid contains 180 copies of the capsid protein. This structure, about 30 to 42 nm in diameter, is surrounded by an envelope possessing glycoprotein spikes projecting from its surface (Figure 4.2, *B*). The entire virion appears as a sphere about 60 to 65 nm in diameter. The RNA genome of alphaviruses like Sindbis is a single molecule around 4.3×10^6 daltons (about 11,700 bases) in size, while that of the rubella virus is somewhat smaller (about 10,000 bases). All togavirus genomes have the typical features of positive-strand RNA: The purified RNA is infectious and has a 5′ cap and a 3′ poly-A tail.

The final group, the *Coronaviridae* (*corona* is Latin for "crown") is named for the very prominent, petal-shaped spikes on the viral envelopes (Figure 4.2, *C*). *Murine hepatitis virus* (MHV) will be the representative of this group. These viruses have helical nucleocapsids that are flexible enough to form a sphere that is then surrounded by the envelope. The genomes of the coronaviruses are the largest single RNA molecules found in viruses. These genomes are about 5.4 to 6×10^6 daltons, equivalent to 27,000 to 30,000 bases (four times the size of the picornaviruses and two and a half times that of the alphaviruses). The single molecule of the genome has the 5′ cap and 3′ poly-A tail characteristic of eukaryotic mRNAs, indicating that the genome is positive-strand RNA.

The characteristics of all of these groups are summarized in Appendix B; page 373.

WHAT STRATEGIES DO THE ANIMAL POSITIVE-STRAND VIRUSES USE TO EXPRESS AND REPLICATE THEIR POLYCISTRONIC GENOMES?

The genomes of the picornaviruses, togaviruses, and coronaviruses are single molecules of RNA that carry the codons for all the viral proteins. To the translation system of a eukaryotic cell, this molecule is therefore the message for a *single* protein. How then do the genomes of positive-strand RNA viruses produce the individual proteins necessary to generate new virions?

Applying Occam's razor* to that question yields the answer, "If all that can be made is one protein, then one protein must be all that is made." The question then becomes, "Is one protein enough?" Is there really no problem at all since a single monocistronic message would suffice? Recall that every RNA virus must encode at least one protein for its capsid but must also encode the RNA-dependent–RNA-polymerase necessary to reproduce its genome. In the simplest hypothetical virus, it is possible that the same protein could fill both these roles.

Nature is not so simple, however. Analysis of the virion proteins of poliovirus by polyacrylamide gel electrophoresis demonstrates that there are four different capsid proteins. Comparison of the number of amino acids in these four proteins (879) to the total coding length of the genome (equivalent to 2477 amino acids) suggests that these capsid proteins are only a fraction of the possible proteins encoded by the virus. In Sindbis virus the situation is even more compelling. Not only must capsid proteins be encoded by the genome, but viral-specific glycoproteins must be synthesized for insertion into host cell membranes to form the spikes of the virion envelope. Thus, it is clear that multiple proteins are indeed encoded by the genomes of positive-strand viruses and that these single RNA molecules must therefore be polycistronic mRNAs.

Picornaviruses

Nevertheless, the Occam's razor answer is correct! The entire genome of a poliovirus is translated from a single initiation site 740 bases from the 5′ end to generate a *single* **polyprotein** of 247 kdaltons. In poliovirus-infected cells it is possible to find enormous polysomes of more than 60 ribosomes as the entire genome is being translated at the same time. The polyprotein is cleaved into somewhat smaller polyproteins even as it is being synthesized (these are sometimes called "nascent" cleavages since they occur as the polyprotein is being "born"), so isolation of the entire 247 kdalton molecule is usually not possible.

Figure 4.3 presents the scheme by which the picornavirus genome becomes expressed as a set of individual proteins. Following a leader sequence at the 5′ end of the molecule are the genes for the four capsid proteins arranged as a block. The nascent cleavage of the original polyprotein generates P1, a 97 kdalton polyprotein containing these four proteins (1A, 1B, 1C, and 1D). The remainder of the genome contains two sets of genes whose protein products are responsible for further cutting of the polyproteins and for RNA synthesis. Cleavage of this 150 kdalton protein yields a 65 kdalton polyprotein P2 (containing 2A, 2B, and 2C) and an 85 kdalton polyprotein P3 (containing 3A, 3B, 3C, and 3D).

The nascent cleavage that separates P1 from P2–P3 appears to be autocatalytic. The 2A protein to the right of the junction between P1 and P2 is a protease that recognizes the amino acid pair tyrosine-glycine as its cutting site, and these are the amino acids at that same junction. The other cleavages, with the exception of the final separation of capsid proteins 1A and 1B during maturation of the capsid, are accomplished by the glutamine-glycine specific protease, 3C.

---

*The philosophical principle propounded by William of Occam (who died about 1349) that assumptions introduced to explain something must not be multiplied beyond necessity. In other words, keep it simple!

(1) Translation of genome into polyprotein

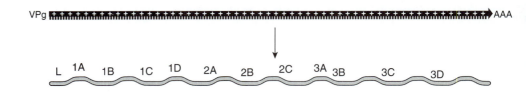

(2) Nascent cleavages by 2A and 3C proteases

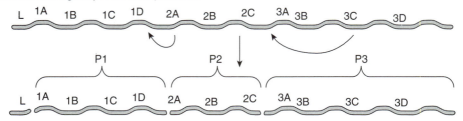

(3) Maturation cleavages by 3C protease

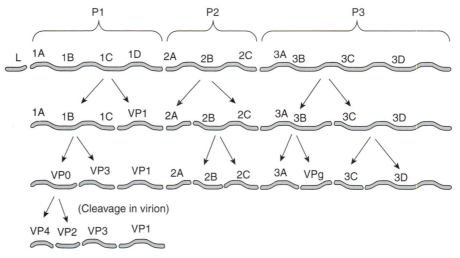

FIGURE 4.3 The genome and translation products of poliovirus. Nascent cleavage of the polyprotein by the 2A protein occurs immediately upon its synthesis. The next maturation cleavages are catalyzed by proteins 2A or 3C, while the final cleavage of VP0 into VP4 and VP2 occurs after genomic RNA has been packaged in the capsid.

This cascade of cleavages can be demonstrated in pulse-chase experiments such as that presented in Figure 4.4. (The principle of a pulse-chase experiment is illustrated in Appendix A, Figures A1–A5.) The upper frame shows the pattern of proteins resulting from a 10-minute pulse with [^{35}S] methionine given to poliovirus-infected HeLa cells. The high molecular weight proteins P1, P2, and P3 are clearly evident. Some cleavage occurs during even a brief pulse period, so some smaller molecular weight proteins are also evident. After

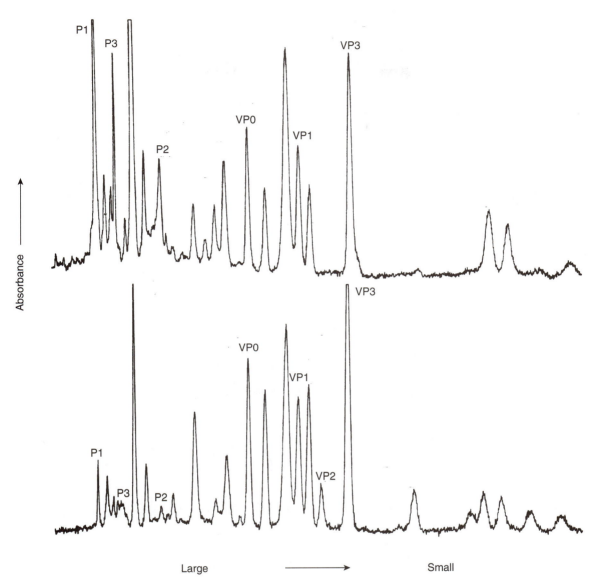

Large ⟶ Small

FIGURE 4.4 A pulse-chase experiment showing the cleavage of the poliovirus polyproteins into individual structural and nonstructural viral proteins. At 3.5 hours after poliovirus infection, HeLa cells were exposed to [^{35}S] methionine for a 10-minute pulse period *(upper panel)* or for a 10-minute pulse followed by a 90-minute chase. The cells were lysed and their proteins separated by SDS-PAGE. Only the proteins formed by the nascent cleavage and their ultimate products are identified in this figure; the other peaks represent cleavage intermediates. Note that the large polyprotein molecules P1, P2, and P3 nearly disappear while the amounts of VP0, VP1, VP2, and VP3 all increase during the chase period. These molecules are identified in Figure 4.3.

a 90-minute chase *(lower frame)* the amounts of the P1, P2, and P3 proteins are greatly reduced, while the amounts of their cleavage products have increased.

Figure 4.5 shows a series of polyacrylamide electrophoresis gels from a similar experiment, this time examining the *in vitro* translation of poliovirus genomic RNA. It is clear that the initial products of the pulse-chase are predominantly higher molecular weight proteins, but as the chase lengthens, the amount of those proteins decreases and new smaller molecular weight proteins appear.

How Are Host and Poliovirus mRNAs Distinguished?

One of the striking effects of poliovirus infection is a rapid shutdown of host cell macromolecular synthesis, including protein synthesis. This raises the intriguing question of how translation of the poliovirus genome can occur when host translation is blocked. How are host

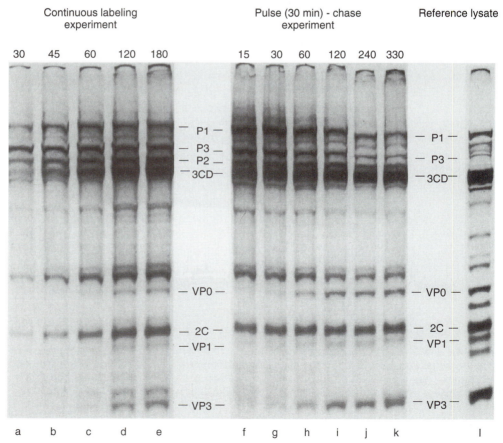

FIGURE 4.5 *In vitro* translation of poliovirus genomic RNA. A 30-minute pulse with [^{35}S] methionine was followed by chase periods as indicated, the products separated by SDS-PAGE, and the results visualized by autoradiography. The reference lysate was formed by electrophoresing materials from [^{35}S] methionine-labeled poliovirus-infected HeLa cells (15-minute pulse). The molecules indicated are identified in Figure 4.3.

and viral mRNAs distinguished? The answer lies in the structure of the picornavirus genome itself. Recall that the 5' terminus of the poliovirus genome has the viral protein VPg attached to it rather than the 7-methyl-guanosine cap characteristic of eukaryotic mRNA. When translation of a eukaryotic mRNA begins, the cap is recognized by a "cap-binding complex" of cellular proteins, one of which is a 220 kdalton protein called eIF4G. In a poliovirus-infected cell, this protein is cleaved, and therefore inactivated, in a series of reactions. The poliovirus protein 2A responsible for one of the nascent cleavages of the viral polyprotein also appears to activate a host cell protease. The activated protease then cleaves eIF4G, preventing translation of host cell mRNAs since their cap structures cannot be recognized.

HOW IS TRANSLATION OF POLIOVIRUS MRNA BEGUN?

As is often the case in science, answering one question raises another. In this case, if poliovirus blocks translation of host cell mRNAs by degrading the cap-binding protein needed to recognize the 5' end of host messages, how does translation of the poliovirus genome initiate? Sequencing of the genomes of a number of picornaviruses provided a clue to the answer to this question. The genomes contain long (741 bases in the case of poliovirus) untranslated regions (leaders) at their 5' ends. As Figure 4.6 illustrates, the RNAs of these leaders can form a series of hairpin loops of various sizes even though there is little similarity in their sequences except for a set of bases upstream of the actual initiation codon. The hairpin loops in the leader region are thought to block the standard "ribosome scanning" mechanism that sets up the translation machinery at the first AUG codon of a eukaryotic mRNA. The conserved sequence at the end of the leader, termed an IRES for "*internal ribosomal entry site*," appears to facilitate formation of the translation initiation complex at an internal location. The process that permits the small ribosomal subunit to attach to the

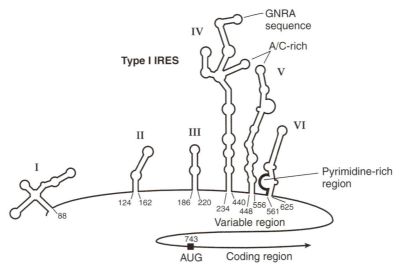

FIGURE 4.6 Hairpin loops in the 5' leader region of poliovirus genomic RNA. The structures labeled II through VI function as an internal ribosomal entry site or IRES to facilitate identification of the initiation codon at position 743.

genomic RNA molecule and locate the next AUG involves a number of cellular proteins. For poliovirus, one of these is the C-terminal fragment of the cleaved eIF4G cap-binding protein. Thus, poliovirus uses the same reaction—cleavage of a protein used in the normal cellular process of initiating translation—both to shut off host cell translation and to facilitate translation of its own viral mRNA.

How Does Poliovirus Replicate Its Genome?

As discussed in Chapter 3, viruses with single-stranded genomes, whether DNA or RNA, replicate in a two-stage process. First, a molecule complementary to the genome is synthesized. This molecule (the replicative intermediate, or RI, form), in turn serves as the template for the synthesis of numerous new genomic molecules simultaneously.

Synthesis of picornavirus RNAs begins as soon as the first polyproteins have undergone their nascent and early maturation cleavages to provide the enzymes and other factors necessary for the process. Just as we saw earlier regarding translation, synthesis of viral RNA molecules requires the presence of hairpin loops and other three-dimensional structures at the ends of the template molecules. The proteins involved in replicating the viral RNA are contained in segment P3 of the poliovirus polyprotein (refer to Figure 4.3).

Figure 4.7 diagrams the steps of synthesis first of a negative-strain poliovirus antigenome and then a new positive-strand molecule. The 3AB protein has a hydrophobic region that causes it to become associated with host cell membranes. The 3CD protein and a host protein called PCBP (*polyC binding protein*) interact with components of a cloverleaf structure at the 5′ end of the viral positive strand that will serve as a template. After 3CD transfers UMPs to 3AB, that molecule is cleaved by 3C to form a new VPg-pUpU-OH that will be the primer for negative-strand synthesis. 3CD is cleaved at the same time to remove the viral RNA polymerase 3D from the protease 3C. The VPg primer and 3D polymerase then bind to the poly-A tail of a positive-strand RNA molecule to initiate negative-strand synthesis.

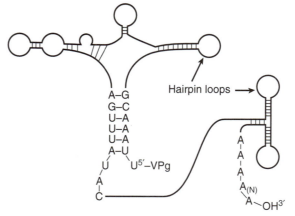

A Sequences at the ends of poliovirus genomic RNA

FIGURE 4.7 Synthesis of negative-strand antigenomes and positive-strand genomes by poliovirus. **A,** The sequences and structures at the ends of the poliovirus genomic molecule. Note that both ends have a series of adenines. *(Continued.)*

B Negative-strand synthesis

Adding UMPs to the 3AB protein to make primer

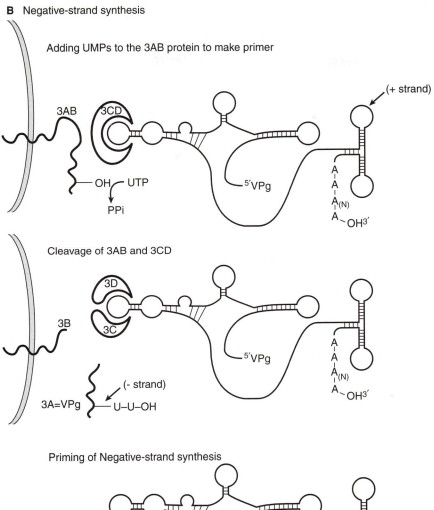

Cleavage of 3AB and 3CD

Priming of Negative-strand synthesis

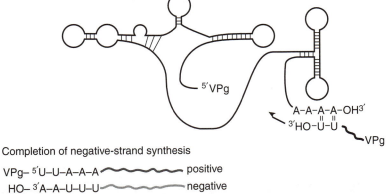

Completion of negative-strand synthesis

VPg– ⁵′U–U–A–A–A positive

HO– ³′A–A–U–U–U negative

FIGURE 4.7, CONT'D. B, Negative-strand synthesis begins with addition of UMPs to 3AB. 3C then cleaves both 3CD and 3AB to generate a primer molecule and free 3D polymerase. After the primer binds to the 3′ end of the positive-strand template, negative-strand synthesis occurs. Note that the 3′ end of the new antigenomic molecule has a pair of adenine bases. *(Continued.)*

C Positive-strand synthesis

Synthesis of primer

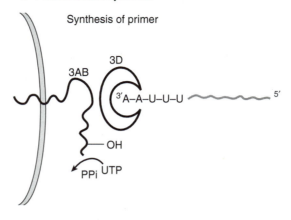

Cleavage of 3AB to release primer

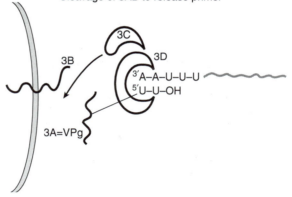

Synthesis of positive-strand RNA

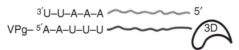

FIGURE 4.7, CONT'D. C, Positive-strand synthesis also begins with creation of a primer molecule by 3D. After cleavage to release the primer, genome synthesis occurs.

This binding reaction appears to be facilitated by the folded structures at the 3′ end of the positive strand.

Positive-strand synthesis is similar but does not require 3C or PCBP. A 3D polymerase that has bound the 3′ end of an antigenome begins synthesis of a new positive strand by adding two UMPs to a membrane-bound 3AB molecule and cleaving 3AB to release a VPg-pUpU-OH primer. This primer base pairs with a pair of AMPs at the 3′ end of the antigenome and elongation commences. Once again, a stem-loop structure in the RNA template appears to facilitate this process.

Togaviruses

The togaviruses also translate their polycistronic genomes into polyproteins, but in a process more complex than that of the picornaviruses. As we have seen, the picornaviruses translate the entire genome at one time, beginning with viral structural proteins and leading into the nonstructural proteins responsible for cleavage of the polyprotein and synthesis of new RNAs. In contrast, Sindbis virus, a togavirus with a genome of about 11,700 bases (49S), initially translates only the 5′ end of its genome directly into a polyprotein containing four nonstructural proteins (Figure 4.8).

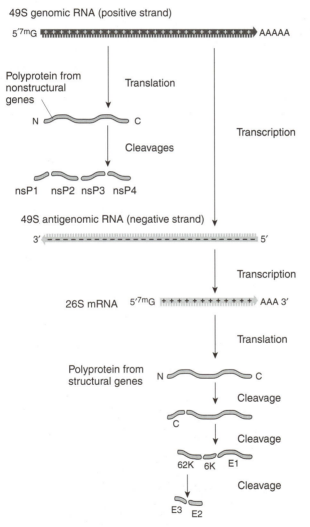

FIGURE 4.8 Scheme for the expression of the Sindbis virus genome. The 5′ end of the genome is translated into a polyprotein that is cleaved to yield nonstructural proteins, including an RNA polymerase. Expression of the 3′ end requires formation of a negative-strand intermediate from which an mRNA equivalent to the 3′ end of the genome can be transcribed. A series of cleavages of the polyprotein translated from this mRNA yields viral structural proteins. Virion proteins are C (capsid) and E1, E2, and E3 (envelope).

One of these is a protease that identifies the alanine-alanine cleavage sites occurring in the Sindbis polyproteins, so cleavage is again autocatalytic. Expression of the remainder of the viral genome appears to require new RNA synthesis, so formation of these enzymes permits the shift from early to late gene expression.

Expression of the 4100 bases on the 3' end of the Sindbis virus genome requires two RNA synthesis steps. First, another 49S RNA molecule is made, complementary to the genome, resulting in a full-length (49S) minus-strand RNA or **antigenome.** This molecule is the template for the synthesis of a shorter (26S) positive-strand molecule that is equivalent to the 3' end of the genome. After this 26S positive-strand RNA is capped and tailed, it is translated into a polyprotein containing the structural proteins needed for the nucleocapsid and envelope proteins of the virion. The capsid protein (the C protein) is the first to be translated (Figure 4.9). The Sindbis early gene encoded protease cuts the C protein from the polypeptide chain while it is still in the polysome. This nascent cleavage, in turn, exposes a signal sequence on the N-terminus of the remaining polyprotein that targets the rest of the translation process to occur in association with the host cell's endoplasmic reticulum. Thus, the final three protein products (envelope glycoproteins E1, E2, and E3) of the Sindbis genome become associated with ER vesicles where they are glycosylated before being transported for insertion into the plasma membrane.

As noted previously, togaviruses must synthesize several different RNA molecules in order to express their entire genomes (see Figure 4.8). Replication of the genome and its expression are therefore interconnected processes, all mediated by the same set of non-

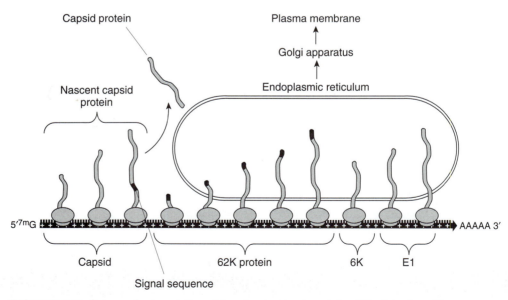

FIGURE 4.9 Scheme for translation of the structural polyprotein of Sindbis virus. The nascent cleavage of the structural polyprotein releases the capsid protein and exposes a signal sequence that causes further translation to occur in association with the endoplasmic reticulum. The envelope proteins extruded into the lumen of the ER are glycosylated in the Golgi and then become inserted into the plasma membrane.

structural proteins. Figure 4.10 illustrates how negative- and positive-strand RNA synthesis in Sindbis virus is regulated by the degree of cleavage that has occurred in the polyprotein translated from the 5′ end of the genome. The formation of the replicative form with a full-length negative-strand RNA begins with the autocatalytic cleavage of the initial polyprotein P1234 between 3 and 4. The resulting RNA-dependent–RNA-polymerase, consisting of P123 and nsP4, becomes attached to host cell membranes and synthesizes a full-length negative-strand complement to the genome.

This negative-strand RNA molecule can then serve as a template in two competing reactions. The first is the synthesis of new full-size genomic RNAs, and the other the synthesis of shorter positive strands (mRNAs) equivalent to the 3′ end of the genome. Synthesis of these positive-strand RNAs is driven by additional cleavage reactions. Cleavage between 1 and 2 significantly reduces the ability of the enzyme complex to recognize the initiation site on positive strands, and consequently reduces antigenome synthesis and shifts the complex of nsP1-P23-nsP4 to synthesis of new, full-length positive strands. The final cleavage

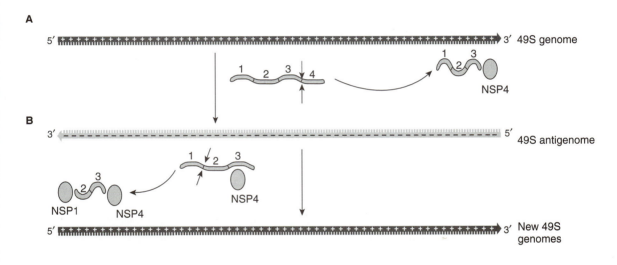

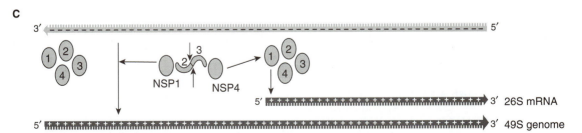

FIGURE 4.10 Sindbis virus positive- and negative-strand RNA synthesis is regulated by cleavages of polyprotein P1234. **A,** The initial cleavage of P1234 releases nsP4 to begin synthesis of the negative-strand antigenome. **B,** The second cleave releases nsP1 from P123 to begin synthesis of new full-length positive-strand molecules. **C,** The final cleave of P23 allows synthesis of both full-length genomic RNAs and 26S mRNAs for the viral structural proteins.

between 2 and 3 generates a complex of four separate, nonstructural proteins that synthesize both new genomes and the 26S RNA needed as mRNA for structural proteins.

Figure 4.11 summarizes the three possible fates of new genomic positive strands. Early in the infective process, when relatively few structural proteins have been synthesized, these RNAs may be used in an amplification process as they serve as mRNAs for the synthesis of more of the initial polyprotein containing the nonstructural proteins. A second use is as templates for the synthesis of additional negative strands. Finally, when capsid proteins accumulate, the full-length positive-strand RNAs are packaged into nucleocapsids. This sequestering of the full-length positive strands acts to facilitate the translation of the smaller structural protein mRNAs since the genomic molecules are no longer available to compete for ribosomes.

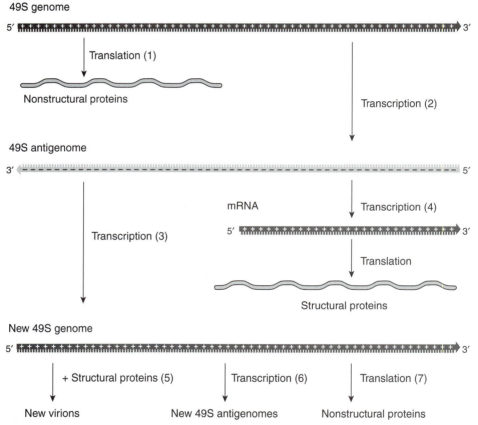

FIGURE 4.11 The possible fates of Sindbis genomic RNA (+49S), its complement (−49S), and the segment encoding the viral structural proteins (+26S). The genome can be partially translated to form nonstructural proteins (1) or copied into its complement (2). This molecule, in turn, can be transcribed into another genome (3) or into the mRNA for viral structural proteins (4). The newly synthesized genomic RNA can follow three pathways: incorporation into new virions (5), serve as a template for additional minus-strand RNA synthesis (6), or serve as mRNA for additional nonstructural protein synthesis (7).

Coronaviruses

Expression of the coronavirus genome introduces several novel strategies found in neither the picornaviruses nor the togaviruses. Although the evidence is not conclusive, initial expression of the very large coronavirus genome may be similar to the pattern seen in the togaviruses. The 5′ end of the genome is translated to form a large protein, which is then immediately cleaved to yield the RNA replicase necessary for synthesis of complementary antigenomic RNA from the genomic template, as well as other nonstructural proteins. *In vitro* translation of the 5′ end of the murine hepatitis virus (MHV) genome, for example, produces a 250 kdalton protein that is cleaved to yield 220 kdalton and 28 kdalton proteins. The smaller of these is also detected in infected cells and may be the replicase protein. An interesting feature of the RNA sequences of the 5′ ends of a number of coronaviruses is that they actually encode several polyproteins rather than just one. Moreover, these initial polyprotein genes appear to contain a characteristic also seen in some bacteriophages: *overlapping open reading frames.* The overlap between the polyproteins in avian infectious bronchitis virus is only 42 bases, which suggests that the two proteins would be translated by a ribosomal frame-shifting process similar to that used by the bacteriophage MS2 in the synthesis of its lysis protein (see pp. 108–12).

As with the togaviruses, expression of the remainder of the coronavirus genome requires synthesis of a negative-strand complement of the entire genome, followed by synthesis of new positive-strand RNA molecules, which have both caps and tails. In MHV one of these positive-strand molecules is a full genome in length, but the others are of seven smaller sizes, ranging from 1.9 to 10.8 kb long. Analysis of the sequences of these seven RNAs has revealed a number of features that suggest that the coronaviruses use a novel method of expressing their late genes. First, the 5′ end of each molecule bears the same "leader" sequence of about 60 to 70 bases that also occurs at the 5′ end of the entire genome. Second, following the leader sequence on each molecule is a seven-base series (UCu/cAAAC) that occurs at five different locations within the genomic RNA molecule. Finally, the 3′ ends of all the subgenomic molecules are the same.

Figure 4.12 diagrams the relationship of the subgenomic RNAs to the MHV genome that these observations suggest. The subgenomic molecules appear to be a **nested set** of mRNAs that share their 3′ ends, but differ by the sequences that follow the leader sequence at their 5′ end. Since translation of a eukaryotic mRNA takes place only from the first initiation codon at the 5′ end to the first termination codon in the proper reading frame, these nested mRNAs would each be translated to yield a different protein product, even though their 3′ ends are the same. The smallest molecule, mRNA 7, encodes a component of the nucleocapsid (the N protein), while mRNAs 6 and 3 encode glycoproteins called M (for "membrane"; also called E1) and S (for "surface"; also E2) that are part of the envelope. mRNAs 2, 4, and 5 appear to encode proteins of 20 kdalton, 14 kdalton, and 13 kdalton sizes, respectively, that are not found as part of the nucleocapsid and so are not structural genes. In MHV and some other coronaviruses, mRNA 2 may be translated using two different reading frames. Although the first appears to encode a nonstructural protein, the second, mRNA 2a, encodes an envelope glycoprotein with hemagglutinating activity.

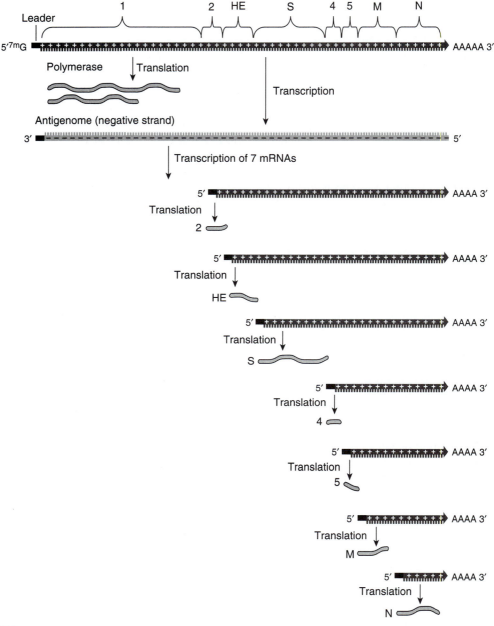

FIGURE 4.12 Relationship between the murine hepatitis virus genome and the nested set of subgenomic RNAs corresponding to its 3′ terminus. The genomic-length molecule is translated to form a protein that is probably the RNA polymerase. After formation of a negative-strand antigenome, a set of seven nested mRNAs is synthesized, each beginning with a short leader sequence read from the 3′ terminus of the antigenome. The unique portion at the 5′ end of each mRNA is translated to synthesize individual viral proteins. Numbers are given to genes whose proteins have not been identified; HE = hemagglutinin, S = peplomer envelope protein, M = virion membrane-associated protein, N = nucleocapsid protein.

HOW IS A NESTED SET OF MRNAS SYNTHESIZED?

An intriguing question is presented by the fact that each subgenomic mRNA has the same leader sequence, which is encoded only at the 3′ end of the template negative-strand RNA molecule, while the rest of the mRNA is transcribed from nearer the 5′ end of the template strand. Since posttranscriptional splicing is a hallmark of eukaryotic transcription, the sequence between the leader and the coding region for each subgenomic mRNA might simply be spliced out of a longer transcript. *In vitro* transcription assays indicate that this is not the case, however. Instead, synthesis of mRNAs appears to be discontinuous, with the viral polymerase beginning synthesis at one location and shifting to another to complete the new molecule. Two different schemes, diagrammed in Figure 4.13, have been proposed for this process. In the first, the leader region is transcribed from the 3′ end of the antigenome and then the polymerase with the nascent RNA molecule still held in place on the enzyme shifts to reinitiate synthesis from one of the seven-base-long internal regions of homology. Transcription of the new mRNA then continues to the end of the template molecule. In the second model, the genome is used as template to begin synthesis of a

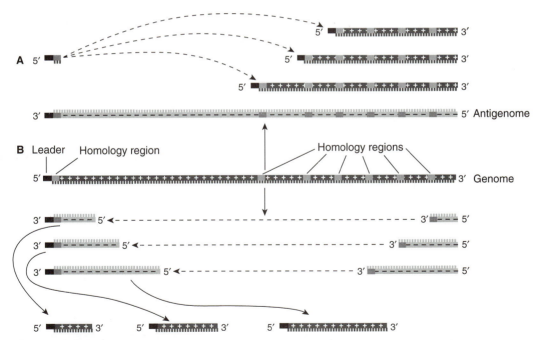

FIGURE 4.13 Two models for discontinuous synthesis of a nested set of mRNAs in coronaviruses. **A,** RNA polymerase carries a newly synthesized primer from the 3′ end of the antigenomic template RNA to one of the internal homology regions and resumes synthesis. **B,** Synthesis of negative-strand RNAs begins at the 3′ end of the genomic RNA and halts at one of the homology regions. The negative-strand molecule is then shifted to the leader region of the genome to complete synthesis of an "antisense RNA" molecule. This molecule, in turn, serves as the template for the synthesis of its corresponding mRNA.

nested set of antigenomic molecules. In this case, the shift moves the nascent molecule from one of the homology regions to the leader-encoding region of the genome. Each of the resulting antigenomic molecules then serves as template for the synthesis of its corresponding mRNA.

Support for these models of discontinuous RNA synthesis comes from genetic studies in which a host cell is infected with two different strains of the same coronavirus. As many as 10% of the progeny viruses produced may have recombinant genomes, indicating that the leader-polymerase complex has begun synthesis on the antigenome of one strain and then shifted to a homology region on the antigenome of the second strain to complete the full-length mRNA. In addition, this mode of replication may account for the relatively high frequency of **defective interfering particles** seen with coronaviruses. Defective interfering particles are viruses that cannot reproduce themselves but can interfere with the replication of normal viruses in the same host cell. Coronavirus defective interfering particles often contain deletions in their genomes, possibly the result of mistakes in the shifting process. Details of the regulation of synthesis of the nested sets of mRNAs as well as new genomic molecules are not yet known.

Figure 4.14 summarizes the strategies used by positive-strand eukaryotic viruses to produce individual proteins from a single RNA genome. The basic theme appears to be the syn-

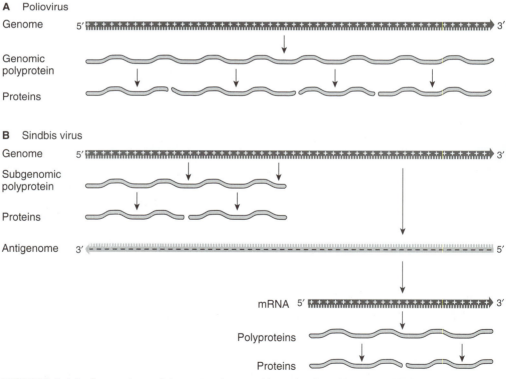

FIGURE 4.14 Comparison of the strategies used by animal positive-strand RNA viruses to produce individual proteins from a single genomic molecule. **A,** Formation of a genomic-length polyprotein. **B,** Formation of subgenomic-length polyproteins. *(Continued.)*

thesis of a polyprotein that is autocatalytically cleaved to produce viral proteins. As the viral genome becomes larger, this process is subdivided into early gene expression from the 5′ end of the genome to synthesize the viral replicase, and late gene expression of structural proteins using mRNAs synthesized after the formation of a negative-strand template molecule.

WHAT STRATEGIES DO PLANT POSITIVE-STRAND VIRUSES USE TO EXPRESS AND REPLICATE THEIR GENOMES?

The vast majority of plant viruses have positive-strand RNA as their genome. Unlike the positive-strand RNA viruses of animal cells, however, many of these plant viruses are **multipartite**; that is, they have RNA genomes that are segmented into two or more molecules, *each of which is surrounded by its own nucleocapsid.* This means that the entire virus is subdivided into component parts, *all* of which must be present in the same cell in order to synthesize new progeny. This type of divided organization is possible in plant viruses because most of them are transmitted from host cell to host cell either by direct movement through cytoplasmic connections, or by the mediation of a vector. In either case, the likelihood is high that at least one of each type of virus particle will be introduced into the new host cell. Subdividing the genome into separate virus particles is not found in either the bacterial or animal

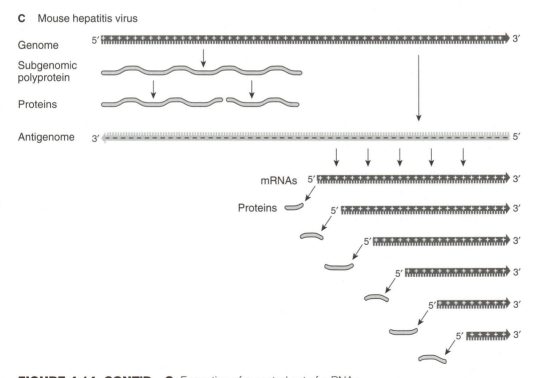

FIGURE 4.14, CONT'D. C, Formation of a nested set of mRNAs.

viruses, which rely on chance encounters between appropriate host cells and virions released into the environment. The probability of the same host cell being infected concurrently by all the necessary viral components would be very low in those circumstances.

Viruses with multipartite genomes may enjoy some selective advantages compared to viruses with monopartite genomes encompassing the same total coding power. The primary advantage may be the ease and efficiency of translation of several small RNAs compared to one large one. Second, multipartite genomes can be encapsidated within smaller particles that may be easier to assemble than larger ones. Finally, multipartite genomes may allow both temporal and physical separation of gene expression to different parts of an infected cell.

The *monopartite* plant viruses have genomes consisting of a single RNA molecule. These viruses are all nonenveloped but may have either helical or icosahedral capsid architectures. Various groups of these viruses use the same strategies as the animal positive-strand RNA viruses for the expression and replication of their genomes. The genome of potato Y virus (the type potyvirus), for example, is a capped and poly-A tailed molecule about 9 to 10.5 K bases long that is translated into a single polyprotein of about 3500 amino acids. As in the picornaviruses, this polyprotein is then cleaved into both structural and nonstructural proteins (Figure 4.15, *A*). Another group of plant viruses, the tymoviruses, use a translation strategy similar to that of the animal alpha togaviruses. The genome of this group has the usual 5' cap, but the 3' end contains the sequence of a tRNA^{val} rather than polyadenosine. The 5' end of the genome is translated into a polyprotein that is cleaved to produce various nonstructural viral proteins. The viral capsid protein is produced from a smaller mRNA molecule that is equivalent to the 3' end of the genomic RNA (Figure 4.15, *B*). Two other groups of monopartite plant viruses, the tobamoviruses and tombusviruses, synthesize proteins in a manner similar to that of the coronaviruses. The genome of tobacco mosaic virus, the type tobamovirus, has a 5' cap and a 3' terminus resembling tRNA^{his}. The 5' end of the genome is translated directly into two polyproteins as a result of readthrough of an amber termination codon. The genes at the 3' end of the genome are then translated from a pair of nested mRNAs that are coterminal with the 3' end of the genome (Figure 4.15, *C*).

Bipartite and *tripartite* viruses have genomes that are subdivided, respectively, into two and three pieces, each of which is separately encapsulated. For example, cells infected with the bipartite cowpea mosaic virus (the type comovirus) produce three particles that can be separated on density gradients: B *(bottom)* particles that contain an mRNA of about 6000 bases; M *(middle)* particles with an mRNA of about 4100 bases; and T *(top)* particles that are simply empty capsids (Figure 4.16). The M mRNA appears to encode a polyprotein containing the two proteins that form the viral capsid. The B mRNA also encodes a polyprotein that contains the viral replicase and a VPg protein that is attached to the 5' end of both mRNAs in place of a cap structure. This virus is thus similar to a picornavirus in both structure and function.

The tripartite viruses use a combination of strategies for production of their proteins. Two of the RNA molecules are monocistronic messages encoding large proteins, usually involved with replication. The third genomic RNA molecule is polycistronic and is translated in a fashion similar to that of the togaviruses (see Figure 4.14, *B*). A nonstructural protein is translated from its 5' end, while the viral capsid protein is translated from a subgenomic mRNA equivalent to the 3' end of the genomic molecule.

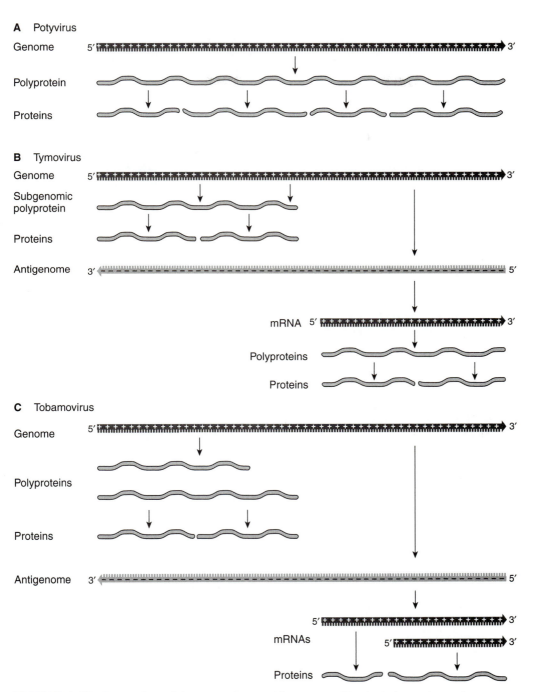

A Potyvirus

Genome

Polyprotein

Proteins

B Tymovirus

Genome

Subgenomic
polyprotein

Proteins

Antigenome

mRNA

Polyproteins

Proteins

C Tobamovirus

Genome

Polyproteins

Proteins

Antigenome

mRNAs

Proteins

FIGURE 4.15 Comparison of the strategies used by monopartite plant viruses to produce individual proteins from their single genomic RNA molecule. The potyvirus and tymovirus genomes are translated to form one or two polyproteins respectively **A, B,** while the tobamovirus genome is expressed by means of both polyproteins and nested mRNAs **C.**

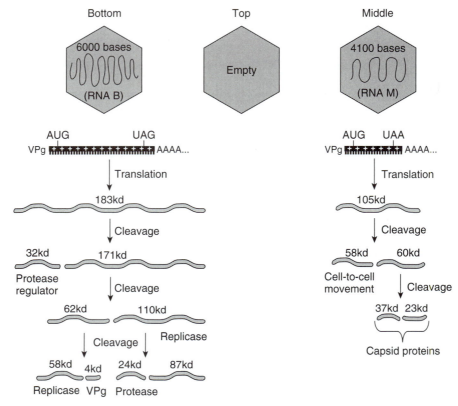

FIGURE 4.16 Genome organization and expression in cowpea mosaic virus, a bipartite plant virus. Each RNA is translated into a polyprotein that is cleaved into individual viral proteins.

WHAT STRUCTURAL FEATURES DISTINGUISH THE NEGATIVE-STRAND RNA VIRUSES?

The negative-strand RNA viruses show a number of common structural features that suggest they have evolved from an ancestral "progenitor" virus. They all have flexible nucleocapsids with helical symmetry surrounding the single-stranded RNA genome. The capsid is composed largely of a single species of protein, with minor amounts of other proteins. The nucleocapsid is surrounded by an envelope whose surface has prominent spikes or protrusions that are virus-encoded glycoproteins involved in attachment and penetration. Their genomes range in length from about 10 to 15 kilobases, giving them about the same coding power as that of the togaviruses just discussed.

The defining characteristic of the negative-strand RNA viruses is that their genomes cannot serve directly as the template for translation of viral proteins. The genomes have neither of the structural features (the 5′ cap and 3′ poly-A tail) found in eukaryotic mRNAs. Moreover, purified genomic RNAs from which *all* viral proteins have been removed have no activity in *in vitro* translation systems.

We shall examine the mechanisms of gene expression and replication in six groups of negative-strand viruses. The family *Rhabdoviridae* (*rhabdo* is Greek for "rod") take their name from their striking morphology. The helical nucleocapsid is coiled into a tube that often tapers at one end. When surrounded by an envelope with prominent spikes, the arrangement produces a bullet- or rod-shaped virion (Figure 4.17, *A*). Although rhabdoviruses infect a wide variety of animal, plant, and insect hosts, the best understood viruses are in the two animal virus genera. Vesicular stomatitis virus (VSV) is the type virus of the vesiculoviruses, while rabies virus is characteristic of the lyssaviruses.

The *Paramyxoviridae* have larger genomes than the rhabdoviruses (about 15 and 11 kilobases, respectively), but much more restricted host ranges, infecting only mammals and birds. The flexible nucleocapsid inside a loose envelope creates virions that appear very irregularly shaped in electron micrographs (Figure 4.17, *B*). The paramyxoviruses are divided into three genera based on, among other things, differences in two biological activities of their envelopes, discussed earlier in Chapter 2. These activities are hemagglutination, or the ability to bind to specific erythrocytes, and cleavage of *N* acetyl-neuraminic acid residues from glycoproteins or glycolipids by the enzyme neuraminidase. Sendai virus, representative of the *paramyxoviruses,* has spikes with both hemagglutination and neuraminidase activities, while the spikes of respiratory syncytial virus, a *pneumovirus,* have neither activity. Measles virus, a member of the *morbilliviruses,* has hemagglutinin spikes but no neuraminidase activity.

The third group of negative-strand RNA viruses came to public awareness in a very dramatic fashion in 1967 when a number of laboratory workers preparing African green monkey kidney cells for culture suddenly developed a new form of hemorrhagic fever. The diseases caused by the Marburg virus that caused this outbreak (named for Marburg, Germany, where the first case appeared) and the Ebola viruses that produced similar highly lethal outbreaks in Zaire and the Sudan have led to these viruses becoming the stuff of potboiler books and films. They are placed in the family *Filoviridae* (*filum* is Latin for "thread") since their virions are about 80 nm in diameter and up to 14,000 nm in length (Figure 4.17, *C*). Their genomes, about 19 kilobases, are larger than either those of the *Rhabdoviridae* or the *Paramyxoviridae.*

The three remaining families of negative-strand viruses are distinguished by having *segmented* genomes within an enveloped virion. Each RNA molecule is surrounded by its own flexible helical nucleocapsid. Unlike the multipartite plant viruses, however, a single virion contains copies of all the segments comprising the entire genomic complement. The *Orthomyxoviridae* viruses influenza A (Figure 4.17, *D*), influenza B, and influenza C all have eight RNA segments in their genomes, although there are significant differences in their host ranges and other biological features. The *Bunyaviridae* (named for a virus first isolated at Bunyamwera, Uganda) are a large group of viruses that have three RNA molecules for their genomes (Figure 4.17, *E*), while the *Arenaviridae* have two genomic segments. These viruses are named for a granular appearance in electronmicrographs (as seen in Figure 4.17, *F*) that is created by ribosomes included within the envelope (*arena* means "sand" in Latin). The presence of these host cell ribosomes means that the virion contains all the usual species of rRNAs (28S, 18S, and 4S to 5S) in addition to the two virus genomic molecules (31S or about 6.3 to 8.4 kb, and 23S or about 3.3 to 3.9 kb).

The characteristics of all of these families of negative-strand viruses are given in Appendix B.

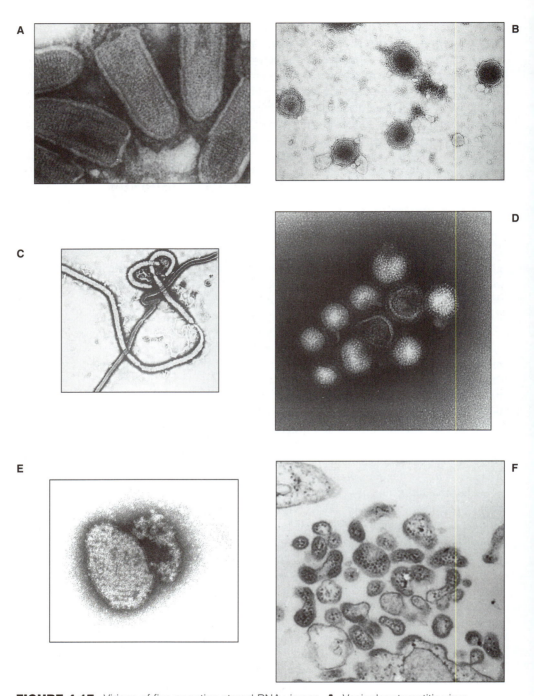

FIGURE 4.17 Virions of five negative-strand RNA viruses. **A,** Vesicular stomatitis virus *(Rhabdoviridae)*, **B,** Sendai virus *(Paramyxoviridae)*, **C,** Ebola virus *(Filoviridae)* **D,** Influenza A virus *(Orthomyxoviridae)*, **E,** Hantaan virus *(Bunyaviridae)* **F,** *(Arenaviridae)*.

WHAT STRATEGIES DO NEGATIVE-STRAND VIRUSES USE TO EXPRESS AND REPLICATE THEIR GENOMES?

We have seen that the positive-strand RNA viruses have solved the puzzle of translating their polycistronic genomes in host cells that only recognize monocistronic mRNAs by employing variations on several basic strategies: polyproteins and/or synthesis of subgenomic molecules that can be treated as monocistronic mRNAs. The negative-strand RNA viruses have a puzzle of their own. If their genomic RNAs cannot be translated directly (as can be demonstrated by introducing the purified RNAs into appropriate cells, or in *in vitro* translation systems), how can they express their genomes at all? In order to be translated, the viral genomes must be copied into their complementary molecules, but this reaction requires an RNA-dependent–RNA-polymerase, which is only encoded by the (untranslatable) virus genome itself—a true "Which came first, the chicken or the egg?" paradox!

The solution to this seeming paradox is straightforward. Since all negative-strand RNA viruses require the presence of an RNA replicase before they can express their genomes, they all package their genomes together with the necessary RNA replicases. This can be demonstrated in two ways. If instead of naked RNA, nucleocapsids (the genome and the proteins associated with it) are introduced into a host cell, the viral genome is expressed. Likewise, *in vitro* transcription systems that copy viral genomes into positive-strand molecules can be constructed from components purified from detergent-disrupted virions.

The replicases of negative-strand viruses must catalyze three different reactions during the viral replication cycle: (1) synthesis of mRNAs from the genome, (2) synthesis of an entire positive-strand RNA antigenome complementary to the genome, and (3) synthesis of new negative-strand genomic RNAs using the antigenomic RNA molecule as template. Although many details concerning the recognition events that occur in each of these processes remain to be described, similarities in the organization of the all negative-strand viral genomes suggest that common strategies may be involved. These common structural features are both large-scale (the same gene order within nonsegmented viral genomes) and small-scale (the nature of the termini of various RNA molecules).

Rhabdoviruses

Although rabies virus is the best known of the rhabdoviruses, vesicular stomatitis virus (VSV) has been the most widely used model system to study gene expression and replication in this family. Inactivation studies using ultraviolet irradiation, together with sequencing data, indicate the genomic organization presented in Figure 4.18. Each of the five VSV genes is transcribed into its own mRNA, to which is then added a 7mG cap at the 5′ end and a poly-A tail at the 3′ end. The protein products of all these genes are found in the complete virion (also shown in Figure 4.18), although in quite different amounts. The sequences labeled T at the 3′ terminus and R at the 5′ terminus are largely complementary to each other.

The first step in expression of the VSV genome after infection is transcription from the negative-strand genome to form mRNAs for the five viral genes. The replicase activity for this synthesis resides in the L and P proteins that together with RNA and the N protein form the nucleocapsid. Polymerizing activity probably is a function of the L protein, with the P protein serving to facilitate both the binding of L to the RNA-N protein complex and its movement down the template strand (perhaps by displacing N). P must be phosphorylated

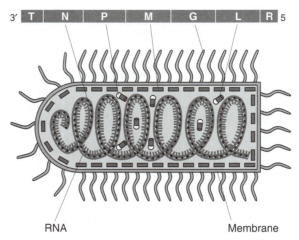

RNA   Membrane

FIGURE 4.18 *Genomic organization of vesicular stomatitis virus (Rhabdoviridae) and location of each gene product in the VSV virion. N = nucleocapsid protein, P = phosphoprotein, M = matrix protein, G = spike glycoprotein, L = RNA polymerase. T and R are inverted terminal repeat sequences and do not encode proteins.*

(hence its present designation) and form a complex of two to four peptides with one L protein in order for transcription to occur.

VSV mRNAs are synthesized in the same order that they occur in the genome (N, P, M, G, L), but each is made in only about two-thirds the amount of the one preceding it in the series. Several theories have been advanced to explain this attenuation. In the first, initiation occurs only at the 3′ end of the genome with transcription of a 46-base "positive-strand leader" from the T sequence. Transcription continues down the genome but stalls at the poly-U sequences following each gene so that "chattering" or repeated use of the same template bases creates the poly-A tail on each message. Synthesis may then cease or may continue to produce polycistronic molecules that must be cleaved to produce individual mRNAs.

The second model suggests that the enzyme binds independently at the beginning of each gene to initiate synthesis of each mRNA separately. The final model, for which the best experimental support exists, is called the stop-start model. This model proposes that each gene is separately transcribed into a monocistronic message, but that separate binding reactions are not needed. The stop-start model says that the polymerase binds first to the 3′ terminus of the template and transcribes the N gene. The chattering reaction that generates a poly-A tail on the new RNA at the end of that gene also causes transcription to halt. About one-third of the time the enzyme then falls off the template. The remaining times the polymerase releases the N mRNA but stays on the template to reinitiate transcription of the P gene. Each succeeding gene is thus transcribed in a lesser amount that those before it.

The process of genome replication in the rhabdoviruses is less well understood than that of synthesis of mRNAs. The same proteins (L, P, and N) catalyze both processes, so why does synthesis of five mRNAs occur in one case and a complete antigenomic molecule occur in the other? Ongoing translation of viral proteins, especially the N protein, appears to be necessary for replication, but not transcription. Several models suggest that the conformations of the template molecules used in transcription and replication are different, perhaps due to different interactions of either the N and P proteins or the N, P, and M (for matrix) proteins with RNA. It is likely that newly synthesized N proteins quickly bind to nascent RNA molecules and hinder recognition of the transcription termination sequences at the end of each of the viral genes. This permits synthesis of a complete positive-strand antigenomic molecule rather than individual mRNAs. It also appears that the P protein that participates in replication is not phosphorylated and only a single copy of it interacts with an L protein. Thus, the conformation of the P protein may be important in regulating the processes of both transcription and replication.

Paramyxoviruses

Although paramyxoviruses are larger and thus have several more genes than the rhabdoviruses, their evolutionary kinship is shown by several features of their genomes. Both families of viruses have the same five genes (called N, P, M, G, L in the rhabdoviruses and NP, P, M, HN, L in the paramyxoviruses) in the same order in the genome. In addition to their shared genes, however, paramyxoviruses like Sendai encode several other proteins.

A variety of other proteins are produced from the P gene sequence. One of these, the C protein that may be involved in inhibiting host cell mRNA synthesis, is translated from a set of bases near the beginning of the P gene mRNA using a reading frame that is shifted one base downstream from the reading frame for the P protein. Another protein of unknown function called V is the result of *RNA editing* of the P gene transcript. RNA editing can take two different forms: the cotranscriptional insertion of nontemplated bases in the molecule being made, or posttranscriptional modification of the templated base. The paramyxoviruses appear to use the former method. In Sendai virus, the mechanism is similar to that used to form the poly-A tails of mRNAs. The transcribing enzyme appears to pause and "stutter" very briefly at a short run of cytosines in the template. This stuttering adds an additional guanosine to the mRNA molecule that causes a translational frameshift at that point. The F_0 or fusion protein is located between the M protein gene and the hemagglutinin (HN) protein gene, the paramyxovirus homolog of the rhabdovirus G protein (Figure 4.19). Finally, the 3′ and 5′ termini of the Sendai genome are complementary to each other, as was also seen in VSV.

Many paramyxoviruses reproduce very poorly in tissue cultures, so details of their mechanisms of transcription and replication are not as well understood as those of the rhabdoviruses. They appear, however, to use the same strategies described previously to produce monocistronic mRNAs for each gene (except P/C) and to make full-length copies of their genomes.

Filoviruses

The evolutionary relatedness of the filoviruses to other negative-strand RNA viruses is also evident from their genetic organization of seven genes ordered in the same fashion as those of the rhabdoviruses and paramyxoviruses. As the Ebola virus map shown in Figure 4.20 indicates, this group of viruses has short overlaps between the endings of several genes and those that follow them. The function of these overlaps is unknown, but they do not appear to cause attenuation of downstream genes as was seen in the rhabdoviruses and paramyxoviruses. A second interesting feature of the Ebola genome is that the major envelope

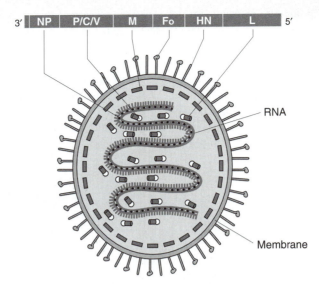

FIGURE 4.19 Genomic organization of Sendai virus *(Paramyxoviridae)* and the location of each gene product in the virion. NP = nucleocapsid protein, P = phosphoprotein, M = matrix protein, F_0 = fusion protein, HN = hemagglutinin/neuraminidase protein, L = RNA polymerase.

glycoprotein gene GP is expressed in two forms. The primary protein expressed from this gene is not a virion protein at all, but is a 60 kdalton protein called SGP since it is secreted from infected cells. The function of SGP is thought to involve interfering with the host immune response to Ebola infection. Synthesis of the actual envelope glycoprotein subunit protein requires RNA cotranscriptional editing of the GP mRNA. A single adenosine is added when the polymerase stutters on a run of seven uridines in the template. This addition produces a frameshift that results in translation to yield a GP molecule that is about twice as long as the SGP molecule.

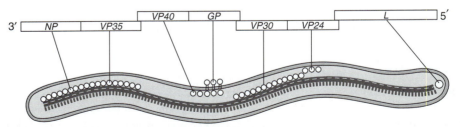

FIGURE 4.20 Genetic map of the filovirus Ebola. Eight proteins are encoded in four sets of overlapping reading frames. Seven proteins are found in the virion. The nucleoprotein NP, the polymerase L, and two structural proteins VP30 and VP35 interact with the viral genome to form the nucleocapsid. The matrix protein VP40, the peplomer or spike protein GP, and VP24 are parts of the envelope. The eighth protein is a soluble form of the GP protein called SGP.

Orthomyxoviruses

Orthomyxoviruses like influenza A virus differ from the rhabdoviruses, paramyxoviruses, and filoviruses both in the architecture of their genomes and in the mechanisms by which that genome is transcribed into mRNAs and replicated to form new genomes. Influenza A virus's genome consists of eight different segments of negative-strand RNA, ranging in size from 890 bases to 2341 bases. All these segments have been sequenced, and the viral proteins that each encodes have been determined by genetic means. These assignments have been confirmed by the technique of hybrid-arrested translation; translation of a viral mRNA in an *in vitro* translation system becomes blocked when that mRNA hybridizes with its genomic complement, so genes are determined by finding which peptides are not synthesized when individual genomic segments are added to the reaction mixture. Table 4.1 presents the gene assignments in influenza A virus.

Table 4.1 | Gene Assignments for Influenza A Virus Segments

Segment Number	Protein	Name Function
1	PB2	Polymerase-7mG CAP binding
2	PB1	Polymerase-elongation
3	PA	Polymerase
4	HA	Hemagglutinin
5	NP	Nucleoprotein
6	NA	Neuraminidase
7	M1, M2	Matrix (membrane) proteins
8	NS1, NS2	Nonstructural proteins

Unlike the negative-strand viruses discussed thus far, the orthomyxoviruses require host cell enzymatic activities to supplement viral enzymes during expression of the viral genome. This was first suggested more than 25 years ago, when it was shown that influenza A virus required a functioning nucleus in which to replicate, in contrast to all other RNA viruses that replicate in the cytoplasm of their host cells. More recently it has been shown that the host enzyme responsible for transcription of cellular mRNAs (RNA polymerase II) is involved in viral RNA synthesis, since the RNA polymerase II inhibitor α-amanitin also blocks viral nucleic acid synthesis. The actual host role in the process appears to be to provide a primer molecule for viral RNA synthesis.

As Figure 4.21 illustrates, the three viral polymerase proteins (PB1, PB2, PA) encoded on segments 1, 2, and 3 form an enzyme complex that functions in both transcription and replication. In transcription, PB2 first binds the 5′ end of a host cell mRNA while the 5′ end of a viral genomic molecule binds to a site on PB1. When the 3′ end of the viral RNA attaches to another site on PB1, the host mRNA is cleaved about 10 to 12 bases from its end to create a primer for positive-strand RNA synthesis on the negative-strand genomic template. Elongation continues until a poly-A tail is made by "chattering" on a run of five to seven uridines about 15 to 20 bases from the 5′ end of the genome segment.

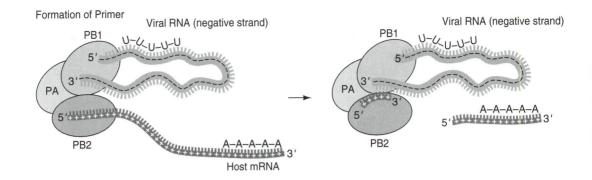

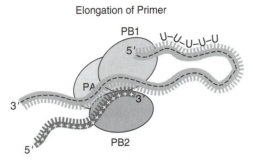

FIGURE 4.21 Model for synthesis of mRNA in influenza virus. A primer is created by cleavage of a host cell mRNA by PB2 while both ends of a viral negative-strand genomic molecule are held on sites on PB1. The primer is then elongated by the polymerase activity of PB1. PA does not function in mRNA synthesis.

The switch from transcription to replication appears to involve a switch in which two of the three components of the complex are active. PA does not play a role in transcription, while PB2 is inactive in replication. In addition, replication requires that newly synthesized NP (nucleoprotein) proteins be available to bind to either negative-strand or positive-strand RNAs that are to be used as template for full-length copying. The details of the process are still sketchy, but it is evident that the presence of NP proteins removes the necessity for a primer and permits *de novo* initiation of RNA synthesis.

The RNAs transcribed from the six largest segments are translated into individual proteins, while those from the two smallest segments give rise to two proteins each. The transcripts from these two segments are synthesized in the nucleus, have the usual eukaryotic 7mG cap and poly-A tails, and possess splice sites similar in sequence to those found in eukaryotic hnRNAs. It is likely, therefore, that the mechanism for producing two proteins from the same genetic information is that of differential splicing of the transcript to form two different final mRNAs. Figure 4.22 illustrates the differential splicing of transcripts from segments seven and eight.

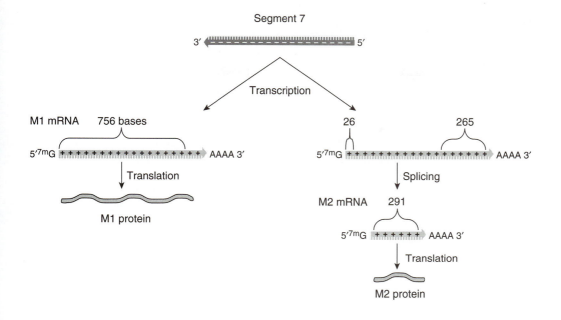

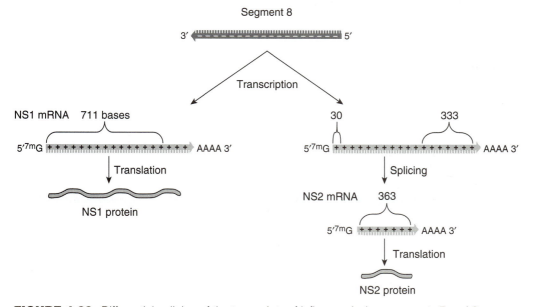

FIGURE 4.22 Differential splicing of the transcripts of influenza A virus segments 7 and 8 to create different mRNAs. In both cases a larger protein is encoded in a nonspliced mRNA (equivalent to the entire genomic fragment) while a smaller protein is read from a spliced mRNA.

CAN A VIRAL GENOME HAVE BOTH POSITIVE- AND NEGATIVE-STRAND RNA?

The *Arenaviridae* and *Bunyaviridae* have segmented genomes that can be shown to contain negative-strand RNAs. mRNAs taken from polysomes in infected cells will hybridize with virion genomic RNAs, indicating that the genomes are not themselves messages. What little is known about the transcription and replication of these two groups of viruses suggests that they employ strategies similar to those used by other negative-strand viruses. The *Arenaviridae* and the genus *phlebovirus* of the *Bunyaviridae* do appear to have one novel strategy, however, for synthesizing some of the proteins. The smallest of the three genomic RNA molecules of the phleboviruses and both the arenaviral RNAs appear to be **ambi-sense,** that is, to be both positive-strand and negative-strand RNA *at the same time* (*ambi* is Latin for "on both sides"). For example, the early gene N at the 3′ end of the phlebovirus small (S) RNA molecule is transcribed into an mRNA for the N protein (Figure 4.23). The viral RNA appears to form a hairpin loop at the junction of the two genes that blocks transcription at that point. After replication of the S RNA to form an antigenomic S RNA molecule, the 3′ end of this molecule (corresponding to the 5′ end of the genomic strand) is transcribed into an mRNA for the NS$_s$ protein. The genomic S RNA molecule thus encoded proteins in opposite directions at its two ends.

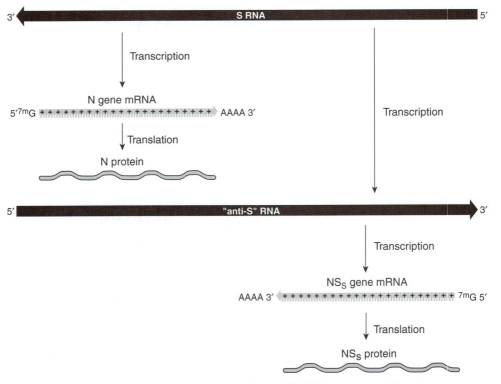

FIGURE 4.23 Expression of the "ambi-sense" small (S) RNA strand of phlebovirus. An mRNA for the N gene is transcribed from the 3′ end of the S RNA, while the mRNA for the NS$_s$ gene is transcribed from the opposite end of the "anti-S" complement of the genomic molecule. The genomic strand thus encodes proteins in opposite directions at its two ends.

HOW DO DOUBLE-STRANDED RNA VIRUSES EXPRESS AND REPLICATE THEIR GENOMES?

The *Reoviridae* are a most unusual group of viruses whose distinctive features were recognized only gradually. These viruses first came to the attention of virologists as "viruses in search of a disease," and were therefore called reoviruses for *r*espiratory, *e*nteric, *o*rphan viruses (recall the echoviruses in the *Picornaviridae* discussed in Chapter 2). The reoviruses were found to be nonenveloped viruses with a *double capsid* structure and, most unusually, to have a genome consisting of 10 to 12 segments of *double-stranded RNA*. Six genera of reoviruses are now recognized: three that infect animals (reovirus, orbivirus, and rotavirus) and three that infect only plants and insects (cypovirus, phytoreovirus, and figivirus). The human reovirus, typical of this family, will be discussed here.

As shown in Figure 4.24, the virion of human reovirus appears round in electron-micrographs and clearly has two layers. The outer capsid, composed of three different proteins, is studded with short spikes that have hemagglutinating activity. The inner capsid, or core, is also composed of three major protein species, but there are several minor proteins in it as well. The core also has spikes (the $\lambda 2$ and $\sigma 1$ proteins) that penetrate the outer capsid and underlie the hemagglutinin proteins. The core surrounds 10 segments of RNA that can be shown to be complementary double-stranded molecules by their base ratios, resistance to enzymes that degrade single-stranded nucleic acids, and other biochemical characteristics. The segments can be separated by electrophoresis into size groups of three *l*arge, three *m*edium, and four *s*mall molecules that encode the *l*ambda, *m*u, and *s*igma groups of proteins, respectively. In addition to the genomic molecules, a large number of short, single-stranded RNA molecules are found in each core. Since most of these oligonucleotides have sequences that are the same as the 5′ end of the positive strand of the genome, they are probably the result of incomplete transcription to create mRNAs.

The reproductive cycle of reovirus is diagrammed in Figure 4.25. The cycle begins with attachment of the virion to the cell surface, possibly to a glycophorin A receptor, followed by penetration by receptor-mediated endocytosis. After the endosome fuses with a lysosome, the outer capsid is degraded by lysosomal enzymes, but the inner capsid remains as a structural entity. The inner capsid somehow then escapes the vesicle and begins the process of expressing the reovirus genome. Unlike the infective process in other viruses, however, in which the virus as a structural entity disappears, the inner capsid of the double-stranded RNA viruses remains intact throughout the entire infective process.

The polymerase(s) necessary for transcription of the genome appears to be an integral part of the inner capsid, residing at the vertices where five capsomers join. This polymerase is activated by the proteolytic digestion of the outer capsid, and by a second event that probably involves cations. Activation has been shown to involve a conformational change in the enzyme, since inactive and active inner capsids have different patterns of contacts between proteins and RNA.

The first transcriptional event is the synthesis of viral mRNAs using the negative strand of the genomic molecules as template. This reaction is like normal transcription (DNA $\longrightarrow$ mRNA) in that the double-stranded template remains intact while the newly synthesized mRNAs are released. This is termed *conservative* transcription to distinguish it from a reaction in which the positive strand of the template duplex would be displaced by a newly synthesized strand ("semiconservative" transcription). The mRNAs are capped by virion core enzymes but do not have poly-A tails. The mRNAs are extruded through the vertices of the core particle into the host cell cytoplasm (Figure 4.26), where translation occurs.

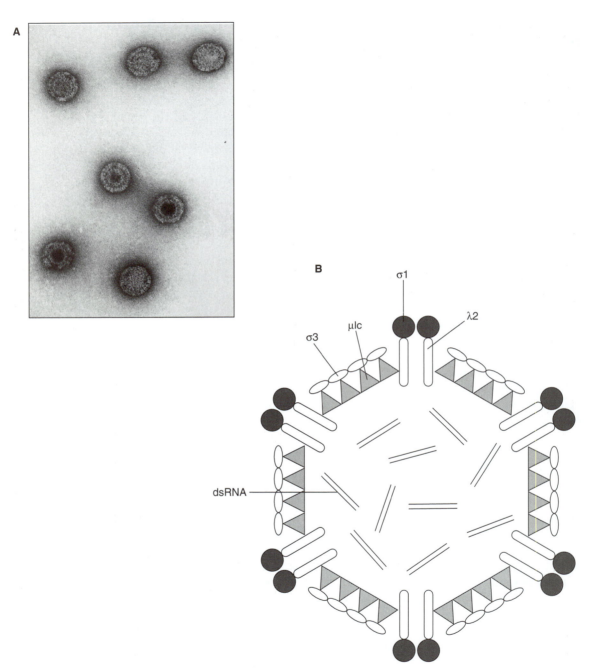

FIGURE 4.24 Electronmicrograph and diagram of the "double" capsid architecture of human reovirus. **A,** Negatively stained reovirus virions. **B,** Diagram of the reovirus virion. The double capsid is composed of σ3 and μ1c proteins. This capsid is pierced by spires made of λ2 and σ1 proteins. The double capsid encloses 10 segments of double-stranded RNA.

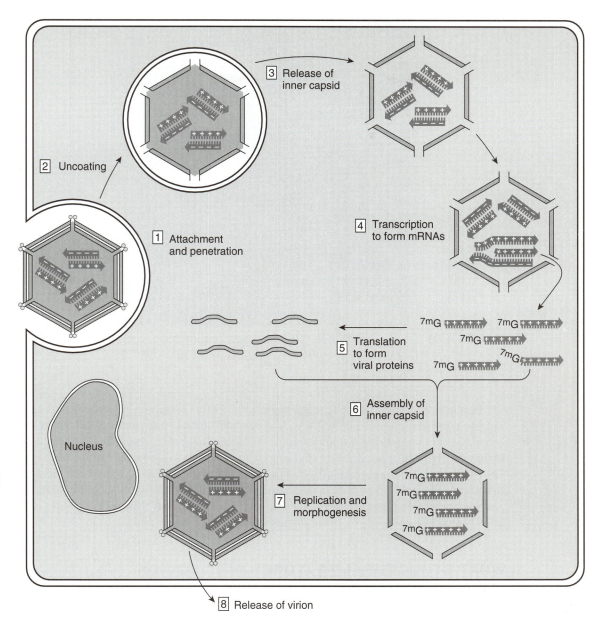

FIGURE 4.25 Diagram of the reovirus reproductive cycle. (1) Attachment and penetration encloses the virion within a vesicle. (2) Uncoating removes only the outer capsid, leaving the inner capsid still enclosing the 10 pieces of double-stranded RNA. (3) The inner capsid is released into the cytoplasm where (4) the negative strands of the genome serve as template for synthesis of positive-strand mRNAs. All RNA synthesis occurs within the inner capsid. (5) After translation of the viral proteins, new inner capsids are assembled around the released mRNAs (6). (7) Each positive strand in the new inner capsid serves as template for synthesis of its complementary negative strand and the outer capsid is assembled around the inner capsid. (8) Mature virions are then released.

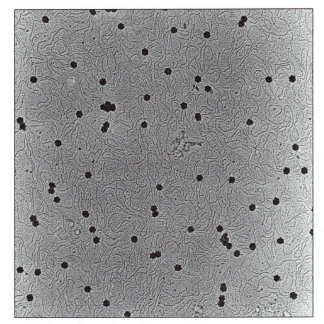

FIGURE 4.26 Electronmicrographs of mRNAs exiting the intact inner cores of reoviruses. The staining procedure used to visualize the RNA molecules causes them to appear far larger than they actually are in relation to the core itself.

Replication of the genome requires the assembly of new virion core particles that contain one copy of each of the 10 mRNAs. The mechanism that ensures that all 10 RNAs are included is not known, but a single-stranded RNA binding protein has been implicated by studies on temperature-sensitive mutants of reovirus. The newly assembled core has an active RNA polymerase that uses each mRNA as a template for the synthesis of their complementary negative strands. These remain associated with the mRNAs to form the double-stranded genomic molecules. The new cores can then synthesize more mRNAs or, if the outer capsid is assembled around them, cease further activity.

WHAT STRUCTURAL FEATURES DISTINGUISH THE RETROVIRUSES FROM OTHER POSITIVE-STRAND RNA VIRUSES?

Although retroviruses have been studied since the turn of the twentieth century (avian leukemia virus was discovered in 1908), the present pandemic of acquired immune deficiency syndrome (AIDS) has focused enormous attention on this group of viruses. Since all the first viruses belonging to this family to be studied were found to cause cancers of some sort, the original name for this group was "RNA tumor viruses" or oncornaviruses (*onco* is Greek for "tumor" + RNA). The discovery that their mechanism of replication requires the *conversion of their RNA genomes into DNA,* so-called *reverse transcription,* however, has led to their current name, the *Retroviridae* (*retro* is Latin for "backward").

The family *Retroviridae* is divided into three subfamilies. The *oncovirinae* contain a large number of viruses that cause leukemias, sarcomas, and mammary carcinomas in a number of different types of animals. Other viruses that do not cause tumors are also included in this group, based on their morphological and molecular characteristics. The oncoviruses are separated into genera on the basis of their morphologies and features of their maturation process during replication. The second subfamily, the *spumavirinae* (*spuma* is Greek for "foam"), is named for the extreme vacuolization that occurs in infected tissue culture cells, producing a bubbly appearance in the cytoplasm. The third group, the *lentivirinae* (*lentus* is Latin for "slow"), take their name from the long incubation period that characterizes the diseases they cause. These diseases include immune deficiencies like AIDS, encephalopathies, arthritises, and pneumonias, but not cancers. The human immunodeficiency viruses HIV-1 and HIV-2 are lentiviruses.

Figure 4.27 diagrams the architecture of HIV-1 as an example of retroviral structure. The retroviruses have icosahedral or wedge-shaped nucleocapsids surrounded by envelopes bearing spikes responsible for attachment of the virion to target cell surfaces. The shape and location of the capsid within the envelope are both traits used in assigning retroviruses to particular species. The genome is a single-stranded RNA that has both a 5′ cap and a poly-A tail, characteristics that suggest a positive-strand RNA. Each virion, however, contains two identical genomic molecules, so the virion is genetically diploid. In addition to the diploid genome, virions also contain other small RNA molecules, particularly tRNAs, which are essential to the infective process.

The genetic structure of all retroviruses follows a similar organization, although numerous additions to the basic structure occur. Fully infective viruses have three structural genes: (1) the *gag* gene (from *g*roup-specific *a*nti*g*en) that encodes the set of proteins that form the nucleocapsid, (2) the *pol* gene that encodes the viral enzymes required for replication, and (3) the *env* gene that encodes glycoproteins that become part of the viral envelope (Figure 4.28). In many oncovirinae, an additional gene whose generic name is *onc* (for "oncogene") is also present. The nature and activities of specific oncogenes will be discussed in Chapter 8. In most cases the *onc* gene interrupts or replaces one of the other structural genes, leading to defective viruses that require the presence of another replication-competent retrovirus in order to reproduce themselves. Although it is the oncogenes that have stimulated much of the research on the oncovirinae, other retroviruses, particularly the lentivirinae, have no *onc* genes. In addition to the three major genes, these viruses contain several small genes that regulate expression of the viral genome. Those of HIV will be discussed later in this chapter.

As we have seen earlier in this chapter, many viruses have distinctive sequences at the ends of their linear genomes, and the retroviruses are no exception. As noted in Figure 4.28, the same sequence of bases is repeated at both termini of the typical retrovirus. Adjacent to those terminal repeats are sequences designated U5 and U3 for "unique 5′" and "unique 3′," respectively. Next to U5 is a "primer binding site" (PBS) about 16 to 18 bases long that is complementary to the 3′ ends of the tRNA species that will serve as a primer molecule during copying of the genome from RNA into DNA. Similarly, next to U3 is a primer site (+P) used when a new positive-strand DNA is made from the newly synthesized negative-strand DNA complement of the genome. The final structural features of the 5′ end are the "dimer linkage site" where the two copies of the genome found in virions attach to one another, and the "packaging signal," which is involved in maturation of the virion and overlaps a portion of the dimer linkage site.

A

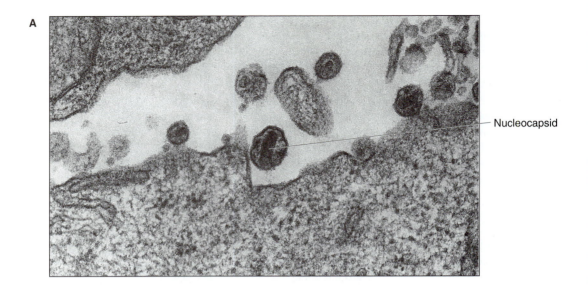

Nucleocapsid

B

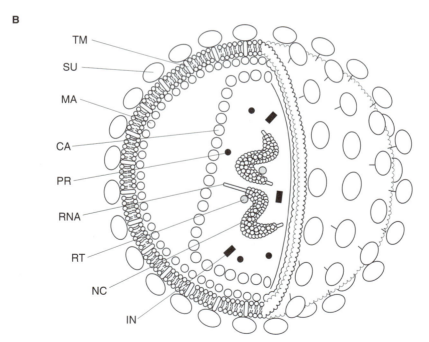

TM
SU
MA
CA
PR
RNA
RT
NC
IN

FIGURE 4.27 A, The virion of human immunodeficiency virus, showing the barrel-shaped nucleocapsid. **B,** Diagram of the architecture of human immunodeficiency virus. Other types of retroviruses have spherical rather than wedge-shaped capsids that may be located centrally or toward the edge of the envelope cavity. TM = transmembrane, SU = surface, MA = matrix, CA = capsid, PR = protease, RT = reverse transcriptase, NC = nucleocapsid, IN = integrase.

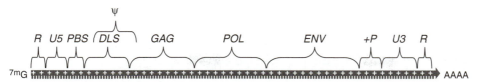

FIGURE 4.28 The genomic organization of a retrovirus. R = terminal redundancy, U5/U3 = unique 5′ and 3′ regions, PBS = negative-strand primer binding site, DLS = dimer linkage site, ψ = packaging signal, +P = positive-strand primer site. The viral genes GAG, POL, and ENV encode polyproteins and are not shown to scale. See text for explanation.

HOW DOES "REVERSE TRANSCRIPTION" WORK?

The "core" of molecular biology is the Central Dogma: DNA is transcribed into RNA that is, in turn, translated into protein. In the 1960s, however, Howard Temin argued that the RNA tumor viruses must commit "heresy," that is, their life cycles had to include the reverse reaction, which copied their genomes from RNA into DNA. Several lines of evidence supported this hypothesis. Although these viruses had RNA genomes, their multiplication was inhibited by agents that interfered with either *DNA* replication or transcription. In addition, nucleic acid hybridization studies showed that infected cells contained DNA sequences in their genomes that were complementary to viral genomic RNAs. These observations suggested that a DNA copy of the viral RNA genome was inserted into the host's chromosome and then transcribed like other "cellular" genes.

In 1970 Temin and David Baltimore, who was working independently, isolated and characterized the enzyme necessary for this scheme to occur: *RNA-dependent DNA polymerase,* now usually called **reverse transcriptase,** the term coined by the journal *Nature.* Although they can use RNA as well as DNA as a template, retroviral reverse transcriptases are similar to other DNA polymerases. They cannot begin synthesis *de novo,* but rather have only the ability to elongate a primer (either RNA or DNA) using the information from a template molecule. Unlike DNA polymerase, reverse transcriptase first uses RNA and then DNA as the template molecule. In addition to their polymerizing activity, reverse transcriptases have an endonuclease activity (termed RNase H) that degrades the RNA strand of RNA-DNA hybrids, producing oligonucleotides with 5′ phosphate and 3′-OH termini. Both the polymerizing and endonuclease activities are required for reverse transcription of the viral genome.

The process of reverse transcription is complex. This complexity results from the problems inherent in making a complete DNA copy of a linear molecule. DNA synthesis in the 3′ to 5′ direction (lagging strand synthesis) requires repeated rounds of priming, elongation, and replacement of the primer by DNA. This full set of reactions is impossible at the 3′ end of the template molecule, however, so the new DNA strand will be incomplete at its 5′ end. As we saw in Chapter 3, bacteriophages with linear DNA genomes solve this problem in a variety of ways. T7 concatemerizes the incomplete genomes into large molecules, completes DNA synthesis through the resulting gaps, and then cleaves the concatemer. λ virus simply circularizes its genome before replication and thereby avoids the problem altogether.

Reverse transcription uses a novel "jumping" strategy to solve the problem of making full-length DNA copies of a linear template. The major features of the process, which occurs in the cytoplasm even though it involves DNA synthesis, are illustrated in Figure 4.29. The first set of reactions synthesizes a negative DNA strand complementary to the positive-strand RNA genome. The primer for the first reaction is a tRNA that was included in the virion itself

(1) tRNA binds to PBS.

(2) Reverse transcriptase elongates 3′ end of tRNA making DNA complementary to U5 and R.

(3) RNase H degrades R and U5 sequences in genomic RNA.

(4) The first "jump": negative-strand DNA pairs with 3′-R sequence in genomic RNA.

(5) Reverse transcriptase elongates negative-strand DNA to create a DNA/RNA hybrid molecule.

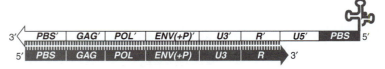

(6) RNase H degrades all genomic RNA molecule except (+P) sequences.

FIGURE 4.29 For Legend see opposite page.

(7) Reverse transcriptase begins synthesis of positive-strand DNA using (+P) RNA as primer.

(8) RNase H degrades all remaining RNA.

(9) The second "jump": the positive-strand PBS sequence pairs with its complementary sequence in the negative strand.

(10) Reverse transcriptase elongates both strands to complete synthesis of a double-stranded proviral DNA molecule.

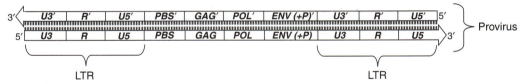

FIGURE 4.29 The process of retroviral reverse transcription. See text for detailed explanation of each step. Note that the process creates a DNA molecule with additional sequences (long terminal repeats = LTR) not found in the genomic RNA molecule.

(tRNA^{trp} for avian viruses, tRNA^{pro} for mouse viruses, and tRNA^{lys} for HIV). This tRNA is bound to the primer binding site (PBS) adjacent to U5 near the 5′ end of the genomic RNA (step 1). Reverse transcriptase first elongates the tRNA primer to generate a short negative-strand DNA molecule complementary to U5 and the terminal repeat (R) at that end (step 2). Reverse transcriptase then uses its RNase H activity to degrade the R and U5 sequences from the genomic RNA molecule. This weakens the binding of the nascent DNA strand to its template, setting the stage for the first "jumping" reaction (step 4). The growing negative-strand DNA molecule moves from the 5′ end of the genome to become paired with the 3′-R sequence. Note that the 5′ tRNA primer is now dangling off the 3′ end of the template molecule so that all the binding is between newly synthesized DNA and the viral genome. This arrangement permits the negative-strand DNA molecule to be elongated until the entire viral RNA strand (including its termini) has served as template (step 5).

The next set of reactions removes the RNA from the RNA/DNA duplex and synthesizes a positive-strand DNA molecule. Reverse transcriptase's RNase H activity degrades all but a short section of the genomic RNA (labeled as +P) that then serves as the primer for positive DNA strand synthesis (step 6). Elongation of the positive-strand DNA continues until synthesis reaches the tRNA that served as primer during negative-strand synthesis (step 7). The tRNA cannot serve as template because its bases are occluded by involvement in intrachain base pairs. All RNA in the complex, including the primer tRNA, is then degraded by RNase H (step 8). This leads to the second "jumping" reaction that moves the new positive-strand DNA molecule to the opposite end of the negative-strand DNA template (step 9). Both DNA strands now can serve as primers and as templates to complete the process of reverse transcription (step 10). The negative-strand DNA is elongated using the U3-R-U5 portion of the positive-strand DNA as template, while at the same time the positive-strand DNA itself is elongated using the rest of the negative-strand DNA as template. This double-stranded DNA molecule is referred to as a **provirus** to distinguish it from the RNA molecule that is contained in the virion.

Note that the proviral DNA that has been synthesized by reverse transcription is not exactly the same as the original viral RNA molecule. The "jumping" steps that permit synthesis of the DNA termini during reverse transcription create a molecule that has *identical* ends consisting of U3-R-U5. These sequences are termed the **long terminal repeats** (LTR). Each LTR contains the sequences that signal a transcriptional unit; the U3 segment has the promoter elements and a cap addition site, while the U5 segment includes a poly-A site (Figure 4.30). Both LTRs could serve to initiate transcription, but usually only the 5′ one does so, while the 3′ LTR provides the necessary termination sequences. Each LTR also contains a transcriptional **enhancer sequence** that interacts with viral or host factors to control the rate of transcription of contiguous DNA sequences. The function of the TAR region of the LTR will be described in a later section of this chapter. The LTRs play a central role in the insertion of the viral genome, as double-stranded DNA, into a host cell chromosome.

FIGURE 4.30 The genetic elements of the HIV-1 long terminal repeat. Transcription is initiated at position +1 at the beginning of the R sequence, while the U3 sequence contains promoter and enhancer sequences. The TAR sequence is involved in regulation of proviral gene expression.

HOW DOES RETROVIRUS PROVIRAL DNA BECOME INCORPORATED INTO THE HOST CELL'S GENOME?

The next steps in the retrovirus life cycle occur in the nucleus. The first of these, integration of the viral DNA molecule into the host genome, is unique to the retroviruses. Other viruses, as we shall see, have a reverse transcription step in their reproductive cycles or are capable of inserting their DNAs into their host's genome, but only in the retroviruses is insertion of the viral DNA into a chromosome *required* for expression of viral genes.

   Several structural features of the provirus and its neighboring host sequences suggest that the mechanism of integration is like that found in the movement of transposons like the yeast's *Ty* or *Drosophila's copia* elements. The proviral DNA is shortened by four base pairs during insertion, and its ends are always 5'TG....CA3' (Figure 4.31, *A*). In addition,

A The organization of an integrated retroviral provirus.

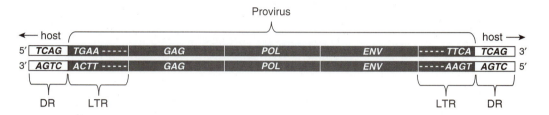

B The process of integration.

1) Creation of staggered nicks

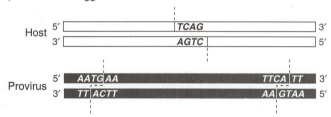

2) Ligation leaving 6-base gaps

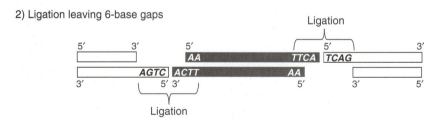

3) Gap-filling to produce direct-repeat sequences in host DNA

FIGURE 4.31 Integration of retrovirus proviral DNA into a host DNA molecule. **A,** The long terminal repeat sequences of the integrated provirus are flanked by four-base direct repeat sequences. **B,** The process of integration. (1) Both the host and proviral DNAs are cut to produce four-base-long staggered nicks. (2) Blunt-end ligation reactions between the host and proviral DNAs produce 6-base long gaps at each junction. (3) DNA synthesis fills the gaps, creating a short direct repeat of host cell base pairs on each side of the provirus.

a short set of bases from the host has been duplicated, so that the provirus is flanked by direct repeats in the host DNA. These features suggest that the proviral DNA is integrated into a host site by a process illustrated in Figure 4.31, *B*. In the first step (*1*) staggered cuts are made in both molecules. The 5′ ends of the proviral DNA strands are then joined to the 3′ ends of the host DNA molecules (*step 2*). A gap-filling reaction using host cell DNA as a template then produces the direct repeats that flank the inserted proviral DNA (*step 3*). Finally, ligation completes the insertion. A viral protein called the "integrase," which in addition to the RNase H and polymerase is encoded by the *pol* gene, probably mediates the staggered cleavage step, while cellular DNA repair enzymes are responsible for the gap-filling reactions.

HOW ARE RETROVIRUS GENES EXPRESSED?

Reverse transcription and insertion of the resulting proviral DNA into the host cell's chromosomes are distinctive retroviral characteristics. In its integrated state, the provirus is expressed entirely by the normal cellular enzyme systems for making mRNAs. As noted previously, the LTR regions contain promoter sites, and enhancer sequences that may be subject to activation by cellular regulating molecules. Gene expression in mouse mammary tumor virus (MMTV), for example, can be regulated by glucocorticoid hormones; genetic engineers have taken advantage of this to "design" hormone-responsive genes by placing a target gene downstream from the MMTV LTR.

The strategies used by the retroviruses to create individual viral proteins are all ones also used by other RNA viruses. Gene expression begins by transcription of the entire provirus sequence into a 35S RNA molecule, which is then capped and tailed in the usual fashion for eukaryotic transcripts. Thus, the virus is, in effect, back to where it began the entire infective process: a polycistronic mRNA. As illustrated in Figure 4.32, a combination of three different strategies is then used to generate the separate viral proteins. The first strategy is at the level of RNA processing. Differential splicing of the 35S transcript produces a 24S mRNA encoding only the final gene in the series *(env)*, while the full-length mRNA is used to produce the *gag* and *pol* proteins. The second strategy involves the translation of the *gag-pol* mRNA. Most of the time, translation of this message stops at a nonsense codon terminating the *gag* polyprotein gene so that the *pol* gene is not translated. Even if simple read-through of the *gag* gene termination codons were to occur, the *pol* gene information in the *gag-pol* mRNA would still not be translated since it is out of phase with the *gag* gene. About 5% of the time, however, a ribosome makes a frameshift after translating the *gag* gene, with the result that a "super" polyprotein containing both the *gag* and *pol* proteins is synthesized. Since the translation products of all three genes are polyproteins, cleavage to generate individual protein molecules is the final retrovirus strategy for gene expression. Cleavage of the *gag* and *pol* polyproteins is mediated by a viral enzyme that is encoded at the junction of the two genes (recall a similar organization in picornaviruses shown in Figure 4.3), while cleavage of the *env* polyprotein is done by a cellular enzyme in the rough endoplasmic reticulum during maturation. Maturation will be discussed more fully in Chapter 6.

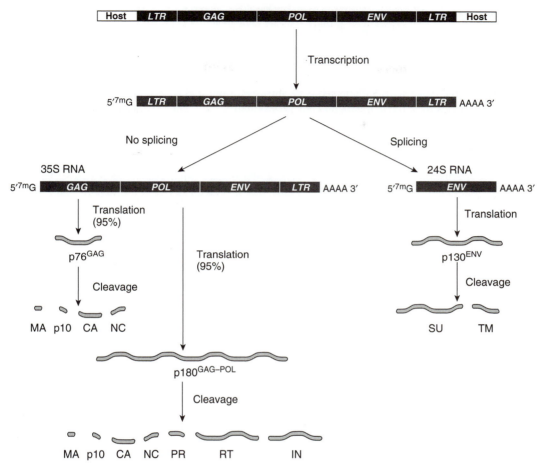

FIGURE 4.32 Gene expression in retroviruses. The entire genome is transcribed into a 35S RNA molecule. A splicing reaction produces the 24S mRNA for the *ENV* polyprotein, while the entire genomic molecule is used for translation of the *GAG* polyprotein, and by a frame shift, the *POL* polyprotein. Each polyprotein is then cleaved into its individual protein components whose identities are given in Figure 4.27. (The sizes given in this example are those for a sarcoma virus.)

HOW DO RETROVIRUSES REPLICATE THEIR GENOMES?

Since the retrovirus genome is positive-strand RNA, synthesis of new genomes from the integrated provirus is simple and straightforward. Normal host cell RNA polymerase II transcribes the entire provirus into RNA and the usual host enzymes modify that transcript to include a 5′ cap and 3′ poly-A tail.

HOW DO THE LENTIVIRUSES DIFFER FROM THE ONCOVIRUSES?

The lentiviruses, as typified by human immunodeficiency virus, differ from other retroviruses both in their genomic organization and in their virion morphology; they have barrel-shaped rather than spherical nucleocapsids as shown in Figure 4.27. Although HIV and the other lentiviruses have the *gag, pol,* and *env* genes, there is no homology between those sequences and those of the other retroviruses. In addition to those polyprotein genes, HIV has six small genes, at least some of whose products influence the efficiency of viral infection and gene expression.

Figure 4.33 shows the location of the genes *tat, rev, nef, vif, vpr,* and *vpu* in HIV-1. The virus appears to make very economical use of its genome since all of these genes are encoded by sequences that overlap those of other genes. Two of these genes (*tat* and *rev*) are encoded by dispersed sequences within the provirus. The *tat* and *rev* genes are read from the same mRNA using different reading frames. That message is created by splicing sequences from the region between the *pol* and *env* genes to those of a region that is part of the 3′ end of the *env* gene itself. The other four genes are encoded as complete units. The *vpu* gene is encoded entirely within the 5′ end of the *env* gene, while the *nef* gene begins in the 3′ end of *env* and continues into U3. The beginning of the *vif* gene overlaps the 3′ end of the *pol* gene. The *vpr* gene does not overlap one of the major HIV genes but does share some bases with *vif.*

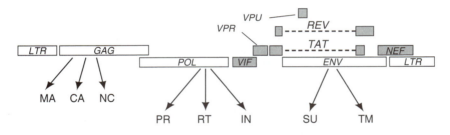

Structural Genes:

gag: MA – myristylated matrix protein
CA – major capsid protein
NC – RNA-binding nucleocapsid protein

pol: PR – protease
RT – reverse transcriptase
IN – integrase

env: SU – envelope glycoprotein (extracellular)
TM – envelope glycoprotein (transmembrane)

Regulatory Genes:

tat: transactivating protein (transcription)
rev: regulator of viral mRNA synthesis and export

Accessory Genes:

nef: "negative factor" (multiple functions)
vif: virion infectivity factor
vpr: cell cycle regulator
vpu: (HIV-1 only) promotes release of virions

FIGURE 4.33 The genomic organization of human immunodeficiency virus-1. The regulatory and accessory genes are indicated by shaded boxes. The dashed lines indicate the splicing reactions needed to produce the *tat* and *rev* mRNAs.

Functions of the HIV-1 Regulatory and Accessory Gene Products

Analysis of HIV-1 mutants has yielded possible functions for most of these genes. The proteins encoded by *tat, rev,* and possibly *nef* all play roles in regulation of viral gene expression. Mutations in *tat* and *rev* strongly inhibit gene expression, so the *tat* and *rev* proteins appear to be positive regulators. Studies have shown that some mutations in *nef* enhance viral gene expression while others decrease it, so the *nef* protein may be both a positive and a negative regulator.

The small (86 amino acid) *tat* protein is a **transactivator** that acts together with a cellular protein to enhance viral transcription. The tat-responsive (TAR) region of the provirus is located between bases +1 to +44 (+1 is the base in the viral transcript to which the 5′ cap is attached). Directed mutagenesis studies indicate that the *tat* protein does not interact directly with the TAR DNA bases themselves, but rather with the RNA molecule that is transcribed from them. The TAR RNA molecule can form a stem with a three-base bulge at position 23 to 25 and a loop at position 30 to 35 (Figure 4.34).Binding of the *tat* protein and a 68 kdalton host protein to the TAR RNA may serve to bring the *tat* protein into position to interact with transcription factors bound at the enhancer and promoter regions of the HIV-1 long terminal repeat. In that position the *tat* protein appears both to stimulate initiation of transcription and to prevent premature termination of transcription at about position +59.

While the *tat* protein stimulates expression of the entire viral genome, the *rev* protein is necessary only for the synthesis of the major viral structural polyproteins. In cells infected with HIV-1 mutants lacking a functional *rev* gene, mRNAs for *gag-pol* and *env* are not produced. Instead, the genome-length transcript is spliced into small mRNAs encoding only the *tat, rev,* and *nef* proteins.

The *rev* protein appears to bind to a specific sequence called the *Rev-responsive element* (RRE) in the transcripts of the structural genes to stabilize their larger mRNAs and facilitate their transport from the nucleus and subsequent translation. This, in turn, enhances transport of full-length genomic RNAs to the cytoplasm and the synthesis of structural proteins, both of which are needed to make new virions.

The remaining small genes of HIV-1 are termed *accessory genes* since they become dispensable when the virus is passaged for long periods in tissue culture. Since they are highly conserved in isolates of HIV-1, however, they are likely to have important functions in natural infections and growth of the virus.

The largest of the accessory proteins, the *nef* protein, is modified by addition of myristic acid, a long-chain fatty acid that allows the protein to be anchored in the inner surface of the cytoplasmic membrane where it may interact with cellular protein kinases. Sequence data also indicate that the *nef* protein has sites where it can be phosphorylated as well as a recognition domain for interactions with cellular signaling pathway components. The *nef* (for *negative factor*) protein was named from initial observations that this protein appeared to down-regulate or inhibit transcription of the HIV provirus. Its functions now appear to be more complex. In various cases, the *nef* protein has been shown to activate transcription, to promote increased endocytosis, and to down-regulate the expression of CD4 and histocompatability antigens on the cell surface.

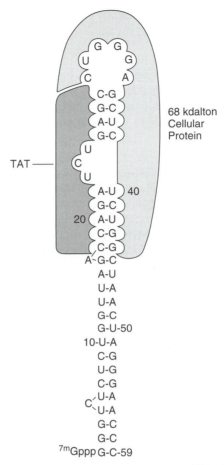

FIGURE 4.34 Interaction of the TAR RNA molecule with the HIV-1 TAT protein and a 68 kdalton cellular protein. The TAT protein binds to the region around the three-base bulge at position 23–25, while the cellular protein binds to the loop at top of the TAR hairpin.

The somewhat smaller *vif* protein is called the *virion infectivity factor* since mutations in this gene cause production of virions that are greatly impaired in their ability to infect new cells. It appears that these virions can enter new host cells and begin reverse transcription of the genome, but that process is not completed and infection is halted. It thus appears that the *vif* protein has a role in some very early step in uncoating or replication.

The *vpr* (*viral protein rapid*) protein is associated with the nucleocapsid in the virion and takes it name from the fact that it appears to affect how quickly HIV-1 infects its host cell. The *vpr* protein causes infected cells to stop in the G_2 stage of the cell cycle. It also binds to the outer side of nuclear pores where it may act to facilitate docking and transport of the DNA provirus into the nucleus.

The *vpu* gene is found only in HIV-1, hence its name for *viral protein unique*. Mutations in the *vpu* gene produce an accumulation of viral cores in intracellular vacuoles and a decrease in extracellular virions. This suggests that the role of the *vpu* protein is in the maturation or release of new virions.

SUMMARY

The RNA viruses of eukaryotic cells must solve two general problems in order to express and replicate their genomes. First, they must synthesize new RNA molecules (either genomes or mRNAs) from information contained in an RNA rather than DNA template. Second, they must synthesize individual proteins from polycistronic genomes (or transcripts made from those genomes) using cellular machinery that reads only monocistronic messages.

The viral solutions to the first problem take two forms. All but one group of RNA viruses encode the information for an RNA-dependent–RNA-polymerase that permits the virus to synthesize new RNA molecules from an RNA template. Negative-strand viruses carry molecules of this enzyme into their host cells at the time of infection, while positive-strand viruses cause it to be synthesized *de novo* as an early gene product. The retroviruses use a very different strategy. These viruses carry a "reverse transcriptase" that creates a double-stranded DNA copy of the viral genome that is then inserted into the host cell's chromosomes to permit expression and replication of viral genes.

The problem of synthesizing individual proteins from polycistronic genomes has been solved in a variety of ways, with particular viruses often employing combinations of the basic themes. The most common strategy for making proteins is the synthesis of a polyprotein that is then cleaved into individual proteins (picornaviruses, etc.). This strategy is frequently combined with the synthesis of subgenomic RNA molecules to permit expression of different sets of genes at different points in the infective process or within different cellular compartments (envelope polyproteins synthesized on rough ER and cleaved in vesicles in the togaviruses, for example). A second strategy is the formation of subgenomic RNAs that are translated to yield only a single protein. These may be nested sets of RNAs in which only the gene encoded at the 5′ end is translated even though other genes are present on the same molecule (coronaviruses). Frequently they are true monocistronic mRNAs transcribed individually from the genome (rhabdoviruses, etc.) or created from longer transcripts by post-transcriptional splicing reactions (retroviruses). In addition, the same mRNA may be translated into several different proteins by using alternative initiation codons (paramyxoviruses), by reading through termination codons to make longer proteins (retroviruses), or by ribosomal frameshifting (coronaviruses, retroviruses). The final strategy is to subdivide the genome so that each segment may encode only a single protein (orthomyxoviruses, reoviruses).

SUGGESTED READINGS

General Reviews

1. "Replication strategies of the single stranded RNA viruses of eukaryotes" by E.G. Strauss and J.H. Strauss, *Current Topics in Micro and Immunol* 105:1–98 (1983).
2. "Evolution of RNA viruses" by J.H. Strauss and E.G. Strauss, *Ann Rev Microbiol* 42:657–683, (1988).
3. "Nonsegmented negative-strand RNA viruses: Genetics and manipulation of viral genomes" by K-K. Conzelmann, *Ann Rev Genet* 32:123–162 (1998).

Positive-Strand Viruses

1. "Protein processing map of poliovirus" by M. Pallansch, O. Kew, B. Semler, et al., *J Virol* 35: 873–880 (1984), provides data indicating that individual poliovirus proteins arise from cleavage of a polyprotein.

2. "Picornaviruses and their relatives in the plant kingdom" by A.M.Q. King, G.P. Lomonossoff, and M.D. Ryan, *Seminars in Virol* 2: 11–17 (1991), discusses the similarities seen in viruses infecting host cells in completely different kingdoms.

3. "RNA signals in entero- and rhinovirus genome replication" by W-K. Xiang, A.V. Paul, and E. Wimmer, *Seminars in Virol* 8:256–272 (1997), reviews the roles of RNA stem-loop and cloverleaf structures in both negative- and positive-strand synthesis.

4. "Restricted initiation of protein synthesis on the potentially polycistronic Sindbis virus 42S RNA" by S. Bonatti, N. Sonenberg, A. Shatkin, and R. Cancedda, *J Biol Chem* 255:11473–11477 (1980), explains how this virus regulates the expression of its nonstructural and structural genes.

5. "The signal for translational readthrough of a UGA codon in Sindbis virus RNA involves a single cytidine residue immediately downstream of the termination codon" by G. Li and C.M. Rice, *J Virol* 67:5062–5067 (1993), describes how the sequence context of the termination codon allows efficient readthrough from the ns3 to the ns4 coding region of the viral genome.

6. "Alphavirus nsP3 functions to form replication complexes transcribing negative-strand RNA" by Y-F. Wang, S.G. Sawicki, and D.L. Sawicki, *J Virol* 68:6466–6375 (1994); "Polypeptide requirements for assembly of functional Sindbis virus replication complexes: A model for the temporal regulation of minus- and plus-strand RNA synthesis" by J.A. Lemm, et al., *EMBO J* 13:2925–2934 (1994); and "Template-dependent initiation of Sindbis virus RNA replication in vitro" by J.A. Lemm, et al., *J Virol* 72:6546–6553 (1998), describe the involvement of the four proteins of the polymerase complex at different stages of RNA synthesis in Sindbis virus.

7. "Genomic organization of murine coronavirus MHV" by S.G. Siddell, In *Positive Strand RNA Viruses,* edited by M.A. Binton and R.R. Reuckert, Alan R. Liss, Inc., 1987, pages 127–135, describes the nested organization of genes in this virus.

8. "Coronaviruses: structure and genome expression" by W. Spaan, D. Cavanagh, and M.C. Horzinek, *J Gen Virol* 69:2939–2952 (1988), reviews genomic organization and expression in the Coronaviruses.

9. "Characterization of an efficient coronavirus ribosomal frameshifting signal: Requirement for an RNA pseudoknot" by I. Brierley, P. Digard, and S.C. Inglis, *Cell* 57:537–547 (1989), shows that translation of two proteins from the large open reading frame at the 5′ end of the genome involves a −1 frameshift induced by a pseudoknot formed in the RNA genome.

10. "Arterivirus discontinuous mRNA transcription is guided by base paring between sense and anti-sense transcription-regulating sequences" by G. van Marie, et al., *Proc Natl Acad Sci USA* 96:12056–12061 (1999), describes experiments using site-directed mutagenesis to study how these viruses produce a nested set of structural gene mRNAs.

Negative-Strand Viruses

1. "Transcription and replication of rhabdoviruses" by A.K. Banerjee, *Micro Rev* 51:66–87 (1987), reviews the organization, expression, and replication of the rhabdovirus genome and the roles of each viral protein in those processes.

2. "Phosphorylation within the amino-terminal acidic domain I of the phosphoprotein of vesicular stomatitis virus is required for transcription but not for replication" by A.K. Pattnaik, et al., *J*

Virol 71: 8167–8175 (1997), presents findings indicating that the level of phosphorylation of the P protein regulates the functioning of the polymerase L protein in this rhabdovirus.

3. "The length and sequence composition of vesicular stomatitis virus intergenic regions affect mRNA levels and the site of transcript initiation" by E.A. Stillman and M.A. Whitt, *J Virol* 72:5565–5572 (1998), supports the scanning polymerase model for mRNA synthesis.

4. "The virion glycoproteins of Ebola viruses are encoded in two reading frames and are expressed through transcriptional editing" by A. Sanchez, et al., *Proc Natl Acad Sci USA* 93:3602–3607 (1996), shows that RNA editing adds a single untemplated adenosine to shift the transcript from one for a secreted protein to that of the viral envelope glycoprotein.

5. "Influenza virus genome consists of eight distinct RNA species" by D. McGeoch, P. Fellner, and C. Newton, *Proc Natl Acad Sci USA* 73:3045–3049 (1976), presents data on the segmentation of the orthomyxovirus genome and what each segment encodes.

6. "Influenza virus, an RNA virus, synthesizes its messenger RNA in the nucleus of infected cells" by C. Herz, E. Stavnezer, R.M. Krug, and T. Gurney, *Cell* 26:391–400 (1981), presents evidence that influenza virus reproduces in the nucleus rather than the cytoplasm like other RNA viruses.

7. "Gene structure and replication of influenza virus" by R.A. Lamb and P.W. Choppin, *Ann Rev Biochem* 52:467–506 (1983) reviews how the influenza virus genome is organized, expressed, and replicated.

8. "Influenza virus nucleoprotein interacts with influenza virus polymerase proteins" by S.K. Biswas, P.L. Boutz, and D.P. Nayak, *J Virol* 72:5493–5501 (1998); "RNA-dependent activation of primer RNA production by influenza virus polymerase: Different regions of the same protein subunit constitute the two required RNA-binding sites" by M-L. Li, C.B Ramirez, and R.M. Klug, *EMBO J* 17:5844–5852 (1998), describe the interactions of the polymerase proteins PA, PB1 and PB2 with the NP protein during transcription and replication of influenza virus.

9. "Sequencing studies of Pichinde arenavirus S RNA indicate a novel coding strategy, an ambisense viral S RNA" by D.D. Auperin, V. Romanowski, M. Galinski, and D.H.L. Bishop, *J Virol* 52:897–904 (1984) describes the unique architecture of small RNA from this virus.

Double-Stranded Viruses

1. "Recent progress in reovirus research" by W.K. Joklik, *Ann Rev Genet* 19:537–575 (1985), reviews reovirus genomic organization and expression as well as the function of various proteins.

Retroviruses

1. "The first human retrovirus" by R.C. Gallo, *Scientific American* 255(12):88–98 (1986), and "The AIDS virus" by R.C. Gallo, *Scientific American* 256(1):46–56 (1987), are a set of articles describing the known human retroviruses.

2. "Reverse transcription" by H. Varmus, *Scientific American* 257(9):56–64 (1987), describes how the retroviruses carry out their "signature" reaction.

3. "Retroviruses" by H. Varmus, *Science* 240:1427–1435 (1988), presents these viruses as more than just pathogenic organisms, and discusses their possible roles as tools in molecular biology and biotechnology.

4. "Human immunodeficiency virus as a prototypic complex retrovirus" by B.R. Cullen, *J Virol* 65:1053–1056 (1991) and "Genetic regulation of human immunodeficiency virus" by K. Steffy and F. Wong-Staal, *Micro Rev* 55:193–205 (1991), review the roles of the HIV regulatory genes on both viral gene expression and the growth of uninfected cells. "The biochemistry of AIDS"

by Y.N. Vaishnav and F. Wong-Staal, *Ann Rev Biochem* 60:577–630 (1991), discusses the organization and expression of the HIV genome.

5. "Two efficient ribosomal frameshifting events are required for synthesis of mouse mammary tumor virus *gag*-related polyproteins" by T. Jacks, K. Townsley, H.E. Varmus, and J. Majors, *Proc Natl Acad Sci USA* 84:4298–4302 (1988); "Signals for ribosomal frameshifting in the Rous sarcoma virus *gag-pol* region" by T. Jacks, H.D. Madhani, F.R. Masiarz, and H.E. Varmus, *Cell* 55:447–458 (1988); "Characterization of ribosomal frameshifting in the HIV-1 *gag-pol* expression" by T. Jacks, et al., *Nature* 331:280–283 (1988), describe how retroviruses utilize different reading frames within the *gag-pol* transcript by a series of −1 frameshifts.

6. "RNA regulatory elements in the genomes of simple retroviruses" by J.D. Banks, K.L. Beemon, and M.L. Linial, *Seminars in Virol* 8:194–204 (1997), reviews the role of RNA folding in regulation of numerous events in the retrovirus life cycle; "Bipartite signal for readthrough suppression in murine leukemia virus mRNA: An eight-nucleotide purine-rich sequence immediately downstream of the *gag* termination codon followed by an RNA pseudoknot" by Y-X. Fong, H. Yuan, A. Rein, and J.G. Levin, *J Virol* 66:5127–5132 (1992); "Two *cis*-acting signal control ribosomal frameshifts between human T-cell leukemia virus type II *gag* and *pol* genes" by H. Falk, et al., *J Virol* 67:6273–6277 (1993); "Structural and functional studies of retroviral RNA pseudoknots involved in ribosomal frameshifting: Nucleotides at the junction of the two stems are important for efficient ribosomal frameshifting" by S. Chen, et al., *EMBO J* 14:842–852 (1995), look at the role of RNA structural features in frameshifting.

7. "HIV-1 regulatory/accessory genes: Keys to unraveling viral and host cell biology" by M. Emerman and M.H. Malim, *Science* 280:1880–1884 (1998), reviews the functions of HIV-1's five accessory genes; "The HIV-1 Rev protein" by V.W. Pollard and M.H. Malim, *Ann Rev Microbiol* 52: 491–532 (1998), reviews the critical role of Rev in HIV-1 infections.

8. *AIDS Update: An Annual Overview of Acquired Immune Deficiency Syndrome* by Gerald J. Stine, Prentice Hall, Englewood Cliffs, NJ (2000), presents the latest information about HIV and AIDS in its yearly editions.

Expression and Replication of the Viral Genome in Eukaryotic Hosts: The DNA Viruses

CHAPTER OUTLINE

- What structural features distinguish eukaryotic DNA viruses?
- What outcomes are possible when a DNA virus infects an animal cell?
- How do animal viruses with double-stranded DNA genomes express and replicate those genomes?
- How do animal single-stranded DNA viruses express and replicate their genomes?
- How do plant single-stranded DNA viruses express and replicate their genomes?
- What features set hepadnaviruses and caulimoviruses apart from other eukaryotic DNA viruses?
- Summary

As we saw in Chapter 4, the RNA viruses of eukaryotic cells have evolved a variety of strategies to overcome the myriad problems they encounter in expressing and copying their RNA genomes in host cells whose own genomes are DNA rather than RNA. The problems facing the DNA viruses of eukaryotes are far less daunting. Since their genomes are of the same molecular type as their host cells, they can simply use the cellular synthetic machinery at hand for transcription of their genes and then for translation of the resulting messages into proteins. In fact, many of the characteristics of these processes were first elucidated in DNA viral systems. We reviewed the steps in these processes in Chapter 3.

If transcription is easier for the DNA viruses than for the RNA ones, replication is more difficult. The first problem in replication is encountered by DNA viruses that rely completely on their hosts' DNA synthesizing systems to copy their own genomes. DNA synthesis in eukaryotic cells does not occur continuously but is confined largely to the S phase of the cell cycle when all the components of the DNA synthesizing systems are present and active. Many of the cells in multicellular organisms rarely divide after they have differentiated, if at all. Small DNA viruses, as well as the RNA retroviruses that use their hosts' systems, therefore must infect cells that have entered S phase naturally. Viruses with somewhat larger genomes have a means of forcing their hosts' cell cycle forward into S phase. Even larger DNA viruses escape the tyranny of the S phase altogether by encoding their own enzymes and accessory proteins for DNA replication.

The other problem arises any time a linear template is used in the synthesis of a DNA molecule, since the priming reactions necessary for DNA synthesis cannot occur at the ends of linear molecules (see Chapter 3). Many viruses solve this problem in the most economical fashion possible: They have circular rather than linear genomes, so the problem simply does not arise. DNA viruses that do have linear genomes, however, have evolved a number of strategies for ensuring that the complete molecule is replicated. These include the use of protein primers, various means of forming concatemers, and reverse transcription.

WHAT STRUCTURAL FEATURES DISTINGUISH EUKARYOTIC DNA VIRUSES?

The DNA viruses of eukaryotes show far more diversity of size and architecture of their genomes than the RNA viruses. The genomes of autonomously replicating RNA viruses range in size from about 1.2×10^6 daltons, or 4 Kb, to about 1.2×10^7 daltons, or 30 Kb, only a tenfold difference. The DNA viruses have the same lower limit as the RNA viruses, about 1.2×10^6 daltons, or 5 Kb, but the upper limit is more than 300 times larger, at around 3.8×10^8 daltons, or 375 Kbp. This larger size is accompanied, of course, by a corresponding increase in the number of proteins encoded by DNA viruses compared to RNA viruses.

The DNA viruses have a number of different arrangements of their DNA molecules. Some of these are straightforward shapes, such as the linear or circular double-stranded DNAs that occur in the genomes of eukaryotic or prokaryotic cells. Other arrangements are unique to the viruses (see Figure 1.2). One group of viruses contains linear single-stranded DNA of either positive or negative polarity. Another group's genome consists of a DNA molecule that is double-stranded with closed ends (visualize a circular single-stranded molecule in which one half of the molecule is complementary to the other half, so that the halves form base pairs). The most unusual organization is that of a negative strand held in the form of a nicked circle by base pairing to a much shorter positive-strand molecule. This genome therefore has characteristics of both linear and circular, as well as single-stranded and double-stranded DNA. The replication of this molecule is correspondingly interesting!

We shall begin our consideration of the DNA viruses with those that infect animal cells, starting with viruses whose genomes are the most similar to those of their hosts and, therefore, rely most directly on host cell processes to express and replicate those genomes. The smallest of these double-stranded DNA viruses are in the family *Papovaviridae,* a name coined from the names of three of its members. *Pa* is taken from the genus *papillomavirus* (*papilla* is Latin for a small, nipple-shaped projection + *oma,* Greek for "swelling" or "tumor"). *Po* is from the genus *polyomavirus* (*poly* is Greek for "many" + *oma*), and *va* from Simian *va*cuolating virus #40 (SV40), a member of the polyomavirus genus. These small viruses have icosahedral capsids about 45 to 55 nm in diameter made from 72 capsomers (Figure 5.1, *A*). None has an envelope. Their genomes are single molecules of circular double-stranded DNA about 5000 (polyomavirus) or 8000 (papillomavirus) base pairs long that encode five to seven proteins.

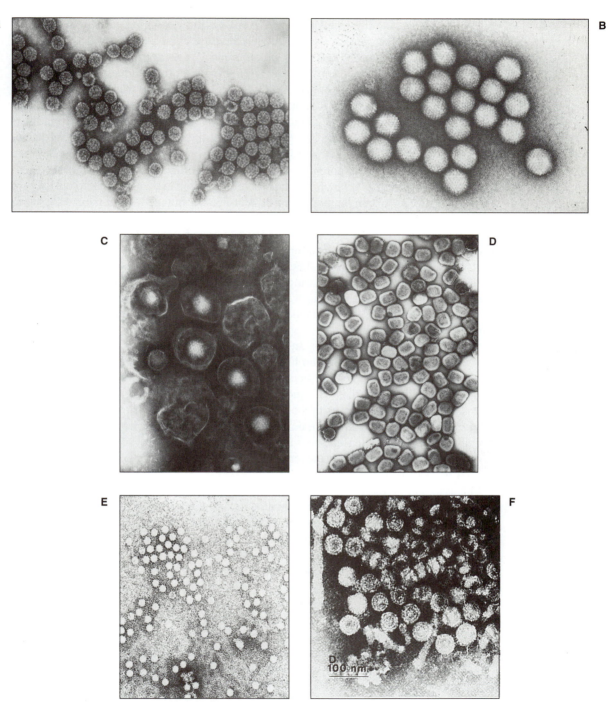

FIGURE 5.1 Virions of various animal DNA viruses. **A,** Papovavirus (polyomavirus); **B,** Adenovirus; **C,** Herpesvirus; **D,** Poxvirus (vaccinia virus); **E,** Parvovirus (B19 virus); **F,** Hepadnavirus (human hepatitis B virus).

The next larger family is the *Adenoviridae,* or the adenoviruses (*adeno* is from the Greek word for "gland"). The adenoviruses are also naked viruses with icosahedral capsids 70 to 90 nm in diameter composed of 240 hexon proteins and 12 penton proteins and their associated fibers (Figure 5.1, *B*). Their genomes are linear rather than circular double-stranded DNA, and are about five times larger (36,000 to 38,000 base pairs) than those of the papovaviruses. They encode at least 10 structural and more than 20 nonstructural proteins.

All the viruses larger than the adenoviruses are enveloped and have complex virion structures. The *Herpesviridae* (Figure 5.1, *C*) have icosahedral capsids about 125 nm in diameter surrounding a ring-shaped core consisting of DNA associated with protein fibrils. Between the capsid and envelope is a compartment called the **tegument** (from the Latin *"tegumentum"* meaning "covering"), which is composed of both structural and regulatory virion proteins. The genome is a single molecule of linear double-stranded DNA about 120,000 to 240,000 base pairs long. The genomes encode on average more than 70 proteins. About half of these are structural proteins, while the rest are enzymatic and regulatory proteins involved in expression and replication of the genome. Human herpesviruses are divided into three subfamilies called alpha, beta, and gamma herpesvirus. The human herpesviruses are herpes simplex viruses 1 and 2, varicella-zoster, cytomegalovirus, Epstein-Barr virus, and Kaposi sarcoma-associated virus, as well as two other viruses that have not yet been linked to a significant human disease. A number of other herpesviruses have been identified in other mammals, in birds, and in fish.

The *Poxviridae,* or poxviruses, are the largest of all viruses known, with brick-shaped virions that measure 300 to 450 nm long by 170 to 260 nm wide (Figure 5.1, *D*). This is the same size as the "elementary body" form of *Chlamydia,* a class of bacteria that are obligate intracellular parasites. This means that poxvirus virions are just visible in the light microscope. The virion consists of a core that is often dumbbell-shaped, two "lateral bodies" next to the thin section of the dumbbell, a lipoprotein coat characterized by ridges, and, finally, an envelope in some poxviruses. The genome is about 130,000 to 280,000 base pairs of double-stranded DNA in the form of a linear molecule with closed ends. The genome encodes more than 100 different proteins.

In addition to these groups of double-stranded DNA viruses, several other families of animal DNA viruses have genomes that consist of either partially or completely single-stranded DNA. The *Parvoviridae* (*parvo* is from the Latin for "small") are named for their tiny virions; their naked capsids made from 32 capsomers are only 18 to 26 nm in diameter (Figure 5.1, *E*). The genomes of parvoviruses are linear single-stranded DNA about 5000 bases long, yet they can encode three to five major proteins. The DNA in virions of members of the genus *parvovirus,* which infects mammalian or avian cells, is negative-strand DNA. The other parvovirus genera appear not to be so careful when individual virions are assembled. Some virions are assembled around negative genomes, while other virions contain positive-strand antigenomes. The *Circoviridae* have circular negative-strand DNA genomes that are about 1700 to 2300 vases long. Some circoviruses appear to encode only a single protein, while others make three small proteins.

The final group of animal DNA viruses we shall consider is the family *Hepadnaviridae* (from *hepatikos,* Greek for "of the liver," + DNA). These viruses have a capsid about the same size as that of the parvoviruses (22 to 25 nm in diameter), but it is surrounded by a tight envelope, so the complete virion is about 45 nm in diameter (Figure 5.1, *F*). The striking feature of hepadnavirus architecture is the organization of the genome: a partially single-stranded circle of DNA. The larger negative-strand molecule, about 3000 to 3300 bases long, is held in the form of a nicked circle by its base pairing with a shorter positive-strand molecule, which varies in length from about 1700 to 2800 bases. This tiny genome, one of the smallest of the animal DNA viruses, is used in a remarkably efficient fashion to encode more than a half dozen proteins.

As we saw in Chapter 4, there are many different types of RNA viruses that infect plant cells. In contrast, only two groups of plant DNA viruses have been discovered thus far. One of these, the *Caulimoviruses* (named for its type species *cauli*flower *mo*saic virus), have nicked circular, double-stranded DNA genomes similar to those of the hepadnaviruses. These 8000 base pairs-long genomes encode seven or eight proteins. The caulimoviruses have icosahedral capsids about 50 nm in diameter, and like all plant viruses, have no envelope (Figure 5.2, *A*). The other group of plant DNA viruses is named the *Geminiviruses* from the Latin word for "twins." Their virions usually are found in pairs with their icosahedral capsids sharing at least one capsomer, resulting in a structure that is about 20 nm in diameter and 30 nm long (Figure 5.2, *B*). The "twin" virions in many cases encapsulate a pair of single-stranded DNAs about 2500 bases long (Figure 5.2, *B*), meaning that the geminiviruses are often bipartite. At least six proteins are encoded by the entire genome.

Some of the largest DNA viruses infect insects and algal cells. The *Baculoviridae* are insect viruses that take their name from their enveloped, rod-shaped virions (*baculum* is Latin for "stick"). Their genomes range from 80 to 230 kilobase pairs of circular double-stranded DNA. Baculoviruses are widely used as expression vectors in insect cell culture systems for eukaryotic genes since they can accept large inserts of foreign DNA and produce copious amounts of proteins that are correctly modified and folded after synthesis. The viruses that infect the unicellular green algae in the genus *Chlorella* have icosahedral virions but share several other structural features with the poxviruses. Their genomes are enormous, ranging from 300 to 380 kilobase pairs, and are in the form of linear double-stranded DNA with closed ends or hairpin loops. Their DNA sequences reveal that like the poxviruses they also have terminal inverted and direct repeat sequences. Viruses that infect species of brown algae also have very large genomes, 160 to 300 kilobase pairs, but unlike the *Chlorella* viruses, their double-stranded DNAs are in the more conventional form of supercoiled circles.

WHAT OUTCOMES ARE POSSIBLE WHEN A DNA VIRUS INFECTS AN ANIMAL CELL?

In the earlier chapters in this book we have explored the strategies by which bacteriophages and the RNA viruses express their genes and synthesize new genomes as steps in a process of **productive infection,** that is, infection leading to the manufacture and release of new virions. Productive infection often leads to the death of the host cell. Productive infection is not, however, the only possible outcome of a virus's entering a host cell. Two other outcomes that occur in certain circumstances, especially with DNA viruses, are *persistent infection* and *transformation* of the infected host cell.

A **persistent infection,** as the name indicates, is one in which the infecting virus remains present in its host cell for long periods without killing or seriously impairing the function of that cell. Persistent infections can be divided into two types. **Chronic infections** are those from which infectious virus can be recovered by conventional laboratory methods.

A

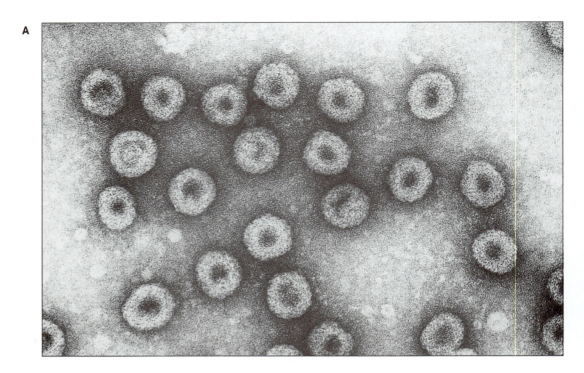

B

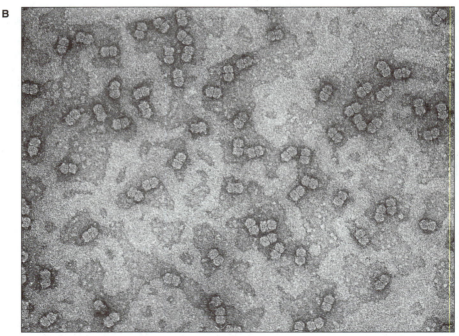

FIGURE 5.2 Virions of various plant DNA viruses. **A,** Caulimovirus (cauliflower mosaic virus); **B,** Geminivirus (maize streak virus).

Latent infections, in contrast, are cases in which the viral genome persists in the infected cell, but infectious virus is present only during "reactivation" episodes.

In **transformation** the infective process changes cell-cell communication and growth regulation in the host cell so that the transformed cell takes on the characteristics of a tumor cell. The viral genome is retained in part or *in toto* in the transformed cell, but infectious virus is usually not produced. Whether a particular virus infection is productive or persistent, or leads to transformation, is determined largely by characteristics of the particular host cell itself. In a **permissive** host cell that has all the functions necessary to support the synthesis of new virions, infection may be productive. Infection of a different, **nonpermissive** host cell that lacks some function critical to viral reproduction may result in transformation.

In this chapter we shall consider the process of genome expression and replication during productive infections by various types of DNA viruses. A detailed discussion of the molecular events that lead to persistent infections and/or transformation will be postponed until Chapter 8.

HOW DO ANIMAL VIRUSES WITH DOUBLE-STRANDED DNA GENOMES EXPRESS AND REPLICATE THOSE GENOMES?

Three families of animal viruses have doubled-stranded DNA in a "standard" conformation (circular or linear) and use primarily host cell biosynthetic systems for the expression and replication of their genomes: the papovaviruses, the adenoviruses, and the herpesviruses. The members of a fourth family of DNA viruses, the poxviruses, have a unique genomic architecture (although it is basically double stranded) and reproduce entirely in the cytoplasm of their host cells using their own enzyme systems.

All these viruses share a common pattern of gene expression and replication. The *early* genes that are expressed immediately upon infection largely produce nonstructural regulatory proteins that promote viral DNA transcription and replication. *Late* genes are transcribed primarily after DNA replication has begun and encode primarily the virion structural components.

The Papovaviruses

The two subfamilies of the papovaviruses, the *polyomavirinae* and the *papillomavirinae,* differ in size of the genome (5000 vs. 8000 base pairs) and in the type of cell that is permissive for a productive infection. Because the papillomaviruses are difficult to culture *in vitro,* the most studied papovaviruses are polyomavirus and SV40 in the polyomavirinae. The genomes of both polyomavirus and SV40 are a supercoiled circle a little more than 5000 base pairs long. Since the viral DNA is complexed with histones, in electron-micrographs it can appear in the beads-on-a-string form characteristic of the nucleosomes seen in eukaryotic chromosomes. Five thousand base pairs is enough coding length for about four average-sized proteins, but both polyomavirus and SV40 encode six proteins, as summarized in Table 5.1. Three of these are structural proteins (labeled VP1, VP2, and VP3), while the others are nonstructural proteins. Most of these are called collectively the **T antigens** since they were first detected by immunological tests using antibodies raised in animals with papovavirus-induced tumors. These viruses must therefore use their small genomes very efficiently.

Table 5.1 | Proteins of Polyomavirus and SV40 Virus

Protein	Polyomavirus	SV40 Virus
Early Gene Products		
Large T antigen	785	708
Middle T antigen	421	—
Small T antigen	195	174
Late Gene Products		
VP1	385	362
VP2	319	352
VP3	204	234
Agnoprotein	—	62

Sizes are given as number of amino acids in each protein.

Transcription.   Polyomavirus and SV40 virions enter their host cells by endocytosis and are transported within endocytic vesicles to the nucleus where uncoating occurs and transcription begins. Since transcription is inhibited by α-amanitin, it appears that host RNA polymerase II is the enzyme responsible. Genetic maps of both polyomavirus and SV40 are presented in Figure 5.3. As can be seen from those maps, each virus has two transcriptional units, corresponding roughly to the halves of the circular molecule taken from a starting point near the viral replication origin. Early gene transcription results in a high molecular weight transcript containing the information for the different T antigens of each virus. Analysis of the amino acid sequences of these proteins indicates that the strategy used by these viruses to produce multiple proteins from the same coding information is *differential splicing of a common high molecular weight RNA to produce different mRNA molecules.* For example, all three T antigens of polyomavirus begin with the same N-terminus sequence, but alternative splicing choices create molecules that differ at their C-termini.

This same strategy is also used in the formation of the mRNAs for late gene products. Transcription of the late gene information, which occurs only after DNA synthesis has begun, generates a high molecular weight RNA containing the codons for all the viral structural proteins. In polyomavirus this transcript is spliced to connect the same untranslated leader sequence to three different sets of exons, yielding mRNAs for VP1, VP2, and VP3.

SV40 uses an additional strategy to generate four different structural proteins instead of three. Differential splicing of the original late gene transcript creates two mRNAs: one for VP1 and a protein of unknown function called the agnoprotein (*agno* is Greek for "unknown") and the other for VP2 and VP3. The two different proteins are translated from each mRNA by initiating at alternative start codons.

The shift from early gene expression to late gene expression is mediated by the large T antigen. In a form of "feedback" regulation, the large T antigen protein binds to the promoter region of its own gene to suppress further transcription from that location.

Replication.   In addition to its role in the regulation of viral early gene transcription, the multifunctional large T antigen protein plays major roles in viral DNA replication and in altering the control of cellular DNA synthesis. Polyomavirus and SV40 virus large T antigens both

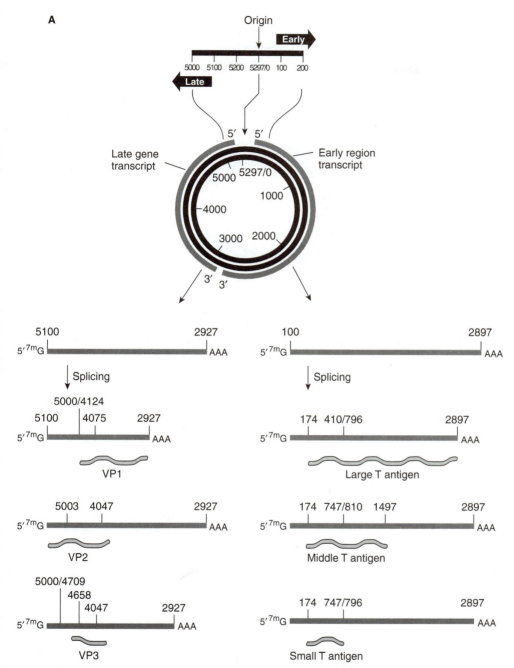

FIGURE 5.3 Genomic maps of two papovaviruses: **A,** polyomavirus; **B,** SV40 virus. Two transcripts are made from promoters on either side of the viral replication origin. These are then spliced as indicated to form individual mRNAs. Splices are indicated by positions of the donor and acceptor bases separated by a slash (e.g., 4917/4572). *(Continued.)*

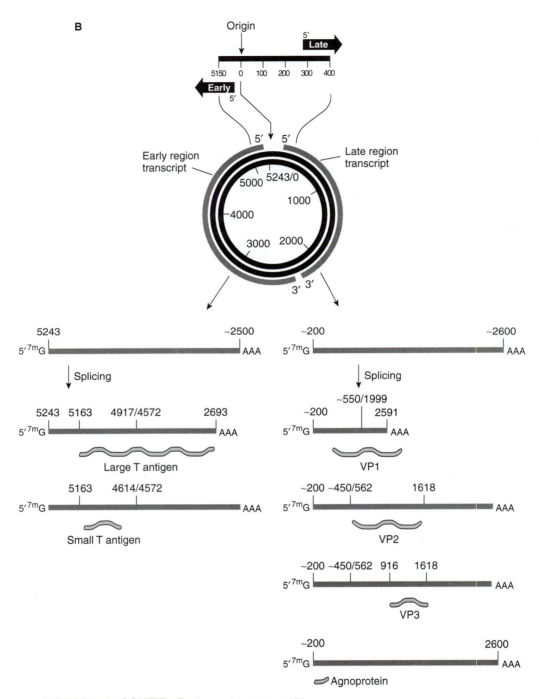

FIGURE 5.3, CONT'D For legend see page 179.

have ATPase activity, while that of SV40 can also adenylate and function as a DNA unwinding helicase. This molecule appears to open the viral genomic DNA at the origin to permit a host primase complex to initiate bidirectional DNA synthesis by host DNA polymerase. Replication of the circular genome produces a *theta intermediate* in which the unreplicated DNA is still supercoiled, but the new daughter strands are relaxed. The result of replication is thus a pair of relaxed molecules, which must then be supercoiled by host topoisomerase II.

The large T antigen protein also appears to interact with cellular regulatory proteins for a number of cellular genes involved in control of the host cell cycle and DNA replication. The ability of large T antigen to induce cellular DNA synthesis is independent of its role in viral DNA synthesis since mutations removing the latter function do not impair the former. In tissue cultures, the presence of large T antigen appears to "immortalize" the infected cells by permitting them to divide indefinitely. This topic will be taken up again in a discussion of transformation in Chapter 8.

The Adenoviruses

The adenoviruses are a homogeneous group of viruses with linear, double-stranded DNA genomes about 36,000 base pairs long, about seven times larger than the papovavirus genomes. Two distinctive features of the adenovirus genome's architecture are (1) the presence of inverted terminal redundancies about 100 to 140 base pairs long, and (2) a 55 kdalton protein attached covalently (serine to dCMP) to the 5′ ends of the linear DNA molecule. These proteins are required for effective replication of the genome.

Transcription.   Adenoviruses enter their host cells by endocytosis, but uncoating does not occur until the viral core reaches the nuclear envelope. At that point the viral genome enters the nucleoplasm while most of its associated proteins remain in the cytoplasm. As might be expected for these larger viruses, their genes are expressed in temporal clusters beginning with the immediate and delayed early genes, which are transcribed before DNA replication begins. Most of the early gene products are involved with regulation of viral and cellular synthetic activities, but some also encode viral structural proteins. Late genes, transcribed after DNA replication has begun, encode the numerous components of the virion pentamers, hexamers, and fibers.

The adenoviruses use the same strategy as the papovaviruses to produce a variety of different proteins from limited genetic resources: alternative splicing of the same high molecular weight transcript to create different mRNAs. In fact, the hypothesis that the mRNAs in eukaryotic cells are generated by removal of sections of the original transcript was based on interpretations of experiments on the synthesis of the adenovirus hexon protein. In 1977, researchers in two laboratories found that when hexon mRNA was hybridized to viral DNA, three loops of single-stranded DNA were formed near the 5′ end of the RNA molecule. These loops were evidence that the hexon mRNA was not colinear with the hexon gene itself.

The genomic map of an adenovirus is extremely complex. The various early genes are transcribed from six different promoters (Figure 5.4). Each transcript is then spliced in at least three different ways to create mRNAs with different sequences. Nearly all the late genes are transcribed from a single promoter, leading to the formation of an RNA

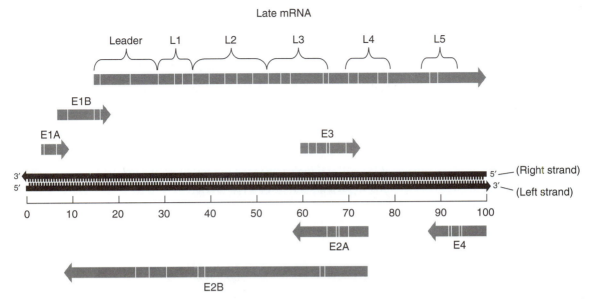

FIGURE 5.4 Genomic map of adenovirus type 2. The six early gene transcripts are indicated by E1A to E4. The positions at which splices are possible within the transcripts are indicated by lines. Three small sequences from the leader portion of the late gene transcript appear in all late gene mRNAs. The labels L1 to L5 indicate regions that share common translation termination codons. Since both strands of DNA are used as templates for transcription, they are not designated "positive" and "negative" but rather "right" and "left" from the direction of transcription on each respectively.

molecule corresponding to about 85% of the length of the genome. This transcript contains the information for components of the virion hexons, pentons, and fibers, as well as other proteins that become associated with the genome in the virion. The number and complexity of the splicing reactions necessary to generate all these different mRNAs are truly remarkable. A sequence that serves as an intron in one splicing reaction becomes an exon in another, and vice versa.

Replication. The problem to be solved by adenovirus during the replication of its genome is that common to all organisms having linear DNA as their genetic material: replication of the ends of the molecule where priming is not possible. Very extensive research using temperature-sensitive mutants indicates that three viral proteins, and at least four host proteins, are involved in the solution to this problem in adenoviruses. One of the viral proteins is the *terminal protein,* which is covalently attached to the dCMP found at each 5′ end of the genome. During the process of DNA synthesis, this molecule is 80 kdaltons large; it is cleaved to 55 kdaltons during packaging of the DNA into virion cores. In addition to the terminal protein, a viral DNA polymerase and viral single-strand DNA binding proteins are required for replication.

Figure 5.5 illustrates the process by which adenovirus replicates its genome. Unlike host cell DNA synthesis, all viral DNA synthesis is continuous and does not involve the formation of RNA-primed Okazaki fragments. Instead, DNA replication is primed by hydrogen bonding of a dCMP attached to an 80 kdalton terminal protein to the 3′ terminal dGTP of the genome. Elongation of that dCMP causes the displacement of the corresponding genomic DNA strand, which is, in turn, stabilized by viral single-strand DNA binding proteins. In *in vitro* assays the displaced strand may assume a "lollipop" or "panhandle" configuration through pairing of the complementary inverted repeat terminal sequences, but the significance of this *in vivo* is unclear. In any event, this single-stranded molecule then serves as the template for a second round of DNA synthesis, again primed by a terminal protein-dCMP molecule, leading to the creation of a complete new genome.

(1)   The adenovirus genome

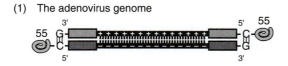

(2)   Primer for "left" strand synthesis is a 80 kd protein

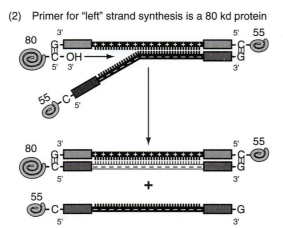

(3)   The displaced "left" strand serves as template for a new "right" strand

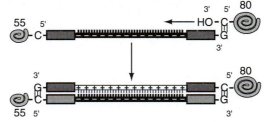

FIGURE 5.5   Model of DNA replication in adenovirus. The genome has inverted terminal repeats (indicated by boxes) and 55 kdalton proteins attached to the 5′ termini. DNA synthesis does not involve Okazaki fragments since both new strands use an 80 kdalton protein as primer. This molecule is cleaved to 55 kdaltons during maturation of the virion.

The Herpesviruses

The herpesviruses are the largest DNA viruses to rely on host cell enzymes for important functions related to expression and replication of the viral genome. The three groups (or sub-families) of herpesviruses were originally differentiated on the basis of biological properties such as host ranges, length of reproductive cycles, effects on their host cells, and types of latent infections established. Some of these features will be discussed in later chapters of this book. Newly identified herpesviruses are now assigned their place based on genome similarities. Table 5.2 presents the characteristics of eight currently identified human herpesviruses. The linear double-stranded genomes of the herpesviruses are about three to six times larger than those of the adenoviruses. Herpes simplex virus, for example, encodes at least 70 proteins in its 152 kdalton genome. Gentle denaturation of this molecule produces more than two strands of DNA, indicating that the genome has nicks or gaps. Nearly all the herpesviruses have sets of repeated sequences, either at the genomic termini or internally.

Transcription.   Possibly because of the coding power available in its large-size genome, herpes simplex virus 1 (HSV-1) does not appear to require special strategies for generating its numerous different proteins. Other herpesviruses like cytomegalovirus and Epstein-Barr virus, however, have been shown to use alternative splicing of high molecular weight transcripts as well as alternate promoter usage during expression of their genomes. What HSV-1 may lack in terms of complicated strategies for encoding information, however, it makes up for in terms of regulating gene expression. Herpesvirus genes can be roughly divided into immediate-early, early, and late categories, but because there is so much overlap in their times of expression, they are instead grouped into sets termed α, β, and γ, respectively. Four of the five α (or immediate-early) gene products are involved in regulation of the expression of all three sets of genes. The β (or early) gene products are primarily involved in nucleic acid metabolism and DNA synthesis. Most of these gene products duplicate cellular enzymatic activities and so are not essential. The γ (or late) gene products include many virion proteins.

Table 5.2 | Characteristics of the Genomes of Human Herpesviruses

Virus	Common Name	Size	Features of Genetic Organization
HHV1	Herpes simplex 1	152	Two unique sections, each
HHV2	Herpes simplex 2	152	flanked by inverted repeats
HHV3	Varicella-zoster	125	Two unique sections, shorter flanked by inverted repeats
HHV4	Epstein-Barr	172	Five unique sections, separated by repeats
HHV5	Cytomegalovirus	240	Two unique sections, each flanked by inverted repeats
HHV6	None	170	One unique section flanked by direct repeats
HHV7	None	145	One unique section flanked by direct repeats
HHV8	Kaposi sarcoma-associated herpesvirus	175	One unique section flanked by direct repeats

The abbreviations HHV1 to HHV8 correspond to human herpesvirus 1 through 8. Genome sizes are given in kilobase pairs.

Expression of the three groups of genes is controlled through a series of feedback loops (Figure 5.6). A γ gene product (α-*trans* induction factor or α-TIF), which is a virion tegument protein that enters the nucleus in association with the viral genome, is the gene activator responsible for transcription of the α genes. The α gene products both autoregulate their own expression and activate the β genes. α and β gene products together regulate γ gene expression by a combination of de-repression and gene activation. The cycle is completed late in the productive infection process when certain γ gene products shut down synthesis from α and β gene promoters.

Replication. Many of the β gene products are involved in DNA replication, either directly or in the creation of nucleic acid metabolites. Only seven of these genes appear to encode required functions, however, including a viral DNA polymerase, helicase/primase and single-strand DNA binding proteins. Since the virus provides its own replicative machinery, it does not require that its host cell be in S phase naturally or be forced into it by viral regulatory proteins.

The simplest model suggested for herpesvirus replication involves first circularization of the linear genome and then rolling circle synthesis to generate multiple copies of the genome. The resulting concatamer is then cleaved into unit genomes during maturation of the virion. Experimental evidence for this model is scant, however. Electron micrographs do indicate the presence of circular DNA molecules in infected cells, but theta

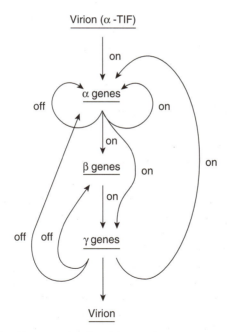

FIGURE 5.6 Regulation of herpes simplex virus 1 gene expression. α genes are activated by a γ gene protein (α-TIF = α-transinduction factor) that is part of the virion. α gene proteins then both autoregulate α gene expression and turn on β genes. α and β gene proteins together turn on γ genes. γ gene proteins are associated with virion assembly and may regulate α and β gene expression as well.

intermediates are not seen. Pulse-label experiments demonstrate that newly synthesized DNA is longer than genome unit length, suggesting rolling circle replication, but sigma intermediates are also not seen. Instead, electronmicrographs show complex networks of DNA. These networks may be the result of homologous recombination, similar to that seen in the bacteriophage T4, since genetic studies of herpesvirus indicate that recombination occurs at rates of as much as 1% to 3% per 1000 base pairs. Such networks do not rule out rolling circle replication, however.

The Poxviruses

The final group of animal viruses with double-stranded DNA genomes is the poxviruses. The virions of some poxviruses are so enormous (at least for viruses) that they are visible in the light microscope. These complex virions (Figure 5.1, *D*) contain genomes ranging in size from 130 to 300 kbase pairs, which encode more than a hundred different proteins. So large is the genome, in fact, that unlike all the other DNA viruses, the poxviruses can carry out all the processes of the Central Dogma entirely in the cytoplasm of their host cells, never entering the nucleus at all.

Smallpox virus (variola virus) is the most notorious member of the poxviruses, but since the united efforts of humankind have led to its eradication as a feral virus, it is not the "type" virus for the family. That honor belongs to vaccinia virus, the closely related virus responsible for cowpox whose use by Jenner to immunize people against smallpox is the source of the term *vaccination*. The architecture of vaccinia's 186 kbase-pair-long genome is typical of the poxviruses. When the double-stranded DNA molecule is denatured, it does not produce two single strands of genomic length. Rather, one molecule of *circular* single-stranded DNA results. These data indicate that the poxvirus genome is in the form of a double-stranded DNA molecule, which has cross-linked ends. It is, in effect, a circle whose two sides can base pair with one another. The "termini" of the genome have inverted repeat sequences about 10 kbase pairs long.

Transcription. A defining characteristic of the poxviruses is their ability to carry out both transcription and DNA replication in the cytoplasm rather than the nucleus of their host cells. They do this by encoding all the necessary enzymes and factors rather than using those in the nucleus of the infected cell. They do, however, rely on their host for the cellular machinery of translation and for the precursor metabolites for all the processes of the Central Dogma.

Transcription of the poxvirus early genes begins as soon as the viral core is released in the cytoplasm of an infected cell. All the enzymes needed to synthesize an RNA molecule and then posttranscriptionally modify it into a eukaryotic mRNA (RNA polymerase, capping enzymes, poly-A polymerase, methylating enzymes, etc.) are parts of the virion itself. About half the genome is transcribed from early gene promoters. Alternative splicing of transcripts does not appear to be a means of producing additional proteins.

Late gene transcription occurs only during and after DNA synthesis. The mechanism by which transcription shifts from early to late gene promoters is unclear, but it certainly involves not only various regulatory proteins but also the configuration of the newly replicated DNA molecule itself. Early genes continue to be transcribed during late gene expression, but they are not translated as efficiently after DNA replication begins. mRNAs are also more rapidly degraded during the later stages of the productive infection process, so regulation of early gene expression may involve a combination of less frequent transcription and faster degradation.

Replication. Early studies of poxvirus replication indicated that concatemers were formed as part of the process, leading to the notion that the genome, whose unique conformation might permit it to open into a single-stranded circular form, was copied by a form of rolling circle replication. Sequence analysis of the concatemers showed that the unit genomes were not linked "head to tail" as rolling circle replication would produce, however, but instead were "head to head" and "tail to tail."

A model accounting for this observation is presented in Figure 5.7. In the initial step a nick near one of the terminal repeats (labelled LTR and RTR for left and right, respectively) creates a free 3′-OH to serve as primer for DNA synthesis (step 2). The template sequence now available to serve as the template contains a set of bases that are self-complementary since they were originally paired with one another. The newly added bases on the 3′ end of the molecule are also self-complementary, so they can fold back on themselves to begin a process of displacement (step 3). The displaced DNA then serves in turn as a template to form a two-genome-long concatemer. At that point the growing point doubles back again to use the newly replicated DNA as a template, forming a concatamer containing four genomes (step 4). The concatamer is then resolved in a two-step process involving staggered cuts (step 5) followed by ligation to complete the formation of the double-stranded genome (step 6).

(1) Nicking

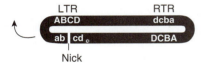

(2) Initial synthesis

(3) Displacement to form 2-genome long concatemer

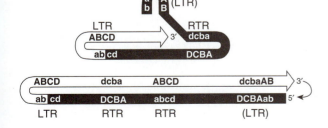

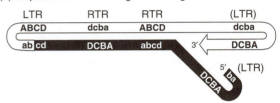
(4) Displacement to form 4-genome long concatemer

(5) Cleavage into genomic-sized molecules

(6) Ligation

FIGURE 5.7 Model for DNA replication in poxvirus. After nicking in one of the inverted repeat sequences to produce a free 3′-OH (steps 1–2), new DNA synthesis generates a four-genome-long concatemer (steps 3–4), which is then cleaved to form individual genomic molecules (steps 5–6).

HOW DO ANIMAL SINGLE-STRANDED DNA VIRUSES EXPRESS AND REPLICATE THEIR GENOMES?

The *Parvoviridae* have single-stranded DNA as their genome. These tiny viruses, with genomes only about 5000 bases long, can reproduce only in cells that are themselves actively synthesizing their own DNA since they have no means to force the cell into S phase. The *Parvoviridae* are divided into three genera. The *parvoviruses* can reproduce themselves in suitable mammalian host cells, while the *densoviruses* reproduce only in insect host cells. The parvoviruses and densoviruses are termed **autonomous** viruses since these viruses have all the information and functions necessary to reproduce themselves in suitable host cells. The third genus, the *dependoviruses,* contains the replication **defective** adeno-associated viruses (AAV). These viruses require that their host cells be infected simultaneously with another DNA virus, such as adenovirus or herpesvirus, to provide some "helper" function(s) necessary for their replication.

The genomes of the autonomous viruses are negative-strand DNA (complementary to mRNAs), while those of the adeno-associated viruses are either negative or positive strand. As is usual in the case of a linear genome, the termini contain special sequences. At each terminus the autonomous parvovirus molecules contain self-complementary sequences that allow the DNA to fold back on itself to form two hairpin loops, giving each end an overall structure with the appearance of a Y or T (Figure 5.8). The sequences at the 3′ and 5′ ends are different from one another, and that at the 5′ end can be in either of two orientations relative to the rest of the genome. The AAV genomes also end in self-complementary sequences, but they are inverted repeats that are the same at both termini.

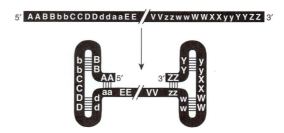

FIGURE 5.8 Formation of hairpin loops at the termini of an autonomous parvovirus.

Although the parvoviruses rely entirely upon the synthetic machinery of their host cells for all the processes of the Central Dogma, they nevertheless encode several regulatory proteins in addition to the three structural proteins of the virion. Production of so many different proteins from such small genomes requires not only employment of a variety of strategies for transcription and translation, but also even posttranslational modification of viral proteins.

All the parvoviruses use two promoters during expression of their genomes (Figure 5.9). The first for the nonstructural or regulatory proteins is at map position 4 near the left end of the genome, while the other for structural proteins is at map position 38. The transcript from the regulatory proteins promoter is first capped and polyadenylated and a small intron between map positions 46 and 48 removed. This molecule can either be used directly as an mRNA for translation of one protein (NS-1) or be spliced again to produce the message for a second protein (NS-2). The same strategy is used for the synthesis of the two larger virion structural proteins, VP-1 and VP-2. The transcript from the map position 38 promoter is

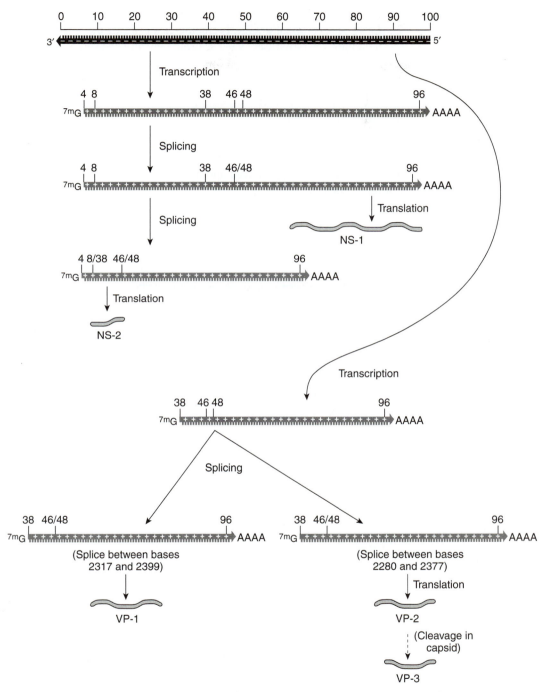

FIGURE 5.9 Expression of parvovirus genes. Nonstructural genes are translated from the 5′ end of a pair of mRNAs transcribed from a promoter at map position 4. NS-1 is translated from the unspliced molecule, while NS-2 has splices between map positions 8 and 38 and 46 and 48. Structural genes are transcribed from a promoter at position 38. The mRNAs for VP-1 and VP-2 have slightly different splice donor and acceptor bases within the 46–48 splice region. VP-3 is a cleavage product of VP-2 generated during formation of the capsid.

spliced in two different ways within the map position 46 to 48 intron to create different mRNAs for VP-1 and VP-2. An additional strategy is employed to generate the smallest structural protein. In the autonomous parvoviruses, VP-2 is proteolytically cleaved after translation to produce VP-3. In AAV the same transcript is translated in two different reading frames to produce VP-2 and VP-3. VP-3 starts at the conventional AUG initiation codon, while VP-2 uses ACG as its initiation codon.

Replication of the parvovirus genomes has several features in common with the mechanism discussed previously for poxvirus replication. The self-complementary sequences at the termini permit "self-priming" to occur so that RNA primers are not required at any point in the process. In addition, all synthesis involves single-strand displacement leading to the formation of concatemers, which are then cleaved into individual genome-length molecules during maturation of the virions.

The simplest example of parvovirus replication is that of the adeno-associated viruses, illustrated in Figure 5.10. With the 3' terminus in its double-hairpin configuration, it serves

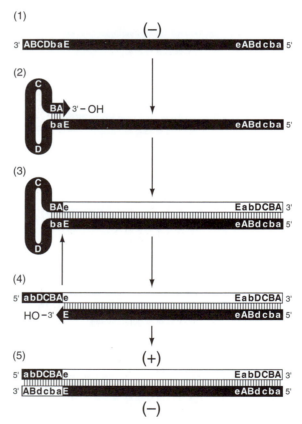

FIGURE 5.10 Model for DNA replication in adeno-associated virus, a defective parvovirus. After folding back upon itself by means of a terminal palindrome (step 2) the 3' terminus of the genome serves as primer for DNA synthesis of a positive strand (step 3). Nicking and unfolding of the negative strand (step 4) permit complete synthesis of a duplex molecule (step 5), either strand of which may be encapsulated within a new virion.

as the primer for elongation of the molecule down the entire length of the genome. A nick in the original strand permits the double-hairpin to "straighten out" and thus serve as template for elongation of the newly created 3′ end. Note that each of the resulting DNA molecules is a hybrid of "old" and "new" DNA. Note also that the sequence of some of the bases within the terminal repeat has been reversed. The linear positive-strand DNA can now serve as template for the synthesis of more negative-strand genomic molecules.

Models describing the replication of the autonomous parvoviruses are *far* more complex than the one presented in Figure 5.10 for AAV. The complexities are introduced by the need to explain how replication yields molecules with different terminal sequences, one of which remains constant (the 3′ end) while the other (the 5′) is inverted every other round of replication. The current model resembles that describing the process used by the poxviruses to replicate their genomes and involves the formation of concatemers by displacement synthesis.

HOW DO PLANT SINGLE-STRANDED DNA VIRUSES EXPRESS AND REPLICATE THEIR GENOMES?

The geminiviruses have both distinctive virions and distinctive genomic architectures. The characteristic fused pair of icosahedral capsids that form the geminivirus virion (see Figure 5.2, *B*) encloses either one or two very tiny molecules of circular single-stranded DNA. The single DNA molecule of maize streak virus (MSV) is only 2687 bases long, making it the smallest known genome of any autonomously replicating eukaryotic virus. Another subgroup of geminiviruses has two genomic DNA molecules, each of which is about the size of the MSV genome; cassava latent virus (CLV), for example, has DNAs 2779 and 2724 bases long in its bipartite genome. Restriction mapping and sequencing have demonstrated that these molecules have no sequence homologies except for a set of about 200 bases that is highly conserved in all the geminivirus genomes.

The DNAs encapsulated in the virions are defined as positive-strand DNAs, but sequencing indicates that the geminivirus genes are transcribed both from that strand and its complementary negative strand. Figure 5.11 presents the location of open reading frames for the MSV and CLV genomes. These ORFs occur in both the positive and negative strands and appear to have numerous overlapping reading frames. A variety of possible promoters and polyadenylation signals occurs in association with these sequences, but it is not clear at present how transcription and translation are regulated. MSV appears to transcribe three mRNAs (two from the positive strand and one from the negative), while CLV makes six mRNAs. Analysis of the RNAs found in infected cells indicates that splicing is not used to create different mRNAs, so it is possible that not all these ORFs are actually used to synthesize viral proteins. The coat or capsid protein ORF has been identified in all the geminivirus genomes sequenced thus far, but the function of other gene products is unknown. Transfection studies have demonstrated, however, that the CLV DNA 1 molecule contains the genes necessary for DNA replication, while the DNA 2 molecule contains those for movement of the virus from cell to cell, either by cellular fusions or via its leafhopper vector.

Few details of the replication of the geminiviruses are known. Bidirectional replication forming theta intermediates is probably not used since a number of researchers have found short genomic concatamers in infected cells. Other lines of evidence also suggest that a form of rolling circle replication occurs. The 200-base conserved region common

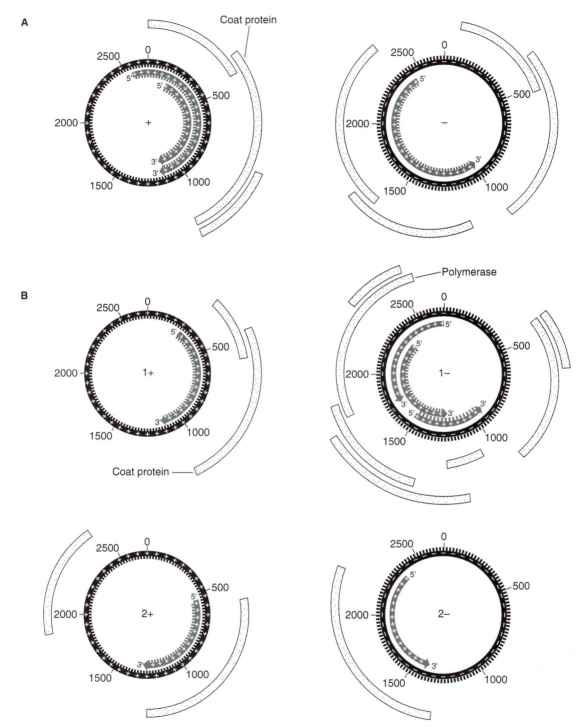

FIGURE 5.11 Genomic maps for two geminiviruses. Although the geminiviruses are single-stranded, open reading frames are found on both the genome and its complement. The form of DNA in the virion is labeled (+). **A,** Maize streak virus. **B,** Cassava latent virus has two different (+) DNA molecules, one in each side of the double virion. The locations of known transcripts and functions are indicated.

to all the geminiviruses is known to be involved in replication, since deletions or other alterations in its structure can interfere with replication. It contains a palindromic sequence capable of forming a hairpin loop, and thus serving as the site of primer synthesis during replication. Within this region is a very highly conserved nine-base sequence that appears related to the bacteriophage φX174 sequence that is cleaved to initiate rolling circle replication in that virus.

WHAT FEATURES SET HEPADNAVIRUSES AND CAULIMOVIRUSES APART FROM OTHER EUKARYOTIC DNA VIRUSES?

This section might also have been titled "Virological Evidence that Nature Can Be Whimsical," given the genomic architectures and modes of replication of the animal hepadnaviridae and the plant caulimoviruses.

The Hepadnaviruses

The hepadnaviruses have the smallest genomes of all the autonomous animal viruses. Despite their very limited apparent coding capacity, however, these viruses are full of surprises. The first of these is that the hepadnavirus virion is not a simple naked icosahedral structure, as are those of all the other viruses with small genomes, but rather has a nucleocapsid (or core) surrounded by a very tight envelope containing three viral glycoproteins.

The second surprise is the architecture of the hepadnavirus genome. In infected cells the genome appears in the form of circular double-stranded DNA about 3200 base pairs long. This is not, however, the form of the genome that is carried in the virion. As Figure 5.12 shows, the virion molecule consists of two *linear* strands of DNA of different sizes. One is about 3200 bases long, and since it serves as template for transcription of all viral mRNAs, it is negative-strand DNA. The other molecule (the positive DNA strand) is of varying length, from 1700 to 2800 bases long. Neither of these DNA strands has a free 5′ end. The larger strand has a protein covalently attached to its 5′ terminus, while the smaller strand ends in

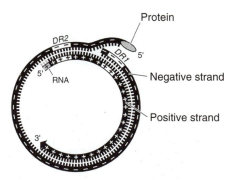

FIGURE 5.12 The architecture of the viral DNA molecule found in the hepadnavirus virion. The genome contained within a virion has a gapped negative strand and an incomplete positive strand that overlap each other. DR = direct repeat sequences.

a 19-base-long RNA molecule. The two DNA strands are base paired in an overlapping fashion to produce a circular double-stranded molecule with breaks in both strands. Immediately upon entering the cytoplasm, the positive strand is elongated until the strands are of essentially equal length. The 5′ ends of both molecules are then unblocked and the genome is converted into a form called *covalently closed circular* DNA (cccDNA).

Two other structural features of the hepadnavirus DNA molecule are important in expression and replication of the genome. The first is a pair of directly repeated 11-base-long sequences (DR1 and DR2) occurring near the 5′ ends of the two strands. These play crucial roles in replication. The other feature is a cleavage/polyadenylation site located about 200 bases from the negative-strand 3′ terminus.

Transcription. Sequencing data indicate the presence of four open reading frames in the genome of the human hepatitis B virus (HBV), the "type virus" of the hepadnaviruses (Figure 5.13). The ORFs are arranged in an overlapping pattern, and since there are no introns present, *every base in the hepadnavirus genome is used to code for at least one protein.* Likewise, every *cis*-acting regulatory sequence is also part of one or more ORFs. This is efficiency *par excellence!*

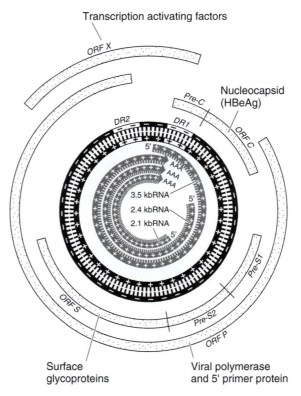

FIGURE 5.13 Genomic map of the hepadnavirus human hepatitis B virus (HBV). After infection of a host cell the genome is made into a complete double-stranded circular molecule. Three mRNAs are transcribed from that molecule, all of which share a common 3′ terminus. The four open reading frames and their products are indicated.

ORF C encodes the major HBV nucleocapsid protein (HBcAg, 21 kdaltons), as well as a smaller (16 kdalton) protein called HBeAg that can be detected in the serum of infected individuals. The ORF P sequence overlaps the end of the ORF C region and continues nearly the rest of the way around the genome. ORF P encodes the virion-associated polymerase (90 kdaltons), as well as the protein primer molecule that is attached to the 5′ end of the negative DNA strand. ORF S occurs entirely within ORF P and encodes glycoproteins (HBsAg) found in the surface shell of the virion. Three in-frame initiation codons permit synthesis of proteins 39, 33, and 24 kdaltons in size from ORF S. The final ORF was designated ORF X since a role for its protein product was not immediately evident. This ORF now appears to encode two, and possibly three, *trans* active regulatory proteins that can facilitate transcription from a number of cellular as well as viral promoter sequences.

Although there are four ORFs in the HBV genome, only three mRNAs are formed. The mRNAs form a nested set with different 5′ starting points but with the same 3′ terminus. The longest of these RNA molecules is, in fact, *longer than the genome itself.* Its promoter is actually upstream from the cleavage/polyadenylation site that, the *second* time it is transcribed, creates the signal for the 3′ end of each mRNA. During translation it appears that multiple initiation sites on the same mRNA are used to synthesize individual proteins.

Replication. Indications that hepadnavirus replication was most unusual came from studies of woodchuck and duck viruses in infected livers. First, viral DNA polymerase was found to be able to make only positive-strand viral DNA, but the only DNA actually found to be synthesized in infected cells was negative strand. Then it was found that actinomycin D, which inhibits DNA-dependent *RNA* synthesis (i.e., transcription), blocked positive-strand but not negative-strand DNA synthesis. These observations, and the discovery of negative-strand DNA/positive-strand RNA hybrid molecules in infected cells, led to the conclusion that hepadnaviruses must replicate their DNA genomes by means of an *RNA intermediate.*

Figure 5.14 presents a scheme for hepadnavirus replication. Many aspects of this model are similar to that given in the previous chapter for replication of a retrovirus (compare Figure 5.14 with Figure 4.29, page 158). Replication begins with the synthesis of a *pregenomic RNA* that is equivalent to the largest of the mRNA molecules (step 1). Note that the pregenomic RNA has terminal repeats since the same set of bases is copied twice during its transcription. The HBV viral polymerase first functions as a *reverse transcriptase* to synthesize a negative-strand DNA copy of the pregenomic RNA molecule (step 2). Synthesis begins at the direct repeat sequence near the 3′ end of the pregenomic RNA molecule (DR2), using as primer the protein that is found still attached to the 5′ end of the completed molecule in the virion. This priming strategy is similar to that seen in adenovirus replication. The pregenomic RNA molecule appears to be degraded by polymerase RNase H activity as DNA synthesis occurs, leaving eventually only a short piece of RNA bearing the 5′ direct repeat sequence (step 3).

The synthesis of positive-strand DNA appears to require a "jumping" reaction like that seen in retrovirus reverse transcription (step 4). The remaining fragment of pregenomic RNA jumps from the DR1 at the 3′ end of the new negative-strand DNA to the DR2 sequence at the 5′ end. Circularization of the template negative strand allows the viral polymerase to begin the synthesis of positive-strand DNA (step 5). Since assembly of the virion core occurs during these steps, synthesis of the positive strand reaches different points in different virions.

(1) Synthesis of "pregenomic " RNA

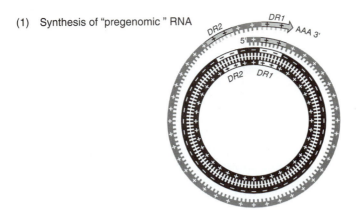

(2) Synthesis of negative-strand DNA begins at 3' DR1 on pregenomic RNA

Primer protein

(3) Degradation of RNA

(4) RNA fragment "jumps" to serve as primer for positive-strand DNA synthesis

(5) Circularization of negative DNA strand and synthesis of positive DNA strand

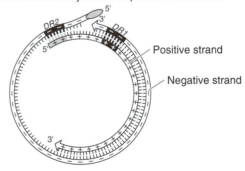

Positive strand

Negative strand

FIGURE 5.14 Model for DNA replication in the hepadnavirus HBV. Replication begins with synthesis of a "pregenomic" RNA molecule, which is somewhat longer than the genome itself (step 1). Reverse transcription of a pregenomic RNA using a protein primer generates a negative-strand DNA molecule (step 2). Nearly all of the pregenomic RNA molecule is degraded during negative-strand DNA synthesis, leaving only a small piece containing DR1, the 5′ direct repeat sequence (step 3). This RNA fragment jumps to the 5′ end of the negative-strand DNA, pairing with DR2, where it will serve as primer for synthesis of the positive DNA strand (step 4). Circularization of the negative strand permits synthesis of positive-strand DNA. Since this reaction occurs within the virion, the length of this strand is variable.

The Caulimoviruses

Although they are much larger viruses, the plant caulimoviruses have many structural and genetic features in common with the hepadnaviruses. Figure 5.15 diagrams the organization of the cauliflower mosaic virus (CaMV) genome. The negative strand of DNA (designated α) is 8031 bases long and has a single nick. There are two positive strands (β and γ) that together equal the length of the α strand. The nicks in the genome are actually locations in which the DNA is *triple stranded* since the 5′ terminus of each strand overlaps the 3′ end of the preceding strand and is therefore displaced from the double helix. Each 5′ terminus has an RNA molecule covalently attached to it.

Transcription. Sequencing data indicate the presence of eight open reading frames in the CaMV genome (Figure 5.15). An important aspect of the arrangement of these ORFs is that the termination codon for one is followed only one or two bases later by the initiation sequence for the next ORF. In several cases there are short overlaps at these junctions, but in only one location is one ORF (VIII) contained within another (IV). Like the hepadnaviruses, the caulimoviruses do not transcribe each ORF into its own mRNA. CaMV, for example, synthesizes only two RNAs: One includes only the ORF VI region of the genome, while the other is 180 bases longer than the entire genome and is thus the CaMV pregenomic RNA molecule.

The mechanism by which these polycistronic RNAs are translated into individual viral proteins is not well understood. As with the hepadnaviruses, neither RNA contains introns,

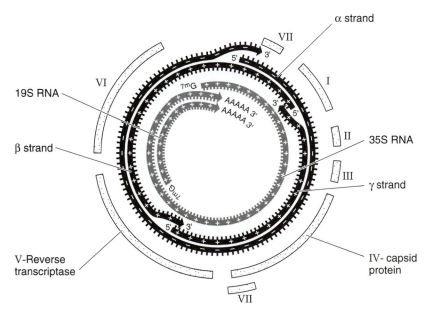

FIGURE 5.15 Genomic map of caulimovirus (cauliflower mosaic virus) showing the architecture of the viral DNA and the locations of transcripts and open reading frames. The genome consists of a full-length α strand and two strands, β and γ, that are complementary to it. All three strands have regions of triple helix where 3′ and 5′ ends overlap for a short distance.

so differential splicing to form alternative mRNAs is not the strategy employed. The juxtaposition of the termination codons for one ORF and initiation codon for the next ORF in these RNAs suggests that the molecules may be translated by ribosomes, which move directly from one ORF to the next without detaching from the RNA molecule.

Replication.   The basic strategy of caulimovirus replication is like that of the hepadnaviruses: synthesis of a new genome from information contained in a pregenomic RNA molecule. Replication of CaMV begins with synthesis of a pregenomic RNA that is 180 bases longer than the genome, and thus contains terminally redundant sequences. As diagrammed in Figure 5.16 (step 1), viral polymerase initiates negative-strand DNA synthesis using as its primer a tRNA^{met} bound about 590 bases from the 5′ end of the pregenomic RNA. In this use of a transfer RNA as a promoter, the caulimoviruses resemble the retroviruses rather than the hepadnaviruses. When the growing point of the new DNA molecule reaches the 5′ end of the pregenome, the polymerase shifts to reading from the 3′ end of the pregenomic

(1)   35S RNA serves as template for synthesis of α-strand of DNA

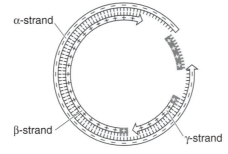

(2)   35S RNA is degraded as DNA synthesis continues

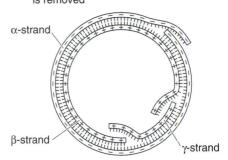

(3)   β-DNA and γ-DNA strand synthesis uses fragments of RNA as primers

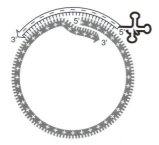

(4)   Synthesis is completed when all RNA is removed

FIGURE 5.16   Model for DNA replication in caulimovirus. Reverse transcription of the pregenomic RNA, primed by a tRNA, shifts over a gap from the 5′ end of the template to the 3′ end (step 1). The pregenomic RNA molecule is degraded as reverse transcription proceeds (step 2). Small fragments of the pregenomic RNA molecule are left at the sites where β and γ strand synthesis begin as α strand synthesis is completed (step 3). Synthesis is completed with the removal of all RNA fragments (step 4).

RNA molecule, probably by means of base pairing in the terminal redundancies. As with both retroviral and hepadnaviral reverse transcription, RNase H activity in the polymerase degrades the pregenomic RNA template as DNA elongation continues (step 2). Positive-strand DNA synthesis begins when the growing point passes the site of the β gap, using as primer some of the RNA template just copied, thus producing the overlapping sequences found at the 5′ end of each strand (step 3). Similarly, a second positive strand is synthesized beginning when the growing point of the negative strand passes the γ nick. Synthesis is completed when all RNA from the pregenomic molecule has been degraded (step 4).

SUMMARY

Since the DNA viruses have genomes that are the same type of nucleic acid as their host cells, they can use their host's normal machinery and processes to express those genomes. For that reason the DNA viruses seem to have fewer novel strategies for gene expression compared to the sizable array of tricks employed by the eukaryotic RNA viruses. DNA viruses are generally transcribed into high molecular weight RNA molecules that host enzymes then modify by adding a 5′ cap and 3′ poly-A tail and by removing any introns present. The resulting mRNAs are translated in the usual fashion by host systems. In some groups of viruses, such as the papovaviruses and adenoviruses, alternative splicing reactions on the same transcript create more than one mRNA from the same initial sequence. In other groups of viruses the same modified transcript may be used to generate several proteins, probably by reinitiating translation at a site immediately adjacent to a termination codon.

The major difficulties encountered by the DNA viruses during this part of their life cycles are in replicating their genomes. DNA synthesis must begin with a primer that provides a 3′-OH to be elongated. While this presents no problem if the genome is circular like that of the papovaviruses, replication of a linear DNA genome requires some "trick" to permit complete copying if elongation proceeds in the 5′ direction.

The DNA viruses have evolved a variety of strategies to ensure complete replication of their linear genomes. The simplest strategy is to circularize the linear genome and then replicate by a rolling circle mechanism; the herpesviruses may use this strategy. The adenoviruses replicate their genomes as linear molecules but use a protein rather than nucleic acid primer, which allows DNA synthesis to begin right at the end of the template strand. Many other viruses have special sequences at the termini of the linear genomes that facilitate their replication by a displacement process. In the poxviruses and parvoviruses, for example, all synthesis is by elongation of a 3′-OH group at the end of a sequence in the parental genome that can fold back and base pair with itself. All DNA synthesis in these viruses is "leading strand" and does not involve Okazaki fragments.

The most unusual strategy employed by the DNA viruses is that found in the hepadnaviruses and caulimoviruses. These viruses use reverse transcription of a longer-than-genome-length RNA transcript to create new DNA genomes. The process is very similar to that seen in the RNA retroviruses and involves both degradation of the RNA template as DNA synthesis proceeds and "jumping" of the primer molecules by means of repeated sequences at the ends of the template strands.

SUGGESTED READINGS

1. "Animal virus DNA replication" by M.D. Challberg and T.J. Kelly, *Ann Rev Biochem* 58:671–717 (1989), reviews the process of DNA replication in the groups of viruses that use their host cells' DNA synthesizing systems.

2. "Adenovirus DNA replication in vitro: Characterization of a protein covalently linked to nascent DNA strands" by M.D. Challberg, S.V. Desiderio, and T.J. Kelly, *Proc Natl Acad Sci USA* 77:5105–5109 (1980), and "Adenovirus DNA replication in vitro: a protein linked to the 5' end of nascent DNA strands" by B.W. Stillman, *J Virol* 37: 139–147 (1981), discuss the use of a protein rather than a nucleic acid as the primer for DNA synthesis.

3. "An amazing sequence arrangement at the 5' ends of adenovirus 2 messenger RNA" by L.T. Chow, R.E. Gelinas, T.R. Broker, and R.J. Roberts, *Cell* 12:1–8 (1977), presents the first description of introns in eukaryotic DNA.

4. "Regulation of adenovirus mRNA formation" by S.J. Flint, *Adv Virus Res* 31:169–228 (1986), reviews the processes of transcription and differential splicing to produce the great number of mRNAs characteristic of adenovirus.

5. "The genomes of the human herpesviruses: Contents, relationships, and evolution" by D.J. McGeoch, *Ann Rev Microbiol* 43:235–265 (1989), reviews the genetic properties of the major human pathogenic herpesviruses.

6. "Herpes simplex virus DNA replication" by P.E. Boehmer and I.R. Lehman, *Ann Rev Biochem* 66:247–384 (1997), describes a rolling circle model for herpesvirus DNA synthesis.

7. "Regulation of herpesvirus macromolecular synthesis. VIII. The transcription program consists of three phases during which both extent of transcription and accumulation of RNA in the cytoplasm are regulated" by P.C. Jones and B. Roisman, *J. Virol* 31:299–314 (1979), reports on how the cycles of transcription that occur in herpesvirus are regulated by feedback loops.

8. "The mechanism of cytoplasmic orthopoxvirus DNA replication" by R.W. Moyer and R.L. Graves, *Cell* 27:391–401 (1981), describes a model for replication of the unusual poxvirus genome, which does not require any Okasaki fragments.

9. "Regulation of vaccinia virus transcription" by B. Moss, *Ann Rev Biochem* 59:661–688 (1990), reviews the details of how the poxvirus vaccinia shifts from early to late gene transcription.

10. "Parvovirus replication" by K.I. Berns, *Micro Rev* 53:316–329 (1990), reviews the genetic organization and regulation found in the parvoviruses.

11. "Hepatitis B virus" by P. Tiollais and M.-A. Buendia, *Scientific American* 264:116–123 (1991), presents a fine introduction to the hepadnavirus's interesting "lifestyle," while "The molecular biology of the hepatitis B viruses" by D. Gamen and H.E. Varmus, *Ann Rev Biochem* 56:651–693 (1987), gives details of their replication and transcription processes.

12. "Biochemistry of DNA plant viruses" by R.J. Shepherd, *The Biochemistry of Plants* 15:563–616 (1989), reviews features of replication and transcription in both the caulimoviruses and geminiviruses.

13. "Nucleotide sequence of cassava latent virus DNA" by J. Stanley and M.R. Gay, *Nature* 301: 260–262 (1983), presents the sequence data that indicated that the two DNA molecules in this geminivirus were different and that both these strands and their complements encoded information.

14. "Giant viruses infecting algae" by J.L. Van Etten and R.H. Meints, *Ann Rev Microbiol* 53: 447–494 (1999), focuses on a prototype *Chlorella* virus and compares it to viruses infecting various brown algae.

Assembly, Maturation, and Release of Virions

We began our examination of the viruses in Chapter 1 by considering how the "one-step growth experiment" gives evidence that the viruses are fundamentally different from all other types of organisms. In the next chapters we examined some of the ways in which viruses enter their host cells, express their own genomes in the synthesis of viral proteins, and create new copies of their genomes. While many of the strategies employed by different viruses to carry out these activities are not those used by the host cells themselves, they are, nevertheless, variations on cellular themes. For example, we have seen that many eukaryotic viruses enter their host cells by binding to a cell-surface receptor that induces the formation of an endocytic vesicle to carry the virus into the cell. This process is not unique, however, since the bacterial intracellular parasites rickettsia and chlamydia are also capable of inducing a eukaryotic cell to take them up in a similar fashion. Likewise, although some of the details may be fascinatingly different, DNA, RNA, and protein synthesis are all carried out in familiar ways.

The final phase of the viral life cycle, however, is *fundamentally different* from that of any other type of organism. A central tenet of biology for more than 150 years has been *omnia cellulae e cellulae* (all cells from cells); that is, cells *divide* to produce new cells. In this chapter we shall examine evidence demonstrating that new viruses do not arise from preexisting viruses, but rather are *self-assembled* from component parts. We shall then consider various questions about the process of assembly. How are proteins assembled in the correct order? What interactions between proteins and nucleic acids ensure that only viral genomes become packaged? How do viruses acquire their envelopes? What are the mechanisms by which viruses are released from their host cells to complete their life cycles?

WHAT IS THE EVIDENCE THAT VIRUSES SELF-ASSEMBLE?

More than 40 years ago genetic studies suggested that bacteriophages such as T4 do not multiply by binary fission like their bacterial hosts, but by some other mechanism. This conclusion was solidified by the Hershey and Chase experiments (described on page 46) that demonstrated that a complete virion did not enter an infected cell. Instead, only phage DNA was injected into the cytoplasm, while virion proteins remained on the cell surface. At the same time electronmicrographs were taken, which showed T4 component parts like empty phage heads and incomplete phage tails accumulating in cells in which phage development was inhibited. Taken together, these observations indicated that the mechanism that produced new virions was not division of an existing virion, but rather *assembly* of new virions from pools of constituent viral proteins and nucleic acids.

Rigorous proof that viruses assemble rather than divide comes from *in vitro* reconstitution experiments in which infective virions are "built" from purified viral component parts. In a series of elegant experiments published between 1955 and 1957, Heinz Fraenkel-Conrat and his coworkers demonstrated that a very simple helical virus (tobacco mosaic virus) could be dissociated into its two parts, RNA and proteins, which could then be combined to recreate infective viral particles. Since virions prepared with RNA from one strain of virus and the proteins from another strain always produced infections characteristic of the RNA donor strain, these experiments also demonstrated that RNA could be the sole genetic material in a virus.

Ten years later, similar experiments indicated that even T4's vastly more complex architecture could be produced *in vitro*. R.S. Edgar and W.B. Wood found that complete virions could sometimes be made by mixing together lysates from cells that had been infected with different mutant strains of T4. For example, the heads from a strain that could not produce tail fibers could be combined with the tails produced by a strain that was unable to make heads in order to make infective virions. Analysis of many such combinations also demonstrated that formation of the head structure requires that components be assembled in a precise order. As Figure 6.1 shows, T4 is actually assembled from subunits created on three subassembly lines. In one, a hollow head is assembled and then filled with DNA. In a second, a tail is built up separately and then attached to the filled head. In the third, tail fibers that were assembled on their own line are then attached to complete the entire virion.

In vitro assembly has now been shown for naked plant and animal viruses as well as a number of bacteriophages. More recently it has been demonstrated that not just assembly, but *all* the steps necessary for the synthesis of poliovirus, can be carried out *in vitro*. Starting with purified genomic RNA, infectious virions of poliovirus can be synthesized in a cell-free extract of HeLa cells that provides all the enzymes and precursors necessary for RNA and protein synthesis. While naked viruses can be assembled *in vitro,* complete, infectious *enveloped* viruses cannot be constructed in this fashion, since their envelopes are derived from their host cells' intact membrane systems, as we shall discuss later in this chapter.

WHAT BASIC STRATEGIES ARE USED TO ASSEMBLE CAPSIDS?

As we saw in Chapter 1, the basic architectures for the virions of all viruses except the complex bacteriophages and the poxviruses are variations on a helix or an icosahedron. Since a viral nucleocapsid, however simple or complex, consists of only proteins and nucleic acids, the assembly of a virion must involve only two types of interactions: protein-protein and protein-nucleic acid. Even in very simple viruses that have only a single species of protein in their capsids, several problems must be solved during the assembly process. What sorts of interactions permit the creation of a helix or icosahedron from identical subunits? How does the virus package only its own proper genomic nucleic acid and not some other molecule available in the cell? For viruses with more complex types of nucleocapsids, there are additional problems. Since viral structural genes tend to be expressed as a group, the "late genes," the capsomers, the proteins necessary to construct the virion, are all synthesized concurrently. What ensures that individual proteins will be incorporated into the virion in the correct order?

Since a viral nucleocapsid consists only of nucleic acid and proteins, a fundamental question is "What is the order of assembly of these components?" Two strategies are possible. In the first, the genomic molecule acts as the focus for the assembly process, so that capsomer proteins are placed *around* the nucleic acid to form the capsid or nucleocapsid. In the second, the process begins with the formation of a hollow capsid that is subsequently *filled* by nucleic acid. As we should have come to expect by this point in our study of the viruses, if two possibilities exist, two possibilities have been used! The "choice" of strategy appears to be a function of the architecture of the capsid rather than the type or size of the nucleic acid forming the genome. Helical capsids and nucleocapsids are formed around their nucleic acids, while icosahedral capsids and nucleocapsids may first be assembled and then filled with nucleic acid.

HOW ARE RIGID HELICAL VIRIONS ASSEMBLED?

The model for assembly of a rigid helical capsid is that of tobacco mosaic virus (TMV). TMV is one of the simplest of viruses, consisting only of an RNA molecule 6400 bases long with

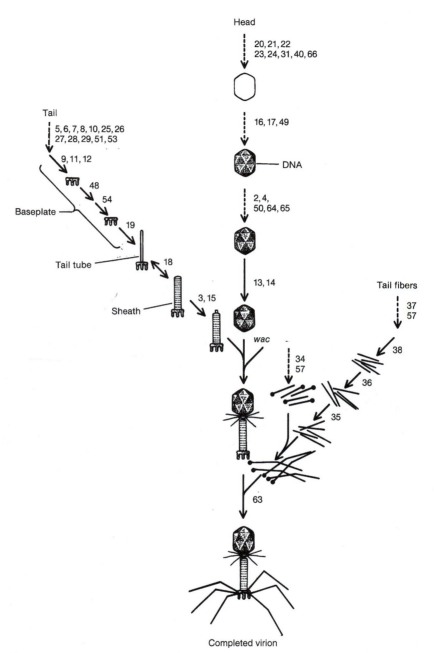

FIGURE 6.1 Assembly of the bacteriophage T4 occurs as the result of three subassembly paths. One path gives rise to the head of the virion and fills it with DNA, while another assembles the tail tube and sheath. After the head and tail are connected, the tail fibers that were assembled on their own pathway are added to complete formation of the virion. Numbers identify the genes whose products enter at each step.

2130 identical capsomers (the A protein) arranged in a helix about it (Figure 6.2). Since it is possible to assemble TMV virions *in vitro* by combining their components, one might be forgiven for assuming that the assembly process must be as simple as its ingredients. If the capsid proteins can interact with each other and with RNA in only one manner, then the individual protein molecules need merely bind to the RNA and to each other to form

A

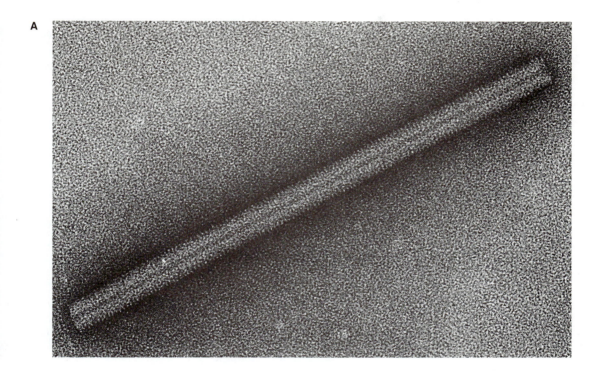

B

Viral RNA

Protein subunit

FIGURE 6.2 Structure of the rigid helical capsid of tobacco mosaic virus. The capsid is constructed entirely from a single type of protein.

the virion. The initial reconstitution experiments indicated that this was indeed the case. First, A proteins were found to be capable of forming a helical structure even in the absence of any RNA, so assembly of the capsid had to be the result of interactions intrinsic to the A proteins themselves. Second, TMV proteins that readily formed a virion around their own RNA molecules were found to recognize other natural or artificial RNAs poorly or not at all. This type of specificity is obviously necessary *in vivo,* so the proper interaction of A proteins with the correct RNA must be "built in" to their structure.

Compared to such a simple model, the actual mechanism of assembly of TMV has an elegant complexity derived from the requirement that only the appropriate RNA be encapsidated in the virion. The process begins without the participation of RNA, as 34 capsid proteins aggregate into a pair of discs. As illustrated in Figure 6.3, the outer portions of these discs interact to hold the discs together, while the inner portions are separated from each

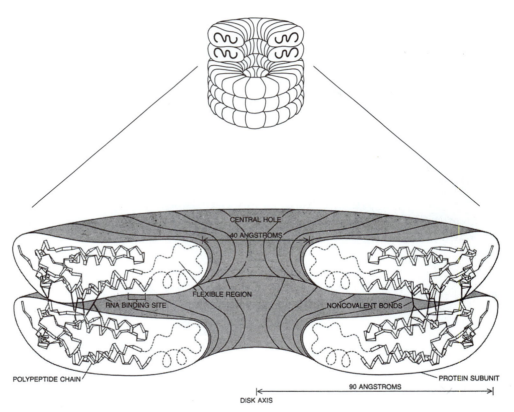

FIGURE 6.3 Cross-section through two discs of the tobacco mosaic virus capsid. The stretches of amino acids in the outer portion of each capsomer that interact with adjoining capsomers are drawn as ribbons. The amino acids in flexible portion of each capsomer that moves to bind the virus's RNA genome are indicated by dotted lines rather than a ribbon. The RNA binding site is shown in the groove between the two discs.

other by a gap that contains the RNA binding site. When RNA enters the region between the discs, the flexible inner portion of the protein chains closes the gap to hold the RNA in place in the helix.

The simplest model for assembly would suggest that incorporation of the TMV genome should begin at one of the termini, but this is not the case. When the genome is allowed to interact with only a small number of A proteins to form the initial aggregation complex, a set of about 500 bases becomes protected from RNase digestion. These bases, called the *pac site* (for "packaging"), are about 1000 bases from the 3' end of the genome. As Figure 6.4 indicates, the sequence at the *pac* site can form a series of hairpin loops. The active residues appear to be the three guanines at the top of loop 1 since any alteration of them or their positions by site-directed mutagenesis blocks normal packaging.

Assembly of TMV is illustrated in Figure 6.5. Initiation requires that the *pac* sequence loop 1 RNA enter a pair of discs (step A) and become intercalated into the gap between the

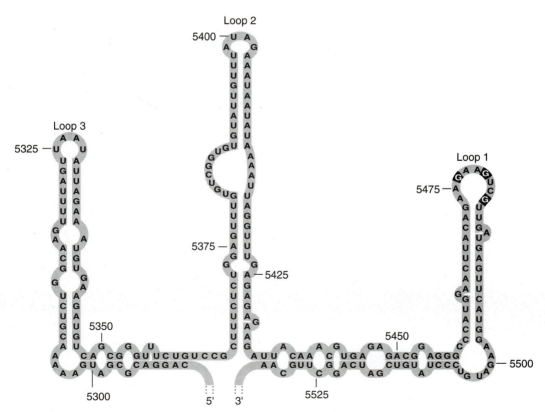

FIGURE 6.4 Stem and loop formations in the region where capsid proteins first interact with the tobacco mosaic virus genomic RNA. Bases 5290 to 5527 of the genome are shown. The three essential guanines on the *pac* site at the top of loop 1 are shown in white type.

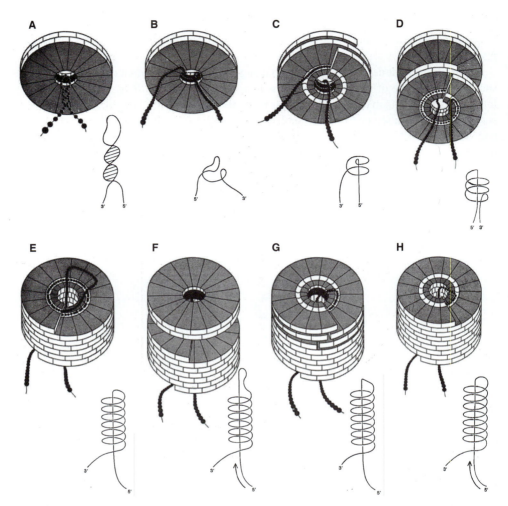

FIGURE 6.5 Model for the assembly of tobacco mosaic virus. Steps **A** through **C** depict initiation of the assembly process. The interaction of a disc with the loop 1 *pac* site (**A** and **B**) causes a shift to a "lock washer" conformation (**C**). Elongation of the helix (steps **D** through **H**) proceeds as new 5′ RNA enters the central hole of the virion and additional discs are stacked on. Each added disc shifts to the "lock washer" conformation to prepare for the next insertion of RNA.

two layers of proteins (step B). When the flexible residues close upon the RNA, the discs shift their conformation to a "lock washer" arrangement, which is the actual beginning of a helical configuration (step C). Overall assembly then proceeds in a 3′ to 5′ direction as RNA is drawn up through the "hole" in the helix and intercalated into additional disc pairs (steps D–H). Note that this scheme permits very rapid assembly since discs do not need to slip down the entire length of the RNA molecule before being added to the growing helix. Extension of the helix in the 3′ direction is much slower and appears to be by addition of small aggregates of A proteins rather than by addition of entire disc pairs.

HOW DOES THE ASSEMBLY OF FLEXIBLE AND RIGID HELICAL CAPSIDS DIFFER?

The highly ordered arrangement of protein and RNA in TMV produces a rigid virion. Many animal RNA viruses also have helical capsids that are surrounded by a membranous envelope; such capsids are generally termed *nucleocapsids* since the complete virion contains additional components. The helical nucleocapsids of enveloped viruses are not rigid, but rather are flexible enough to permit their folding within the confines of the envelope. Although the level of structural detail available for the TMV capsid has not been obtained for any of these animal viruses, such flexibility clearly indicates a different type of organization of the proteins and genomic RNAs in those viruses.

The major nucleocapsid protein in all the helical animal viruses is designated N or NP. These proteins are generally about 50 to 60 kdaltons in size and have a number of basic amino acid residues; they may also be phosphorylated. In the coronaviruses, which have positive-strand (or messenger) RNA as their genomic molecule, the N proteins alone form the nucleocapsid. The rhabdoviruses, paramyxoviruses, and orthomyxoviruses have negative-strand RNA genomes and consequently must bring into the cell the enzymes necessary to replicate their genomes before they can be expressed. These viruses therefore have polymerase components as small parts of their nucleocapsids in addition to the structural N proteins.

Since all the animal viruses with flexible capsids protect those structures with envelopes, the capsomers alone do not have the burden of shielding the viral genome from damage by nucleases or other environmental threats. The proteins surrounding the genome may consequently be organized in a looser, more open fashion requiring fewer capsomers than those in viruses with naked capsids. For example, in TMV's rigid capsid, only three nucleotides interact with each A protein in the formation of the helix. In the flexible helix of the rhabdoviruses and paramyxoviruses, in contrast, that ratio is about six nucleotides per N protein. In the orthomyxoviruses it is three times that, or about 20 nucleotides per protein subunit. Since the RNA in the orthomyxovirus nucleocapsid remains sensitive to RNase, it has been suggested that the arrangement of nucleic acid and protein is actually the reverse of the usual situation, and that the RNA is wrapped around the NP proteins rather than being surrounded by them.

In all of these viruses, the N proteins also serve regulatory functions during replication of the viral RNAs. An important question about the assembly of nucleocapsids in these viruses, therefore, is how aggregates formed of *genomic* RNA and N proteins are distinguished from those formed from nongenomic RNAs. Sequence data suggest that the 5′ ends of genomic molecules may contain *pac* site-like regions, which would function as the origin for assembly of nucleocapsids. The antigenomic RNAs do not contain such sequences. The observation that rhabdovirus nucleocapsids are elongated in the 5′ to 3′ direction supports the notion of a *pac* site at the 5′ end of the genomic molecule.

In other cases, the apparent *pac* site is located internally or may be formed by interactions between the termini of linear molecules. For example, in mouse hepatitis virus, a coronavirus, the 31 kbase-pair molecule that is both a new genome and the mRNA for the polymerase protein contains a *pac* site about 10 kbase pairs from the 3′ terminus. A 61-nucleotide sequence at this location forms a hairpin loop that is necessary (but perhaps not sufficient) for packaging. In influenza virus, all eight different genomic segments contain complementary sequences at their ends that cause these molecules to form a terminal "pan-handle" structure about 15 base pairs long. Since the influenza virus replicase binds to about 12 to 15 nucleotides at the 3′ end of each segment, it is possible that the panhandle serves as a packaging signal if replicase protein binding initiates formation of the nucleocapsid.

HOW ARE THE CAPSIDS OF ICOSAHEDRAL RNA VIRUSES ASSEMBLED?

A 20-sided figure, an icosahedron, is one of the simplest closed, three-dimensional shapes that can be made from identical subunits that interact similarly with one another. Although theoretically only 20 subunits would be needed to construct an icosahedral capsid, even the smallest viruses use 60 identical subunit peptides to do the job. As Figure 6.6, *A,* shows, each face of the icosahedron is composed of three subunits. When these subunits are identical, or are in identical environments, the figure is said to be a T = 1 capsid (the "T" is from "triangle"). Larger capsids are constructed using 60 subunits in multiples of three (180, 240, 360, and so forth). If the subunits are identical but not in the same environment, as in Figure 6.6, *B,* two different organizations occur. Five capsomers meet at each vertex, while six capsomers meet in the center of each face. This arrangement of nine identical capsomers on each face produces a T = 3 capsid. In a T = 4 capsid (Figure 6.6, *C*) each face of the icosahedron can be subdivided into four smaller triangles, each built of three subunits. In this type of capsid all vertices are surrounded by five identical capsomers (pentamers), while the intersections between adjacent faces are formed by six capsomers (hexamers), this time two each of three different types.

The capsomer proteins of a number of icosahedral viruses appear to share certain structural "themes" that can be related to their roles in assembling the entire capsid. Figure 6.7 depicts the "core structure" X-ray crystallographic studies have shown in the subunit proteins from a variety of plant and animal capsids. This core structure contains two important features. The first (Figure 6.7, *A*) is eight regions of amino acids in the β-pleated sheet conformation that are linked by regions of α-helix or random-coil, which can be quite large in some cases. The other feature is an "arm" of variable length that is generally made from the residues at the N-terminus of the peptide. The tertiary structure (Figure 6.7, *B*) of the core is a compact wedge (often called the "barrel") with the "arm" extending away from it. Figure 6.7, *C,* shows the structure of the tomato bushy stunt virus capsid subunit. The N-terminus "arm" enters from the left; the α-helix between the βG and βH regions is mostly hidden at the back of the figure.

Assembly of a structure as complex as an icosahedron appears to require intermediate forms. A general model for the assembly process in RNA viruses suggests the following steps. The first is a **nucleation** reaction in which RNA interacts with a few capsid subunits to form an initiation complex. It is likely that this reaction involves specific sequences in the RNA. Next, dimers formed of individual subunits are added to form a "cap" based on a

T = 1

A

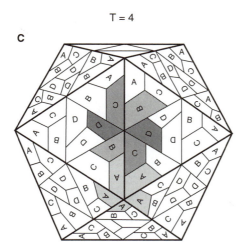

T = 3

B

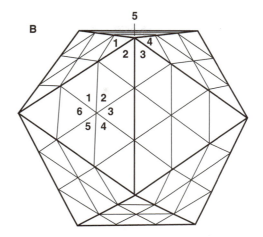

T = 4

C

FIGURE 6.6 Models of icosahedral capsids. **A** shows a T = 1 capsid in which the three capsomers on each face are identical. **B** shows a T = 3 capsid in which each face is constructed of nine identical capsomers. Note that the vertices are surrounded by five identical capsomers, while six capsomers meet in the center of each face. **C** shows a T = 4 capsid. Each face of this capsid contains four triangles built of three subunits each. The central triangle on each face contains three identical capsomers, while the other three triangles contain three different capsomers. Note that all vertices are surrounded by five identical capsomers (pentamers), while the intersections between faces are formed by six capsomers (hexamers), two each of three types.

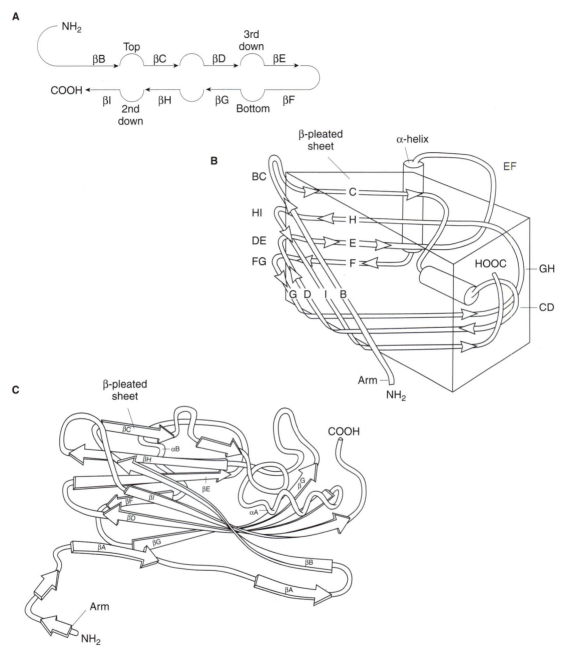

FIGURE 6.7 Three views of core structure for a capsomer subunit protein of an icosahedral RNA virus. **A** shows the conventions for designating eight regions of β sheet (labelled βB through βI) that are connected by loops of α-helix or random coil. The "tail" of the core structure is the run of amino acids at the N-terminus. **B** models the compact barrel or wedge-shaped structure formed by folding of the core structure. The loops are identified by the β-pleated sheet regions they connect. **C** depicts the actual conformation of a tomato bushy stunt virus capsomer, as determined by X-ray chrystallographic analysis. β-pleated sheet is shown as ribbon.

central pentamer vertex. In the remainder of the assembly process, additional dimers are placed in the growing shell.

Several types of alterations in the structure of individual proteins occur during the assembly process in many viruses. The first of these is a conformational change in the orientation of the "arm" sequence during its interactions with other subunits. In some viruses, extension of the "arm" appears to regulate the local curvature of the subunits as they are added to the growing capsid shell. The other form of alteration involves cleavage of the peptide chain of a subunit during the assembly process. These cleavages usually occur after the viral capsid has formed its spherical shape, frequently just before the final encapsidation of the genomic nucleic acid. Since the architectural requirements for a subunit during assembly may be different from those for the same subunit during the infective process (recognition, attachment, penetration, and uncoating), this cleavage may create conformational changes that shift the structure of the subunit from one activity to another. For example, a conformational change in a protein may allow the capsid to open readily when it enters a cell, just as cocking the spring in a Jack-in-the-box prepares the "Jack" to pop up when the top opens.

The Assembly of Poliovirus

The assembly of the picornaviruses like poliovirus has been extensively studied as a model system for naked icosahedral viruses, but numerous unanswered questions remain. The poliovirus virion contains 60 copies each of structural proteins called VP1, VP2, VP3, and VP4. The major structural proteins (1, 2, and 3) all have the basic core structure described previously, but with extensive differences in the sequences of the loops connecting the regions of β-pleated sheet. These loops and the C-termini of the peptides are on the outer surface of the capsid, while the inner surface contains VP4 and the N-termini of VPs 1, 2, and 3.

The major features of poliovirus assembly are illustrated in Figure 6.8. As we saw in Chapter 4, this virus expresses its positive-strand RNA genome as a single *polyprotein* that is then cleaved into both structural and nonstructural proteins. The first posttranslational cleavage of the genomic polyprotein (step 1) yields smaller structural polyprotein (P1) containing VP1, VP3, and VP0. VP0 is itself a polyprotein containing both the VP2 and VP4 peptides. The structural polyprotein P1 is then cleaved (step 2) to separate VP1 from VP3/VP0. This cleavage causes a rearrangement of the peptides, particularly the N-termini and C-termini, and leads in turn to a third cleavage and rearrangement (step 3) to form the basic "building block" or *protomer* of the capsid. The protomer is also called a 5S unit from its sedimentation coefficient.

The final cleavage of P1 allows the N-termini (the "arms" of the core structure) of five protomers to interact to create a pentamer, or 14S unit (step 4). The vertex of the pentamer is formed when the "arms" from five VP3 molecules intertwine (in β-conformations) like the leaves of a microscope's iris diaphragm to form a central cylinder. This cylinder is in turn surrounded by three-stranded β-sheets created from the N-termini of VP1 and parts of what will later become VP4. This arrangement is stabilized by the presence of the fatty acid myristate bound to the N-terminus of VP4.

Twelve pentamers then combine to form a **procapsid** (73S) or empty icosahedral structure ready to receive a genomic molecule (step 5). The next step (6) converts the empty procapsid into an RNA-filled *provirion* (155S), but the mechanism by which RNA enters

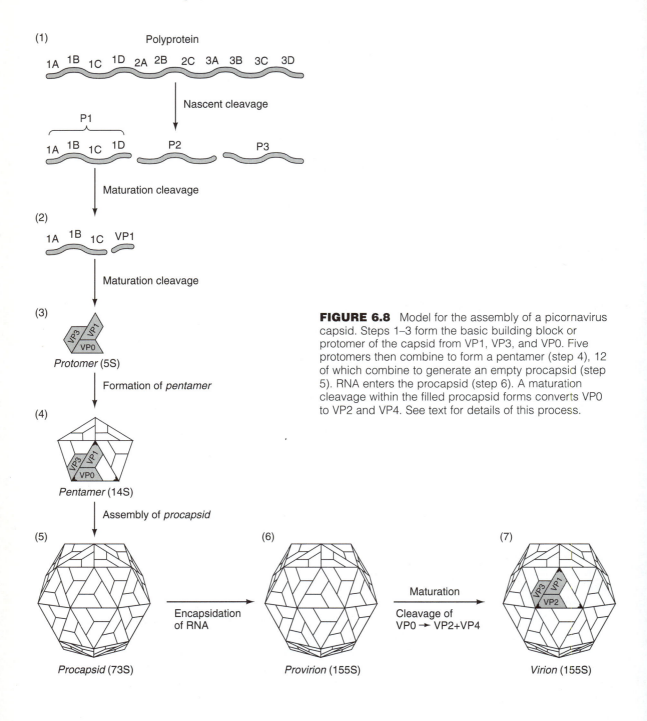

FIGURE 6.8 Model for the assembly of a picornavirus capsid. Steps 1–3 form the basic building block or protomer of the capsid from VP1, VP3, and VP0. Five protomers then combine to form a pentamer (step 4), 12 of which combine to generate an empty procapsid (step 5). RNA enters the procapsid (step 6). A maturation cleavage within the filled procapsid forms converts VP0 to VP2 and VP4. See text for details of this process.

the procapsid is not known. Although exactly the same proteins are present in both the procapsid and the provirion, studies using specific antibodies directed against the amino acids on the surfaces of these structures indicate that different groups are exposed in the procapsid and provirion. These findings suggest that the process of encapsidation involves significant shiftings or rearrangements of the capsomers as a result of packaging the poliovirus RNA.

The final step (7) in the assembly process is a so-called *maturation* cleavage of VP0 to form VP2 and VP4. This cleavage appears to stabilize the mature virion since it becomes much more resistant to degradation by heat, pH, and detergents than is the provirion. This increased stability may be due to the formation of a seven-stranded β-pleated sheet from parts of VP3 and VP1 in one protomer and VP2 from a neighboring protomer.

HOW IS THE RNA ARRANGED IN AN ICOSAHEDRAL CAPSID?

X-ray crystallography and nucleic acid sequencing of structural genes together have permitted construction of a model for the assembly of the protein capsid components of virions of small RNA viruses like poliovirus. The next question, then, is what is the arrangement of nucleic acid within the capsid? In helical capsids the regularity of the interactions of RNA with capsid proteins was quickly evident from structural analysis. This has not been the case for the icosahedral viruses, however. Even high-resolution X-ray crystallographic studies have not been able to define the organization of RNA in most cases. These data suggest that unlike in helical viruses, RNA is not packaged in completely ordered arrays in icosahedral viruses.

This is not to say that there is no organization of the RNA in the capsid, however. About 20% of the nucleotides in the genome of bean pod mottle virus (a comovirus), for example, appear to be associated with hydrophilic pockets formed by the interaction of three neighboring core structures. Figure 6.9 illustrates how the RNA may be threaded in the same polarity across the inner surface of each of the 20 faces of the virion. The organization of the remaining 80% of the genome within the interior of the virion is still unknown.

HOW ARE SIMPLE ICOSAHEDRAL DNA BACTERIOPHAGES ASSEMBLED?

The assembly process for icosahedral capsids of either RNA or DNA viruses appears to be similar in outline. In both cases, assembly of empty *procapsids* begins when a sufficient concentration of structural proteins accumulates in the cytoplasm. Genomic nucleic acids then interact with the procapsid to complete assembly of the virion. A significant difference between assembly of icosahedral capsids in RNA and DNA viruses, however, occurs in the relationship of the assembly process to replication of the viral genome. As noted in Chapter 3, replication of DNA in many bacteriophages leads to the formation of concatemers, which must then be cut into genome-length molecules for insertion into a virion. This cutting process is usually linked to assembly of the virion and is mediated by virion components.

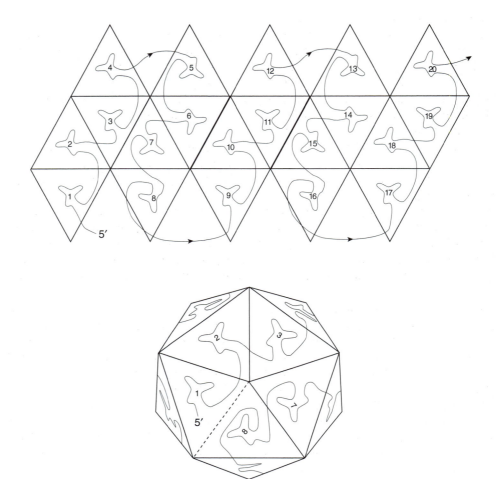

FIGURE 6.9 Model for the arrangement of a portion of the RNA within the bean pod mottle virus capsid. The genomic molecule interacts with the inner surface of all faces of the capsid, but its arrangement in the interior of the virion is not known.

Many bacteriophages have complex architectures that include not only an icosahedral head, but also a helical tail and its associated fibers. As we saw at the beginning of this chapter, these components are assembled in separate series of steps and then combined into the final virion.

The architecturally simplest DNA viruses are bacteriophages like φX174 that have circular single-stranded genomes. φX174's T = 1 capsid consists of 60 copies of F protein forming the faces of the icosahedron and 12 spikes at the vertices, each made of five G proteins and one H protein. The complete virion also contains 60 copies of a very basic protein (J) associated with the viral DNA molecule. The J protein probably helps to condense the DNA by neutralizing its negative charges.

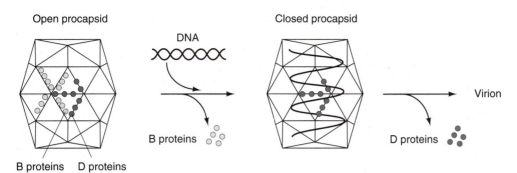

FIGURE 6.10 The bacteriophage ϕX174 procapsid has two sets of scaffolding proteins. Protein B forms an internal scaffold while protein D forms an external one. The inner scaffold is displaced when DNA enters the procapsid. The external scaffold is then removed to complete the process of maturation.

The process of assembly for ϕX174 appears to occur in association with the inner surface of the cytoplasmic membrane where the growing procapsid is held by the very hydrophobic viral H spike protein. A series of intermediate-sized structures can be isolated from infected cells, indicating that assembly proceeds by sequential additions to a growing procapsid. In addition to the proteins found in the virion, two other proteins (B and D) are associated with the procapsid until DNA fills it. As Figure 6.10 illustrates, the empty procapsid has an external **scaffolding** formed by viral D proteins and an internal scaffolding made from viral protein B. As the name "scaffolding protein" indicates, these proteins help to create and maintain the proper overall shape of the procapsid until it is stabilized by being filled with nucleic acid. The internal scaffolding is displaced as DNA enters the procapsid, while the external scaffolding is removed after the procapsid is filled as the final step in the maturation process.

Accumulation of procapsids in the cytoplasm mediates a dramatic shift in the mechanism of viral DNA synthesis in ϕX174. As noted in Chapter 3, ϕX174 initially creates a double-stranded replicative intermediate of its circular genome that replicates in the conventional bidirectional fashion of its bacterial host, producing a θ intermediate. Later in the infective cycle, DNA synthesis becomes coupled directly to packaging of the genome into procapsids.

Figure 6.11 presents a scheme for *rolling circle* replication of ϕX174 that relates DNA synthesis to assembly of the virion. The viral nickase, or A protein, cuts the positive strand of the double-stranded replicative intermediate at a specific site, which is thus both a replication origin and a *pac* site. The A protein remains attached to the 5′ end of the positive strand. DNA synthesis by *E. coli* enzymes, using the new 3′ end of the positive strand as primer, then gradually displaces the original strand from the duplex. It appears that the A protein (attached to the 5′ end of the genome) also binds to an opening in the procapsid, which causes the displaced strand to fold itself into the capsid interior as synthesis proceeds. It is possible that the J protein enters at the same time. When the replication complex has traversed the entire template strand, the A protein cuts the new genomic molecule from the concatemer and ligates its ends together to form a circle.

(1) Nick at the *pac* site to initiate rolling circle replication

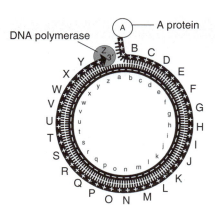

(2) The protein binds to opening in procapsid

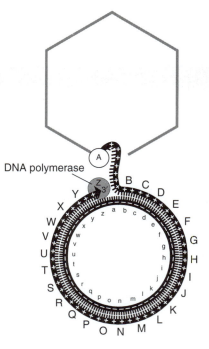

(3) As it is displaced, positive-strand DNA enters the capsid

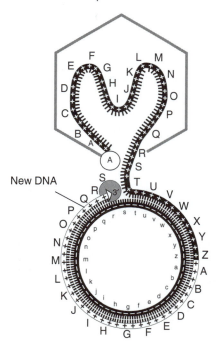

(4) The protein cleaves the positive strand and ligates the ends to form a circular molecule within the virion

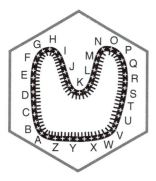

FIGURE 6.11 For legend see opposite page.

HOW ARE THE CAPSIDS OF COMPLEX BACTERIOPHAGES ASSEMBLED?

Many of the double-stranded DNA bacteriophages have complex virions that consist not only of an icosahedral "head" containing the genome, but also a helical "tail," which mediates attachment and penetration. As we noted in our discussion of *in vitro* reconstitution experiments at the beginning of this chapter, the construction of such intricate virions utilizes a series of independent assembly lines to form the major components of the virion that are then combined into the final product. Genetic and biochemical experiments have provided a reasonable picture of how the protein constituents of the head and tail are assembled in viruses like λ and T4.

As with the simple icosahedral DNA bacteriophages, the complex bacteriophages appear to create an empty *procapsid,* or "head" unit, in a series of assembly steps and then fill that procapsid with DNA. Both of these processes require a special structure called the **DNA translocating vertex,** which serves as the initiation complex for assembly of the head and in the movement of DNA both into the head during maturation as well as out of the head during penetration. As its multiple roles suggest, the DNA translocating vertex is located at the junction of the head and tail in the virion. An intriguing feature of that location is that the capsid vertex has *fivefold* symmetry while the tail connected to that vertex has *sixfold* symmetry. A special protein, the **portal protein,** shaped like a donut with 12 knobs surrounding it, occurs at this "mismatch" junction and appears to mediate all the DNA-translocating vertex functions.

Genetic studies indicate that the portal protein in T4 is gp20 since gp20⁻ mutants accumulate the coat protein gp23 but do not form heads. The process of head assembly is initiated by the portal protein, which is bound to the inner surface of the cytoplasmic membrane. The procapsid requires not only the portal and coat proteins, but also scaffolding proteins (gp22) that build an inner shell that is then removed when assembly is complete so that they are not found as part of the mature virion.

Assembly of the procapsid proceeds through several steps. The first constructs a double shell consisting of coat proteins on the outside and scaffolding proteins on the inside (Figure 6.12). This structure, often called the Procapsid I, has a spherical form rather than the icosahedral shape of the capsid and is smaller in dimension. When the Procapsid I is complete, the scaffolding proteins are removed, either by proteolytic digestion as in T4,

---

FIGURE 6.11 Bacteriophage φX174 DNA synthesis by the rolling circle mechanism is directly linked to packaging of the newly synthesized DNA within a procapsid. When the shift from bidirectional to rolling circle replication occurs, a nick at the *pac* site on the positive strand of DNA creates a free 3′ end where elongation occurs (step 1). The 5′ end of the nicked DNA strand has a molecule of the A protein attached to it. This A protein binds to the opening in the procapsid (step 2) so that as DNA synthesis proceeds the displaced strand enters the procapsid (step 3). When a complete genome has entered the procapsid, the A protein cleaves the genome from the concatemer and ligates its ends to form a circular molecule within the mature virion (step 4).

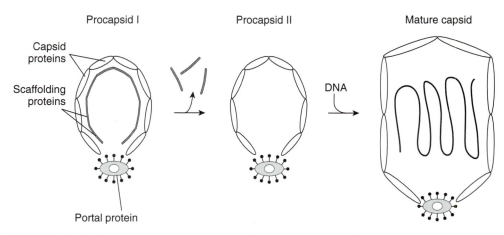

Procapsid I   Procapsid II   Mature capsid

Capsid proteins

Scaffolding proteins

DNA

Portal protein

FIGURE 6.12 Formation of the icosahedral "head" of a complex bacteriophage. The somewhat rounded Procapsid I structure is stabilized by scaffolding proteins that are removed to yield the Procapsid II structure. Conformational changes in the Procapsid II proteins cause the shell to expand and become more angular in shape. DNA filling may occur with or subsequent to this expansion.

or by recycling intact proteins as in the phage P22, to generate an empty procapsid, which may be called the Procapsid II. Conformational changes in the coat proteins, often caused by proteolytic cleavages, cause the shell to expand to its final dimensions and assume the angular shape found in a mature head. This expansion and reorganization of the procapsid prepares it for the final step of the process, packaging the viral genome, or is concurrent with it.

At the same time as these steps are occurring, the helical tail of the virion is being generated on its own assembly line. As we saw in Chapter 3, these tails can be short or long, "contractile" (T4), or flexible (λ). Assembly of the T4 tail begins with the formation of the *baseplate,* which will make the distal end of the tail in the completed virion. The subunits of the tail tube are then polymerized upon the baseplate. The sheath proteins that surround the tail tube in "contractile" tails, as well as minor components that stabilize the entire structure, are then added to complete the tail assembly process. The DNA-filled head is then joined to the tail. If the virion also contains tail fibers, these are assembled on their own line and added to complete the construction of the virion. These steps in the assembly of T4 are summarized in Figure 6.1.

WHAT STRATEGIES DO DNA BACTERIOPHAGES USE TO FILL THEIR CAPSIDS?

Many DNA viruses form *concatemers,* or multigenome length molecules, rather than single genomes when they replicate their DNAs. (You may wish to review bacteriophage DNA replication in Chapter 3.) An important step in the assembly of such viruses is "measuring" the correct amount of DNA for each new virion and then cutting the concatemer at the proper site. Several general strategies have evolved to solve these problems, all of which utilize similar components. Two of these components have already been mentioned: the portal protein, which is part of the DNA translocating vertex, and the major capsid proteins themselves. The third component is a **terminase,** which is a complex of ATP-binding proteins that not only cuts DNA strands but also has a role in translocation of DNA as well.

Figure 6.13 summarizes the different approaches bacteriophages have taken to cutting and packaging concatemeric DNAs. In the first group of viruses, cuts are always made at defined sequences within the concatemer. In λ virus, for example, the terminase complex recognizes an extended *cos site* (from *co*hesive *s*ite) of about 200 base pairs, and then makes single-strand nicks at the ends of a 12-base sequence within the *cos* site. Cutting at two consecutive *cos* sites creates identical genomic molecules with unpaired "cohesive" or "sticky" complementary 5′ ends. These cutting reactions take place at the DNA translocating vertex of the procapsid, so the genome between the first and second *cos* sites is drawn into the virus head. These cutting and packaging reactions can also be induced in an *in vitro* system employing any double-stranded DNA with *cos* sites 37 to 52 kbases apart. In recombinant DNA technology, hybrid plasmids called **cosmids** containing the λ *cos* sites are widely used as vectors for cloning large (45 kbase pairs) fragments of DNA.

Other viruses like T3 and T7 also make staggered cuts at defined sites to produce individual genomic molecules with very long single-stranded stretches at their 5′ ends. These single-stranded sequences are then used as template for elongation of the shorter 3′ strands, leading to the creation of genomic molecules that have blunt ends and are terminally redundant.

Many bacteriophages appear to "measure" the DNA in their concatemers out into "headful" lengths; that is, the amount of DNA that can be contained within the expanded procapsid during packaging. In some viruses like P22, measuring begins at the first encountered *pac* site in the concatemer. The DNA is cut at that location and packaging carried out until a length of DNA equivalent to 104% of the genomic length is inserted into the viral capsid, at which point the next cut is made. Each succeeding genome is packaged beginning where the previous genome ended, so the result is a collection of genomes with permuted terminal redundancies.

Other bacteriophages like T4 simply initiate headfilling at a random location near the end of the concatemer and package "headful" lengths equivalent to 102% of the genome. Note that regardless of the packaging scheme, all of these viruses have a full-length copy of their genome in each capsid. Since any site on the genetic map of a virus like T4 can be the location where an individual genome was cut from the concatemer, it is possible that one or more genes might be separated from the promoter that governs their expression. The "extra" DNA included in the genomic molecules when cuts are made may help ensure that most virions will contain a complete functional set of all transcriptional units.

While the "measuring" mechanisms of many viruses are known, the arrangement of the DNA within the viral head is far less well understood. Early models suggested that the DNA was looped into the head from the inner surface of the capsid to the central axis, somewhat the reverse of the way thread is looped onto a spool. Recent evidence indicates that

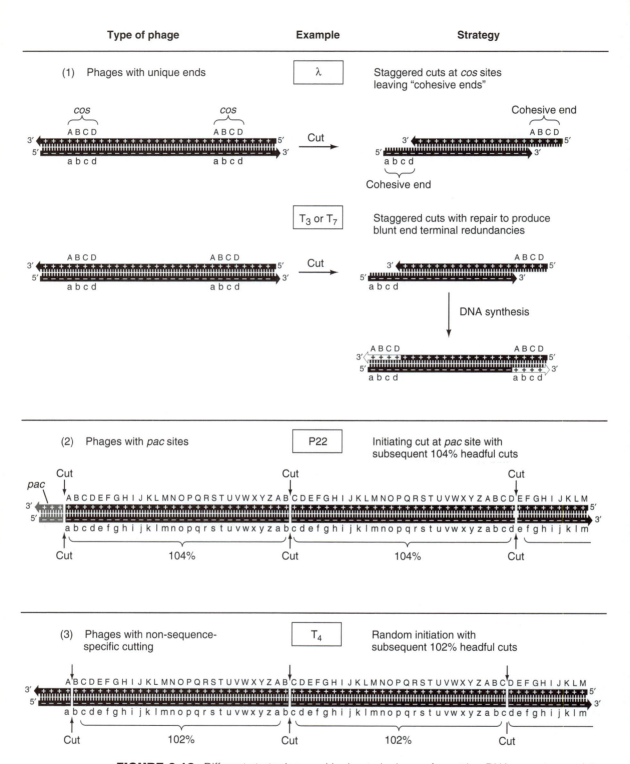

FIGURE 6.13 Different strategies used by bacteriophages for cutting DNA concatemers into sizes that will fill their capsids.

this is not the case, however, and that the DNA is instead folded into rods, which may be packed into the head as arrays parallel to the tail of the virion.

HOW ARE THE ICOSAHEDRAL DNA VIRUSES OF EUKARYOTES ASSEMBLED?

Compared to the information available about the assembly of the DNA bacteriophages, little is presently known about the details of assembly in even the smallest of the icosahedral DNA viruses infecting animal cells. The capsid of polyomavirus, one of the papovaviruses, for example, has an unusual arrangement of subunits that appears to violate the "rules" for assembly of icosahedral shapes. Rather than the usual 12 pentamers and 60 hexamers, all of the polyomavirus capsomers are grouped into pentamers. The virion is thus more spherical than icosahedral. Assembly of this strange virion appears to involve a series of intermediate steps similar to those seen in the small icosahedral RNA viruses. The process probably begins with condensation of the genomic DNA with histones to produce nucleosomes. This forms a compact particle around which the capsid then is built. Histone 1 is lost during formation of the capsid, but there are no covalent modifications of the capsid proteins in the process.

Assembly of the adenoviruses also appears to proceed through a series of stages like those seen in the bacteriophages and RNA viruses. Pentamers and hexamers are assembled as subunits that are then placed in position in the growing capsid, while two other proteins appear to act as scaffolding proteins during the process. The linear DNA of an adenovirus genome has a protein attached to each 5′ terminus, but genetic studies indicate that a *pac* sequence located between 290 and 390 base pairs from the left end of the genome is responsible for initiating entry of DNA into the procapsids. Adenovirus DNA is condensed into a compact core structure by basic proteins encoded by the virus rather than by host cell histones.

The steps in the assembly of capsids in even larger viruses like the herpesviruses have not been identified. These viruses replicate their linear genomes by a process that produces concatemeric molecules. Procapsids are apparently filled by a mechanism that involves recognition and cutting of specific repeated sequences in the concatemer.

HOW DO VIRUSES WITH SEGMENTED GENOMES ENSURE THAT VIRIONS CONTAIN A COPY OF EACH SEGMENT?

For viruses like influenza A that have helical nucleocapsids contained within an envelope, the answer appears to be simple: They don't. Any mechanism that ensured that one copy of each of influenza A's eight segments was included in each virion would necessitate multiple complex recognition events and interactions. It appears to be more efficient simply to package at least eight random lengths of viral RNA. The ratio of virus particles to actual infective units in an influenza A virus stock is comparable to the ratio predicted for such random packaging. There is some evidence, however, from studies in which defective and wild-type viruses were used to coinfect the same host cell, that assembly of new virions may have an element of segment-specific control.

In contrast, at least some viruses with icosahedral capsids appear to have specific mechanisms for packaging their segmented genomes. One well-characterized example is that of the bacteriophage φ6, which has three double-stranded RNA molecules as its genome. Like the animal reoviruses that have segmented double-stranded RNA genomes, positive-strand RNAs serve both as mRNAs and are encapsidated during maturation. The procapsid is also the viral replicase and synthesizes a complementary negative strand for each positive strand. Each positive-strand RNA has its own unique *pac* sequence located between 200 and 300 bases from its 5′ end, so each sequence can be individually recognized during filling of the procapsid. Almost no defective particles are formed during maturation, indicating that the phage accurately packages a copy of each segment into the procapsid. *In vitro* packaging experiments suggest that the procapsid has high affinity sites that recognize each segment. Since packaging of the longest segment is dramatically enhanced by prior packaging of the middle-sized segment, it is likely that the long RNA site is either created or activated by filling of the middle RNA site.

HOW DO VIRUSES EXIT THEIR HOST CELLS?

The final event in the viral life cycle is *release* of mature virions into the extracellular environment. How this release is accomplished is a function of the architecture both of the virion itself and of the host cell. Many animal viruses and virtually all bacteriophages have naked virions. These viruses cause their host cells to **lyse** or break apart, literally spilling the new virions out into the extracellular milieu. Other animal viruses (and a tiny number of bacteriophages) surround their capsids with a membranous envelope. These viruses use a special form of exocytosis that combines the final events of maturation with a mechanism for release of the completed virions.

The plant viruses appear to use neither of these strategies for the release of virions from infected cells. As we noted in Chapter 2, transmission of plant viruses generally relies on direct passage from cell to cell via cytoplasmic connections in the same plant, or on mechanical inoculation or insect vectors between different plants. None of these mechanisms requires that virions be released from infected cells after they have assembled.

HOW DO NAKED VIRUSES EXIT THEIR HOST CELLS?

All but a handful of bacteriophages complete their life cycles by lysing their host cells. This lysis actually involves two different sorts of activities since a bacterial cell is surrounded by a cytoplasmic membrane and then by a complex cell wall (see Chapter 1). The cytoplasmic membrane is first destabilized by the action of viral late gene proteins (gp t in T4 and gp s in λ). In T4 the action of gp t appears to be regulated by proteins encoded by the two genes of the *rII* locus. These genes take their names from the observation that when their gene products are missing or defective, the virus lyses its host cell more rapidly than usual (hence "r" for "rapid lysis"). Disruption of the cytoplasmic membrane then allows other viral enzymes to degrade the peptidoglycan layer of the bacterial cell wall. T4 uses a *lysozyme* to break down the sugar backbone of peptidoglycan, while λ produces an *endolysin* that cleaves the peptide chains that cross link the peptidoglycan backbones. Other bacteriophages like ϕX174 appear to have a gene product (gp E) that activates a set of host cell autolytic enzymes that then degrade the membrane and cell wall.

Naked animal viruses like the picornaviruses, reoviruses, papovaviruses, adenoviruses, and parvoviruses are also released by lysis of their host cells. Although they are not naked viruses, poxviruses also lyse the cells they infect. Lysis of infected cells by these viruses does not appear to be the result of the production of specific viral proteins designed for that purpose as is the case with the bacterial viruses. As we shall discuss in the next chapter, many animal viruses inhibit or block important steps in the metabolism of their host cells. The cumulative damage that accrues to infected cells when their metabolic activities are inhibited seems to lead to cellular disintegration and release of virions. It has been suggested that disruption of lysosomes, which would release lysosomal hydrolytic enzymes into the cytoplasm, plays a role in this process, but no particular viral protein has been shown to mediate such a process. Animal viruses that kill their hosts during release of the mature virions are termed **cytocidal** or **cytolytic.**

HOW DO ENVELOPED VIRUSES ACQUIRE THEIR ENVELOPES?

In many animal viruses the *nucleocapsid* of nucleic acid and protein is surrounded by a membrane called the *envelope*. Envelopes are derived from locations in a host membrane that have been modified to replace cellular proteins and glycoproteins with the virus-encoded glycoproteins that play crucial roles in the attachment and penetration stages of the virus's life cycles (see Chapter 2). The first experiments to demonstrate these alterations in cellular membranes utilized specific antibodies to identify constituents on the surfaces of infected cells. Antibodies against viral envelope proteins or glycoproteins were found to bind to discrete patches on the surface of infected cells, while antibodies to most normal plasma membrane proteins or glycoproteins did not bind to those same patches. Electronmicrographs showed the altered patches to be the sites where virions formed. The lipids of the host membrane and the viral envelope are usually the same types, although their proportional contributions to the membrane may differ.

Maturation of enveloped viruses, illustrated in Figure 6.14, generally requires four steps. The first is the synthesis and insertion of viral glycoproteins and proteins into target host membranes. This process uses the usual protein secretory pathway, but involves translation of viral mRNAs for envelope proteins in place of host cell mRNAs for membrane proteins. Note that the fusion of a vesicle bearing viral glycoproteins in effect creates a new area of membrane and in the process pushes aside the existing membrane components. The reactions of the second step assemble the viral nucleocapsid. The third step prepares for the process

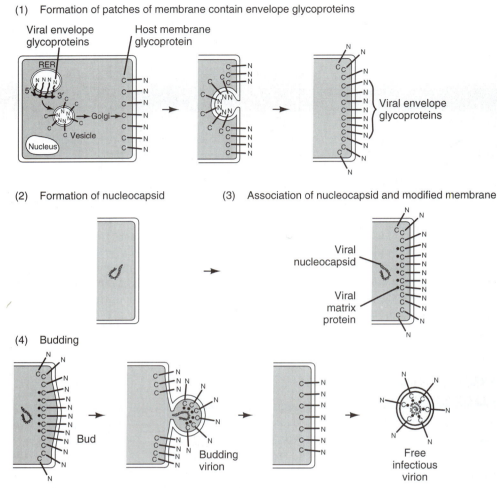

(1) Formation of patches of membrane contain envelope glycoproteins

(2) Formation of nucleocapsid

(3) Association of nucleocapsid and modified membrane

(4) Budding

FIGURE 6.14 Maturation and budding of an enveloped virus from the plasma membrane. Modified patches of membrane are created as a result of fusion of vesicles containing viral glycoproteins with the plasma membrane (step 1). The viral nucleocapsid is assembled (step 2) and becomes associated with the inner surface of the modified membrane patch, perhaps containing viral matrix proteins bound to the C-termini of the envelope glycoproteins (step 3). Budding is a process of evagination or exocytosis that produces a mature enveloped virion (step 4).

of exocytosis by bringing the nucleocapsid and modified membrane together. The final step is the formation of an enveloped virion by **budding.** The nucleocapsid becomes enwrapped in membrane to form a vesicle that is then "pinched off" from the membrane to release the enveloped virion (Figure 6.15)

Budding would seem to offer a chance that an infected cell could survive the production of new virions. In a few instances that is indeed so. Cells infected with retroviruses,

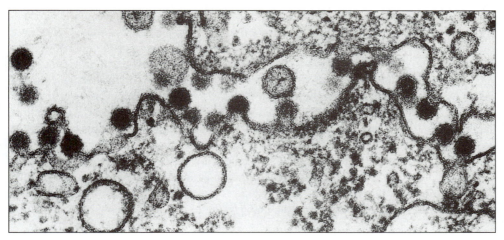

FIGURE 6.15 Semliki Forest virus (Toga) budding from the cytoplasmic membrane.

for example, can bud virions continuously for many years. In most cases, however, even viruses that complete their life cycles by budding are cytocidal.

HOW DO NUCLEOCAPSIDS IDENTIFY MODIFIED MEMBRANE LOCATIONS FOR BUDDING?

The budding process requires that nucleocapsids identify patches of the appropriate membrane that have been modified by insertion of viral proteins. In some viruses, like the retroviruses, formation of the nucleocapsid appears to be linked directly to the budding process, since the first detectable nucleocapsids occur immediately beneath the plasma membrane where budding will occur. The C-terminal domain of the envelope proteins presumably is responsible for directing assembly and budding at that location. In the orthomyxoviruses, paramyxoviruses, and rhabdoviruses, a *matrix* or M protein is required for proper maturation. The M proteins associate with the inner surface of the plasma membrane in proximity to the embedded envelope proteins. Interaction of nucleocapsid proteins with M proteins initiates the formation of the bud. In rhabdoviruses, binding of M proteins to the nucleocapsids is responsible for the formation of the tightly coiled "bullet" shape characteristic of those virions.

HOW DOES A VIRUS TARGET A PARTICULAR MEMBRANE REGION AS THE SITE OF BUDDING?

An aspect of this process that has received much attention is the question of how viral envelope proteins are routed into the proper membrane. Their synthesis appears to proceed by the usual cellular pathway of translation of the envelope mRNAs on rough ER, process-ing and glycosylating in the Golgi, and transport to the plasma membrane via Golgi-derived vesicles. Analysis of envelope mRNAs reveals that they do not have the equivalent of a "signal sequence" to direct their proteins to the appropriate target, however. Studies in which various parts of the envelope proteins are deleted or altered, or in which hybrid or chimeric proteins are created, suggest that the entire conformation of these glycoproteins is involved. For example, orthomyxoviruses like influenza virus bud from the apical surface of epithe-lial cells (toward a lumen) while rhabdoviruses like vesicular stomatitis virus bud from the opposite basal surface (toward the underlying tissue) (Figure 6.16). When chimeric gly-coproteins are made with the N-terminal (extracellular) portion of the influenza virus hemagglutinin (HA) combined with the C-terminal (intracellular) portion of the vesicu-lar stomatitis virus G protein, the chimeric proteins are inserted into the apical mem-brane. The reciprocal combination is inserted into the basal surface membrane, indicating

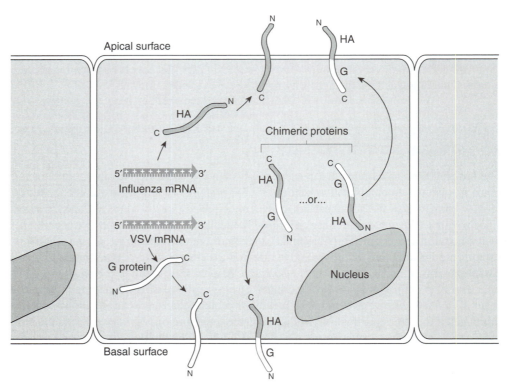

FIGURE 6.16 Sorting of viral envelope glycoproteins is controlled by NH_2-terminal domain. Influenza A virus hemagglutinin glycoprotein (HA) becomes inserted into the apical cell surface, while the vesicular stomatitis virus G glycoprotein enters basolateral membranes. Chimeric glycoproteins whose ends are from different viruses are sorted according to the origin of their NH_2-terminal domains: NH_2-HA/G-COOH glycoproteins are inserted into apical membrane, NH_2-G/HA-COOH glycoproteins into basolateral membranes.

that the extracellular domain of these proteins is responsible for guiding the glycoproteins to the proper location. Deletion of the C-terminal 79 amino acids from the G proteins causes that protein to be secreted rather than membrane bound, however, so both domains are important in the overall process.

DO ALL VIRUSES BUD THROUGH THE CYTOPLASMIC MEMBRANE?

Although the basic process of forming an enveloped virion is the same for all viruses, not all viruses use the cytoplasmic membrane as the source of their envelopes. Table 6.1 summarizes the properties of various groups of budding animal viruses. Although direct budding through the cytoplasmic membrane is the choice of most RNA viruses, the internal membrane systems of the rough endoplasmic reticulum and Golgi and even the nuclear envelope are also used.

Figure 6.17 depicts how five different viruses use internal membrane systems to complete their maturation processes. The helical nucleocapsids of the coronaviruses bud from the cytoplasm into rough endoplasmic reticulum near the *cis* side of the Golgi (Figure 6.17, *A*). As the virus particles then move through the Golgi complex, sugars are added to the envelope proteins to complete the maturation process. The virions are then transported to the cell surface in vesicles that fuse with the cytoplasmic membrane to release the virions. Bunyaviruses bud directly into the Golgi complex and then proceed through the same steps as the coronaviruses.

Table 6.1 | Properties of Budding Viruses

Membrane System	Virus Family	Envelope Proteins	Matrix Protein	Nucleic Acid and Nucleocapsid	Site of Nucleocapsid Assembly
(1) Plasma membrane					
■ Apical surface	Orthomyxovirus	HA, NA	Yes	(−) RNA/helical	Nucleus
	Paramyxoviruses	F, HN	Yes	(−) RNA/helical	Cytoplasm
■ Basal surface	Retroviruses	SU, TM	Yes	(+) RNA/icos.	Cytoplasm
	Rhabdoviruses*	G	Yes	(−) RNA/helical	Cytoplasm
	Alphaviruses	E1, E2	No	(+) RNA/icos.	Cytoplasm
(2) Endoplasmic reticulum**	Coronaviruses	E1, E2	No	(+) RNA/helical	Cytoplasm
	Togaviruses (Flaviviruses)	M, E	No	(+) RNA/icos.	Cytoplasm
(3) Golgi**	Bunyaviruses	G1, G2	No	(−) RNA/helical	Cytoplasm
	Coronaviruses	E1, E2	No	(+) RNA/helical	Cytoplasm
	Poxvirus	Many	No	DNA/complex	Cytoplasm
(4) Nucleus	Herpesviruses	B, C, D, E	No	DNA/icos.	Nucleus

* = except for rabies virus that buds from apical surface of acinar cells in the salivary gland.
** = virions bud into vesicles and are released from cell by exocytosis.

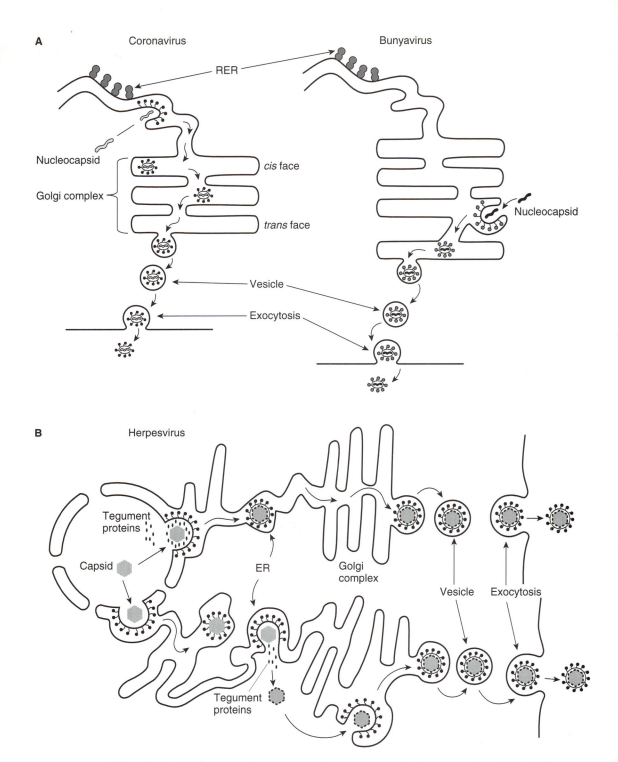

FIGURE 6.17 The use of internal membrane systems during maturation. **A,** The RNA coronaviruses and bunyaviruses bud into the rough endoplasmic reticulum and Golgi complex respectively. **B,** The DNA herpesviruses bud into the inner nuclear envelope membrane. One model of this process suggests that the tegument proteins are added before that budding step, while a different model says that two budding steps are needed and the tegument proteins are added in the cytoplasm between those steps. *(Continued.)*

C Poxvirus

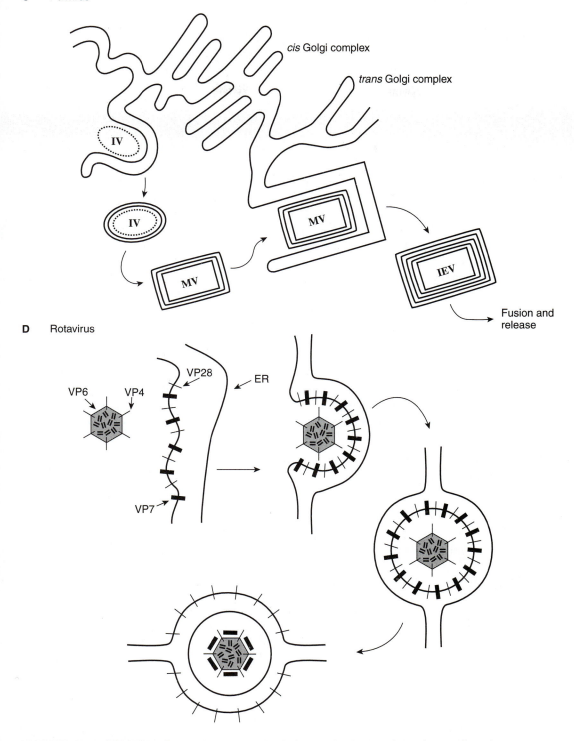

FIGURE 6.17 CONT'D. C, Poxviruses acquire their complex layers of membranes from the *cis* and *trans* faces of the Golgi complex. **D,** The naked double-stranded RNA rotaviruses use budding into the endoplasmic reticulum to create the outer of their two capsids.

Two models have been proposed to describe the process by which the DNA herpesviruses surround their capsids with tegument proteins and an envelope (Figure 6.17, *B*). Both theories recognize that these viruses assemble their capsids within the nucleus and then bud through the inner membrane of the nuclear envelope. One model says that the tegument is added to the virus in the nucleus as part of this budding process, so the virion simply traverses the ER and Golgi and exits the cell via fusion of a vesicle with the cytoplasmic membrane. The other model is more complex. After budding into the space between the membranes of the nuclear envelope, the virion exits that space by fusing its transient envelope with the ER membrane. The tegument proteins are added in the cytoplasm, whereupon the virion buds into the Golgi to acquire its permanent envelope.

The maturation of two other groups of viruses involve cellular membranes but in unusual ways. Poxviruses like vaccinia virus have very complex structures that include several membrane layers (Figure 6.17, *C*). These membranes appear to be derived from the sides of the Golgi complex by a process that does not involve budding. The immature virion (IV) acquires its first membrane from the *cis* face of the complex and changes from a circular to a "mature" brick shape in the process (MV). This particle in turn acquires a second membrane layer from the *trans* face of the Golgi complex to become an intracellular enveloped virus (IEV). This virion may be released by cell lysis or by exocytosis through the cytoplasmic membrane

The most bizarre maturation process to involve the process of budding is that of the *naked* rotaviruses (Figure 6.17, *D*). These viruses have double-layered icosahedral capsids with spikes that surround 11 double-stranded RNA molecules. The inner capsid layer is built from two proteins called VP4 and VP6. This structure buds through the endoplasmic reticulum at locations at which two other viral proteins, VP7 and VP28, are embedded. The transient membrane then dissolves in some way that leaves VP7 forming the outer layer of the mature capsid and removing VP28 entirely. The mature virions are finally released by cell lysis.

SUMMARY

The final steps in a virus's life cycle assemble the virion and in most cases release it from its host cell. Helical capsids are usually assembled around strands of nucleic acid, while icosahedral capsids are generally constructed as empty shells that are then filled by nucleic acid. Subassembly lines are used by complex bacteriophages to generate the heads, tails, and fibers that are finally assembled into the complete virion.

Naked viruses generally lyse their host cells to release their virions. Viruses with envelopes exit their host cells and acquire their envelopes in the same process, a specialized form of exocytosis called budding. Since enveloped viruses also disrupt their host cells' metabolisms, budding viruses may also cause their host cells to lyse.

SUGGESTED READINGS

1. "The assembly of a virus" by P.J. Butler and A. Klug, *Scientific American* vol. 239, pages 62–69 (1978), describes the interactions that occur during the assembly of tobacco mosaic virus. The original work is presented in "Reconstitution of active tobacco mosaic virus from its inactive protein and nucleic acid components" by H. Fraenkel-Conrat and R.C. Williams. *Proc Natl Acad Sci USA,* 41:690–698 (1955) and "Virus reconstitution II. Combination of protein and

nucleic acid from different strains" by H. Fraenkel-Conrat and B. Singer, *Biochem Biophys Acta* 24:540–548 (1957).

2. "Building a bacterial virus" by W.B. Wood and R.S. Edgar, *Scientific American* vol. 217, pages 61–74 (1967), describes *in vitro* reconstitution experiments to construct a T4 virion. The original work is presented in "Morphogenesis of bacteriophage T4 in extracts of mutant-infected cells" by R.S. Edgar and W.B. Wood, *Proc Natl Acad Sci USA* 55:498–505 (1966).

3. The rolling circle-capsid complex as an intermediate in φX DNA replication and viral assembly" by K. Koths and D. Dressler, *J Biol Chem,* 255:4328–4338 (1980), describes the linkage between a shift to rolling circle replication and assembly of the bacteriophage's virion.

4. "The DNA translocating vertex of DS-DNA bacteriophage" by C. Bazinet and J. King, *Ann Rev Microbiol* 39:109–129 (1985) and "DNA packaging in DS-DNA bacteriophages" by L. W. Black, *Ann Rev Microbiol* 43:267–292 (1989), review the formation of phage procapsids and mechanisms for packaging DNA into those procapsids.

5. "Structure of a viral procapsid with molecular scaffolding" by T. Dokland, R. McKenna, L.L. Ilog, B.R. Bowman, N.L. Incardona, B.A. Fane, and M.G. Rossmann, *Nature* 389:308–313 (1997), and "Topologically linked protein rings in the bacteriophage HK97 capsid" by W.R. Wikoff, L. Liljas, R.L. Dida, H. Tsuruta, R.W. Hendrix, and J.E. Johnson, *Science* 289:2129–2133 (2000), discuss the construction of procapsids and their conversion to mature virions.

6. "*In vitro* packaging of the single-stranded RNA genomic precursors of the segmented double-stranded RNA bacteriophage φ6: The three segments modulate each other's packaging efficiency" by M. Frilander and D.H. Bamford, *J Molecular Biol* 246:418–428 (1995), describes how a virus with a segmented genome packages all three segments.

Naked Animal Viruses

1. "Common features in the structures of some icosahedral viruses: a partly historical overview" by S.C. Harrison *Seminars In Virol* 1:387–403 (1990) and "Icosahedral RNA virus structure" by M.G. Rossmann and J.E. Johnson, *Ann Rev Biochem* 58:533–573 (1989), review the architectural similarities in the subunit proteins of icosahedral virion capsids.

2. "Cell-free, de novo synthesis of poliovirus" by A. Molla, A. Paul, and E. Wimmer, *Science* 254:1647–1651 (1991), describes the generation of complete, infectious poliovirus virions using only purified poliovirus RNA and a cell-free extract of HeLa cells.

3. "Three-dimensional structure of poliovirus at 2.9 A resolution" by J.M. Hogie, M. Chow, and D.J. Filman, *Science* 229:1358–1365 (1985), describes the structure of the poliovirus capsid protein "core" and the organization of capsid subunits in the virion.

4. "Conformational changes in poliovirus assembly and cell entry" by O. Flore, C.E. Fricks, D.J. Filman, and J.M. Hogle, *Seminars in Virol* 1:429–438 (1990), discusses the role of cleavages that occur in picornavirus assembly.

Enveloped Viruses

1. "Assembly of animal viruses at cellular membranes" by E.B. Stephans and R.W. Compans. *Ann Rev Microbiol* 12:189–516 (1988), reviews the maturation of enveloped viruses that bud from various membrane systems.

2. "The large external domain is sufficient for the correct sorting of secreted or chimeric influenza virus hemagglutinins in polarized monkey kidney cells" by M.G. Roth, D. Gundrsen, N. Paul, and E. Rodriguez-Boulan. *J Cell Biol* 104:769–782 (1987), and "A sorting signal for the basolateral delivery of the vesicular stomatitis virus (VSV) G protein lies in its luminal domain: Analysis of the

targeting of VSV G-influenza hemagglutinin chimeras" by T. Compton, I.E. Ivanov, T. Gottlieb, M. Rindler, M. Adesnik, and D.D. Sabatini, *Proc Natl Acad Sci USA* 86:4112–4116 (1989), describe analysis of the mechanism by which the envelope glycoproteins of different viruses become inserted into different cell surfaces in polarized epithelial cells.

3. "*cis*-acting genomic elements and *trans*-acting proteins involved in the assembly of RNA viruses" by S. Schlesinger, S. Makino, and M.L. Linial, *Seminars in Virol* 5:39–49 (1994), describes the packaging signals of a variety of enveloped RNA viruses.

4. "Defective RNAs inhibit the assembly of influenza virus genome segments in a segment-specific manner" by S.D. Duhaut and J.W. McCauley, *Virol* 216:326–337 (1996), presents evidence that there may be a mechanism for ensuring that all genome segments are packaged in influenza A virus.

5. "Cell biology of viruses that assemble along the biosynthetic pathway" by G. Griffiths and P. Rottier, *Seminars in Cell Biol* 3:367–381 (1992), describes how five different viruses use internal membrane systems in their maturation processes.

Chapter 7

Effects of Viral Infection on Host Cells: Cytological and Inductive Effects

The first six chapters of this book have focused on the viruses themselves: their architectures, how they enter their host cells, their modes of reproduction, and how they exit their host cells. In the next two chapters, we shall consider the effects that various viruses have on the metabolism and other properties of their host cells. We shall begin by discussing the mechanisms by which cytolytic viruses alter their host's metabolism, particularly protein synthesis, to satisfy their own biosynthetic needs. In animal cells these effects may result in structural damage to the host cell. However, neither prokaryotic nor eukaryotic cells are without means to defend themselves against viral infection. Restriction/modification systems in prokaryotes and the interferon system in animal cells are two examples of such defenses. The viruses, of course, have countermeasures against these defenses.

WHAT MORPHOLOGICAL CHANGES IN HOST CELLS ARE ASSOCIATED WITH VIRAL INFECTIONS?

Often when an animal virus establishes a productive or lytic infection in its host cell, distinctive changes termed **cytopathic effects** (*cyto* is Greek for "hollow vessel" and *pathos* is Greek for "disease"), abbreviated CPE, occur in that cell. These changes are often visible when infected tissues or cell cultures are examined by light microscopy, especially when stains like Feulgen or acridine orange fluorochrome are used to label DNA and RNA concentrations within the cells. Other CPEs are seen most easily in tissue cultures since they involve changes in the overall morphology and/or growth properties of infected cells. The variety of cytopathic effects seen in virus-infected animal cells is summarized in Table 7.1.

Some CPEs are localized alterations at particular sites within infected cells. The most common CPEs of this type are in the form of **inclusion bodies,** which are microscopically visible sites of viral assembly or cellular damage. In the electron microscope, inclusion bodies are often observed as crystals of virions or nucleocapsids, or accumulations of viral components within so-called *factories* where active synthesis is taking place. The rigid helical capsids of tobacco mosaic virus, for example, can form large, platelike arrays of virions aligned in parallel sheets (Figure 7.1, *A*). Icosahedral virions like those of adenovirus can also form "crystals" in the cytoplasm of infected cells (Figure 7.1, *B*).

Table 7.1 | Cytopathic Effects of Viral Infection of Animal Cells

Cytopathic Effect	Examples of Viruses	
Inclusion Bodies		
■ Virions in the nucleus	Adenovirus	
■ Virions in the cytoplasm	Rhabdovirus—rabiesvirus (Negri bodies)	
■ "Factories" in the cytoplasm	Poxviruses (Guarnieri bodies)	
■ Viral proteins and host microtubules	Reoviruses	
■ Clumps of ribosomes in capsids	Arenaviruses	
■ Clumps of chromatin	Herpesviruses	
Morphological Alterations		
■ Nuclear pyknosis (shrinking)	Picornavirus—poliovirus	
■ Proliferation of membrane	Picornavirus—poliovirus	
■ Proliferation of nuclear membrane	Alphaviruses	
■ Vacuoles in cytoplasm	Papovaviruses	
■ Formation of syncytia (cell-cell fusion)	Paramyxoviruses	
	Coronaviruses	
■ Margination and breakage of chromosomes	Herpesviruses	
■ Rounding up and detachment of tissue culture cells	Herpesviruses	
	Rhabdovirus—VSV	
Apoptosis	many	

A

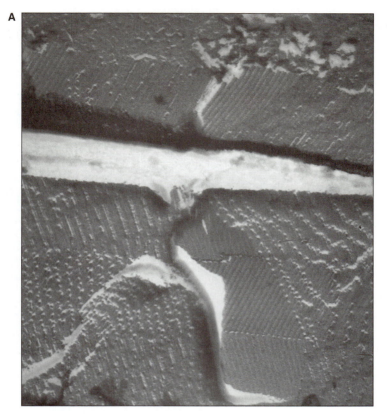

B

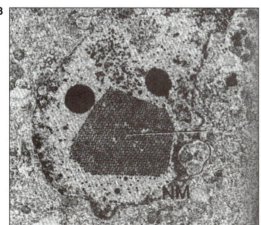

FIGURE 7.1 One type of inclusion body is formed by aggregations of virions. **A,** Freeze-etched preparation of tobacco mosaic virus crystals in a mesophyll cell, demonstrating the regular array produced by stacking of rigid helical nucleocapsids. **B,** Icosahedral virions of adenovirus can also form regular arrays.

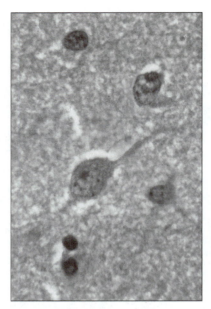

FIGURE 7.2 Negri bodies are cytoplasmic inclusion bodies diagnostic of rabiesvirus infection formed by cytoplasmic masses of rabiesvirus nucleocapsids.

Historically, the presence of particular inclusion bodies in infected cells has been used as a diagnostic tool. For example, the Negri body (Figure 7.2), a cytoplasmic mass of rabiesvirus nucleocapsids that stains cherry red with deep blue granules with a mixture of basic fuchsin and methylene blue dyes, has long been an important indication of rabies infection. Fluorescent antibody staining tests that are specific for individual viral proteins are replacing these types of classical chemical staining techniques in most cases, however.

Although many inclusion bodies are made up of virions or nucleocapsids, others are formed from a combination of viral and host cell components. The reovirus inclusion bodies that surround the nucleus of infected cells, for example, are composed of viral proteins and the host cell microtubules that normally make up the mitotic apparatus.

Numerous other indications of viral infection may also be seen in tissues or cultured cells. Viral infection often produces marked changes in gross morphology of cultured cells, causing them to round up and eventually detach from the substratum upon which they were growing (Figure 7.3, *A*). Striking examples of morphological changes in individual cells are shown in Figure 7.3, *B*. These changes include the development of inclusion bodies in the cytoplasm and the generation of vacuoles that give the cytoplasm a foamy appearance. Enveloped viruses may also promote the fusion of adjacent cells to produce multinuclear syncytia or giant cells. Some viruses produce changes in the appearance of the nuclear contents as chromosomes are broken and the chromatin moves to the surface of the inner nuclear membrane.

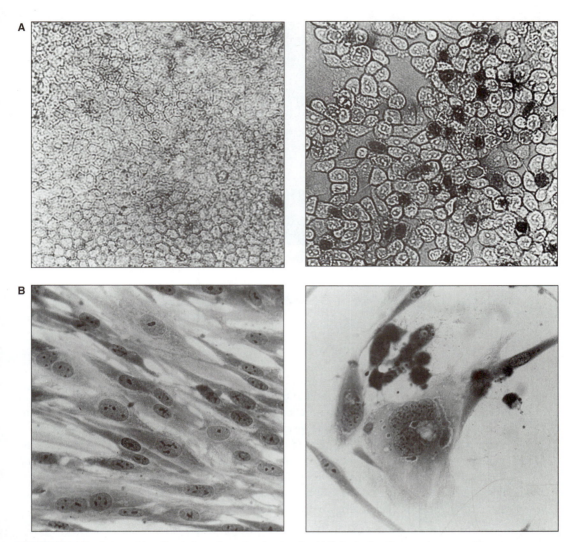

FIGURE 7.3 Cytopathic effects in cultured animal cells. Uninfected cells are shown on the left and infected cells on the right. Panel **A** shows rounding up in monolayer cultures of epithelioid cells Caco-2 infected with rotavirus. Panel **B** shows the effects of infection by the paramyxovirus Newcastle disease virus (NDV) in the fibroblastic cell BHK. Two multinucleated giant cells with dark-staining inclusion bodies and light-appearing vacuoles in their cytoplasms, as well as several rounded cells, are pictured.

The ultimate cytopathic effect of viral infection is **apoptosis,** or programmed cell death. Apoptosis is a genetically controlled process that is crucial to normal tissue development during embryogenesis since it removes cells that are no longer needed. It plays a similar role in maintaining homeostasis and in removing cells that are damaged by viruses or other agents.

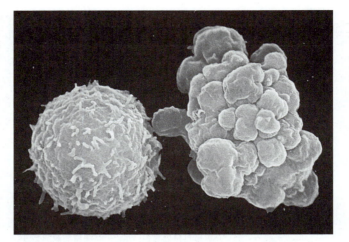

FIGURE 7.4 Apoptosis, or programmed cell death, is characterized by profound structural changes. The normal leukocyte (*left*) is spherical and has only small projections on its surface. The apoptotic cell (*right*) is in the process of breaking apart into small membrane-bound vesicles containing degraded chromatin.

Apoptosis produces very characteristic changes in the dying cell, which are illustrated in Figure 7.4. In cell culture, an apoptotic cell shrinks and partially detaches from its substratum as its plasma membrane forms large evaginations called "blebs." Within the cell the chromatin is first condensed and then cleaved into nucleosome-sized fragments. The process ends with the cell breaking apart into membrane-bound vesicles called "apoptotic bodies." Since the mechanism by which this first line of defense against viral infection occurs is related to how the body defends itself against tumors, we shall defer further consideration of apoptosis until the next chapter where we discuss how viruses cause cancer.

WHAT CAUSES CYTOPATHIC EFFECTS IN ANIMAL CELLS?

Cytopathic effects such as the Negri bodies seen in rabiesvirus-infected neurons have been known for nearly 100 years, but the causes of such effects are only now being discovered. A combination of advances was necessary to permit a shift in focus in the study of CPE away from that of the clinician interested in cellular damage and its relationship to disease and toward that of the virologist interested in the interactions of viruses with their host cells. In the middle decades of the twentieth century, cell culture techniques were developed that permit study of virus-host cell interactions under controlled conditions. More recently a vast armamentarium of molecular techniques to study the processes of the Central Dogma has been created. Together these tools have allowed analysis of the changes that occur in cells when viruses enter them, carry out the expression of their own genomes, synthesize their specific macromolecules, and finally construct and release their virions. In only a few cases are cytopathic effects produced by some harmful protein with no other known purpose in the infective process. Most cytopathic effects are, instead, the *secondary* result of changes wrought in host cell metabolism as viruses reproduce themselves. For example, the rounding up seen in infected cells in tissue culture illustrated in Figure 7.3, *A,* is probably the result of changes in the cytoskeleton, particularly microfilaments. The cytoskeletal changes are, in turn, the result of changes in host cell protein synthesis.

Regardless of the type of nucleic acid in their genomes, all viruses must form mRNA molecules that are then translated by existing host translation systems. While some viruses simply share their host's translation system, many have evolved strategies that enable them to convert that machinery to their own preferential, or even exclusive, use. Cytopathic effects then result in cells infected by such viruses as a secondary effect of inhibition of host cell protein synthesis. Host cell mRNA synthesis is also frequently altered during a productive infection. This alteration is sometimes a direct result of viral infection, but it may also result from inhibition of host protein synthesis. Host DNA synthesis may also be inhibited, usually as the result of interference with protein synthesis.

WHAT STRATEGIES DO VIRUSES USE TO INHIBIT EUKARYOTIC HOST CELL PROTEIN SYNTHESIS?

Protein synthesis is a very complex process requiring the participation of a number of different components in a linked set of three reactions (initiation, elongation, and termination). Initiation is the most complicated of these reactions, requiring an mRNA, large and small ribosomal subunits, a special Met-tRNA$_i^{met}$, and a number of *initiation factors* (eIF$_{1-6}$), as well as GTP and ATP. After the translation system has been assembled, elongation requires only charged tRNAs, two *elongation factors* (EF$_1$ and EF$_{1\beta}$), and GTP, while termination uses a single termination or release factor (eRF) and GTP. If the host's translation system is to be converted to viral rather than host use, altering the process of initiation is obviously the most economical strategy since the infected cell would then not "waste" components like ribosomal subunits or charged tRNAs in making partial host proteins.

Viral takeover of host cell protein synthesis can be nearly total. Inhibition of protein synthesis may occur in stages, however. When protein samples pulse-labelled at hourly intervals following infection with mengovirus, a member of the cardiovirus subgroup of the picornaviruses, are separated on polyacrylamide gels, the patterns show that synthesis of most host proteins is inhibited after the first hour, while the remainder is not blocked until about 6 hours (Figure 7.5). By 7 hours after infection, however, essentially the only proteins being synthesized are viral proteins.

Table 7.2 presents a summary of some of the mechanisms used by animal viruses to inhibit their host cells' translation systems, thereby making the components of those systems more available for viral protein synthesis. Some of these mechanisms could be called "crude but effective," while others involve more subtle alterations in particular factors used at a certain step in the process. As more information has become available regarding these mechanisms, it has become evident that strategies that may appear to be direct and simple actually overlie more subtle mechanisms. The data in this table also make it clear that the same virus can use more than one strategy and that viruses within the same group can employ

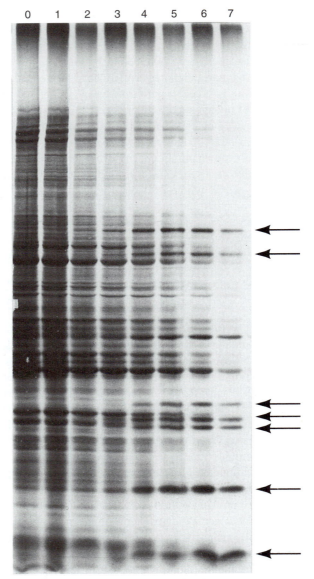

FIGURE 7.5 Two-stage inhibition of cellular protein synthesis in mengovirus-infected mouse L cells in culture. At hourly intervals following infection, cultures were labelled for 30 minutes with [^{35}S]methionine, then lysed and proteins separated on SDS-polyacrylamide gels. The results were visualized by autoradiography. Arrows identify viral proteins. Most host cell protein synthesis ceases after the first hour, while the remainder ceases at about 6 hours after infection.

Table 7.2 | Mechanisms of Viral Inhibition of Eukaryotic Host Cell Translation

Mechanism	Viruses
1. Competition	
■ Based on abundance of mRNAs	Reovirus, VSV, influenza, mengovirus
■ Based on higher initiation efficiency	
2. Failure to transport host mRNAs from nucleus	Adenovirus
3. Degradation of host mRNAs	Bunyavirus, herpes, influenza
4. Blockage of initiation complex formation	Poxviruses
5. Covalent alteration of translation factors	
■ Inactivation by cleavage	Polio
■ Inactivation by phosphorylation	Poxvirus, VSV, reovirus, adenovirus
6. Increased intracellular $[Na^+]$	Encephalomyocarditis virus, Sindbis
7. Inhibition by viral protein	Mengovirus

different mechanisms (poliovirus, mengovirus, and encephalomyocarditis virus are all members of the *Picornaviridae,* for example). Although not shown in the table, the same virus may inhibit synthesis in one type of host cell but not in another.

Competition

The most straightforward method a virus might have of co-opting its host translation system is to produce such an abundance of viral mRNAs that they outcompete the more limited number of host mRNAs for the available ribosomal subunits. This mechanism was advanced, for example, to explain how the rhabdovirus vesicular stomatitis virus (VSV) and several reoviruses inhibit host protein synthesis. As is the fate of many hypotheses that are so attractive in their simplicity, however, newer data suggest that the actual process is more complicated. In both viruses, viral protein synthesis as well as the presence of viral mRNAs is required before host cell protein synthesis is inhibited. Furthermore, studies with recombinant reoviruses map the protein-inhibiting activity of that virus to segment S4, which contains the gene for the major virion core protein σ3. The σ3 protein is capable of binding to RNA, so its presence may tip the balance in favor of the virus by making a portion of the host mRNAs inaccessible to ribosomes.

Although simple competition between viral and host mRNAs based on their relative abundances does not appear to be a strategy for promoting preferential expression of viral information, some viruses do use competition in the formation of initiation complexes as a means of promoting their own protein synthesis at the expense of their host cells. For example, influenza virus mRNAs are somewhat more efficient at competing for eIF-2 than are the host cell mRNAs. Other viruses compete spectacularly well against their hosts. When equal amounts of mengovirus mRNA and a host mRNA (globin mRNA) are added to an *in vitro* translation system, for example, the mengovirus mRNA outcompetes the globin mRNA in forming initiation complexes by *35-fold*. A highly accessible and easily recognized initiation sequence at the 5′ end of the mengovirus mRNA has been suggested as part of the reason for its ability to form initiation complexes so readily.

Inhibition of Host mRNA Transport from the Nucleus

Adenoviruses are DNA viruses that are transcribed and replicated in the nucleus. Like host cell mRNAs, adenovirus mRNAs are posttranscriptionally modified and then transported from the nucleus to the cytoplasm for translation. A striking feature of adenovirus infection is that when host protein synthesis is maximally inhibited late in the infective process, both host and viral genes are still transcribed and their transcripts processed, but only the viral mRNA is found in the cytoplasm. Since no differential degradation of host mRNA in the nucleus is found, it appears that a failure to transport host mRNA from the nucleus to the cytoplasm is responsible. Inhibition of host translation would therefore seem to result from a shift in the relative abundances of host and viral mRNAs in the cytoplasm.

The mechanism that allows selective transport of viral mRNAs while host mRNAs remain in the nucleus appears to involve two viral proteins called E1B-55 KDa and E4 or F6 that shuttle between the nucleus and the cytoplasm. This selective transport alone is not responsible for the dramatic inhibition of protein synthesis that occurs in infected cells, however. This has been shown by experiments indicating that host mRNAs *already in the cytoplasm* are not able to begin new rounds of translation as effectively as viral mRNAs. These experiments suggest that a viral protein acting during the process of initiation is also involved in blocking host protein synthesis.

Degradation of Host mRNA

Infection by a variety of viruses including orthomyxoviruses, herpesviruses, and bunyaviruses can cause a rapid degradation of host mRNAs, with its concomitant effect on host protein synthesis. In all of these cases the resulting inhibition of host cell translation is actually a secondary effect of the particular mechanisms those viruses have evolved to regulate the expression of their own genomes.

"Cap snatching," or cleaving off the first dozen or so nucleotides at the 5' end of a host mRNA in order to use the resulting fragment as the primer for the synthesis of a viral mRNA, is a strategy employed by several groups of viruses with segmented negative-strand RNA genomes. As we discussed in Chapter 4, orthomyxoviruses like influenza A derive their primers from newly synthesized host cell mRNAs that are still in the nucleus. The bunyaviruses cleave cytoplasmic mRNAs for the same purpose. In both groups of viruses, the viral RNA polymerase both creates primers with an endonuclease activity and then elongates them. This strategy serves a two-fold purpose since it reduces competition for host cell ribosomes by reducing the level of functional host mRNAs as well as ensuring that viral mRNAs will have the necessary structural features like the 5' cap to initiate translation efficiently.

The bunyaviruses are transmitted from one vertebrate host to another via a mosquito vector. An interesting feature of this life cycle is that the virus produces a decrease in protein synthesis only in the cells of the vertebrate host and not those of the mosquito vector (which must remain alive and functional in order to transmit the virus to another vertebrate host). It has been suggested that the virion replicase is less active in the invertebrate cells and therefore does not significantly reduce the pool of host mRNAs.

Herpesviruses are large DNA viruses that regulate the expression of their genomes through a series of positive and negative feedback loops. In herpes simplex virus an α (immediate-early) gene stimulates the transcription of the β (delayed-early) genes, and β genes both inhibit the transcription of genes and stimulate the transcription of γ (late or structural) genes, which then in turn inhibit both α and β genes. Host protein synthesis, as well as other macromolecular synthesis, is very rapidly shut off following infection with the human herpes simplex 1 and 2 viruses.

Translation shutoff occurs as a two-stage process. The first stage begins immediately upon infection and is characterized by degradation of host mRNAs, and consequently disaggregation of host polysomes. The second stage occurs at the time that β gene products begin to be expressed and completes the shutdown of host protein synthesis. The first stage, degradation of host mRNAs, occurs even when all viral or host cell transcription and translation are blocked, so this stage must be mediated by a protein that enters with the herpes simplex virion itself. Stage two, complete shutdown of host protein synthesis, requires synthesis of a viral protein.

Since HSV-1 and HSV-2 differ in the rate and completeness of shutdown in stage one, genetic studies with recombinant viruses permitted mapping of a *virion host shutoff* or *vhs* locus in HSV-2. The *vhs* gene, part of the γ1 set, appears to encode a 58 kd protein that like other γ genes becomes a structural component of the virion. An interesting feature of the *vhs* protein is that it appears to actively degrade *both* host and viral α and β gene mRNAs. Since *vhs⁻* mutants of HSV-2 are viable but produce about 10-fold fewer virions than wild-type HSV-2, it has been suggested that the functions of the *vhs* protein are not only to help make pools of host ribosomes and other metabolites available for viral synthesis, but also to facilitate rapid transitions between translation of the viral immediate-early and delayed-early gene mRNAs. Presumably, sequestering of the *vhs* protein into maturing virions allows ample translation of γ gene mRNAs.

Blockage of Initiation Complex Formation

The strategies like pouring out massive amounts of viral message to out-compete the host or degrading mRNAs might be considered as not very subtle means of appropriating a host cell's protein synthetic equipment for viral use. Many types of viruses employ finer-edged tools that modify components of the translation system to accomplish the same end. Poxviruses like vaccinia, for example, use two different mechanisms to alter the specificity of host ribosomes so that viral mRNAs are recognized in preference to host ones. One of these employs a viral protein to prevent the formation of the translation initiation complex. Like the herpes simplex *vhs* protein, the vaccinia inhibitory protein appears to be part of the virion core structure, so its effect does not require any *de novo* protein synthesis after infection. The 11 kdalton phosphoprotein acts on the 40S ribosomal subunit to hinder the association of that subunit with Met-tRNA^met, GTP, and eIF-2 to begin the process of translation. This inhibition applies equally to host and viral protein synthesis, so another virus protein or proteins appears to be required to permit translation of viral messages. This required protein is not a part of the virion since transcription and translation of early vaccinia genes are necessary for continued viral protein synthesis.

Covalent Modification of Translation-Related Components

Numerous viruses have specific reactions that covalently modify ribosomal subunit proteins or the factors required for initiation or elongation. One of the structural proteins of the core of the poxvirus vaccinia, for example, is a protein kinase that sequentially phosphorylates, and thereby inactivates, a set of three ribosomal proteins.

Other viruses such as vesicular stomatitis virus, adenovirus, and reovirus induce phosphorylation of the α subunit of eIF-2. As Figure 7.6 illustrates, when this initiation factor catalyzes the formation of an initiation complex, a GTP bound to the α subunit of eIF-2 is

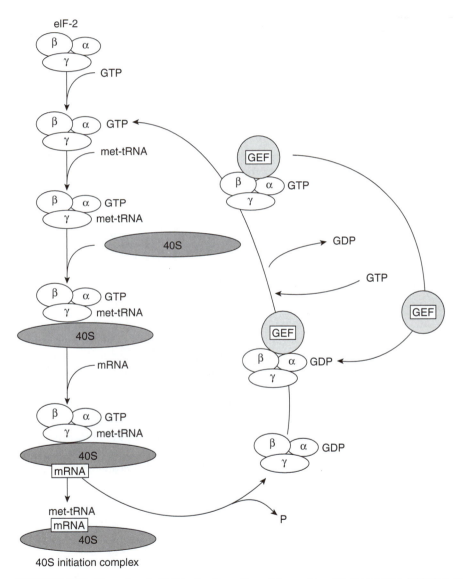

FIGURE 7.6 The role of eIF-2 in the formation of the eukaryotic translation initiation complex. eIF-2 has three subunits: α, β, and γ. The GTP bound to the factor's α subunit is cleaved to GDP during formation of the initiation complex. GTP exchanging factor (GEF) aids in replacing that GDP with a new GTP so that eIF-2 can participate in a new initiation reaction.

hydrolyzed to GDP. A second factor, variously called eIF-2β or GEF (GTP exchanging factor), is then involved in replacing that GDP with a new GTP so eIF-2 can participate in a new initiation reaction. When eIF-2 is phosphorylated, it forms a complex with GEF that is unable to exchange its bound GDP for a new GTP, so the rate of translation initiation is reduced. Since the bulk of mRNAs in the host cell's cytoplasm will be viral rather than host cell, as discussed previously, the balance is tipped further toward committing the cell's protein synthesizing machinery to making viral products.

The enzyme that phosphorylates eIF-2α is called *protein kinase RNA-dependent,* or *PKR.* Figure 7.7 illustrates how PKR functions. PKR is synthesized as an inactive proenzyme

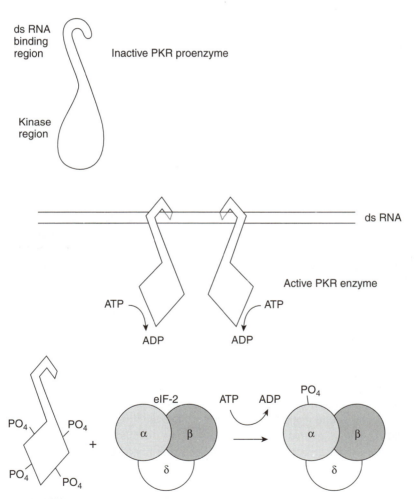

FIGURE 7.7 Protein kinase RNA-dependent, or PKR, is synthesized as an inactive proenzyme. When two molecules of proenzyme bind to double-stranded RNA, each becomes activated and phosphorylates the other. The phosphorylated active enzymes can then release from the RNA molecule and phosphorylate the α subunit of eIF-2.

that itself must be phosphorylated in order to become active. The trigger for that phosphorylation is the presence of double-stranded RNA. Such RNA is produced not only when RNA viruses express or replicate their genomes, but also when DNA viruses transcribe from both strands in the same location in the viral genome, producing complementary RNA molecules that can pair with one another. When two PKR proenzyme molecules bind to a molecule of double-stranded RNA, each undergoes a conformation change that causes it to add phosphates to serines at four different locations on the other PKR molecule. The activated PKR molecules then attach phosphates to eIF-2α proteins.

A wide variety of viruses have some mechanism to alter the synthesis, degradation, or activity of PKR. HIV down-regulates the synthesis of the proenzyme, while picornaviruses increase the rate of its degradation. Many groups of viruses, including the papovaviruses, adenoviruses, herpesviruses, poxviruses, orthomyxoviruses, and retroviruses, encode proteins or RNAs that block the activation process or the activity of the activated PKR enzyme.

By far the best understood mechanism for subverting the host's macromolecular synthetic processes is that employed by poliovirus to inactivate a normal translation initiation factor. Poliovirus is a positive-strand RNA virus, but its genome does not have all the structures characteristic of a eukaryotic mRNA. Of particular interest in the present context is the fact that the poliovirus genome lacks the usual 7-methylguanosine "cap" on its 5′ terminus. During normal host translation initiation this cap is recognized by a "cap-binding complex" (eIF-4F) consisting of three peptides called eIF-4G, eIF-4A, and eIF-4E (Figure 7.8). Early in a poliovirus infection, the eIF-4GI/II complex is cleaved so that host-capped mRNAs are no longer recognized, and host translation ceases. Genetic studies indicate that the viral protease 2A that plays a role in cleaving the initial viral polyprotein is also involved in destroying the cap-binding activity.

In addition to cleaving eIF-4G, several groups of viruses also use phosphorylation as a means of inhibiting the role of the cap-binding complex in initiation of translation. As shown in Figure 7.8, the eIF-4E peptide must be phosphorylated in order to bind to eIF-4G. Both adenoviruses and influenza A virus inactivate the cap-binding reaction by removing the phosphate from eIF-4E. Several picornaviruses accomplish the same end in a more indirect fashion by acting on a small protein called 4E-bp1 that is involved in cell growth regulation by insulin and growth factors. When 4E-bp1 is phosphorylated it competes with eIF-4G for binding of the active form of eIF-4E.

Increases in Intracellular Cation Concentrations

One of the environmental conditions that can influence the rate or efficiency of an enzymatic reaction is the concentration of particular ions, especially cations like Na^+ and K^+. Several decades ago it was found that concentrations of Na^+ that inhibited the translation of cellular mRNAs did not inhibit the translation of mRNAs from a number of different types of viruses. Other observations indicated that changes in the permeability of the plasma membrane resulted in an increase in intracellular Na^+ concentration in cells that were infected by those same viruses. Taken together, these findings led to the hypothesis that viruses could create a favorable environment for differential translation of their own mRNAs by increasing cation concentrations within their host cells. Careful studies of the timing of onsets of inhibition of cellular translation and of increased intracellular Na^+ concentration, however, have demonstrated that in many cases the "cause" appears to *follow* rather than precede the "effect."

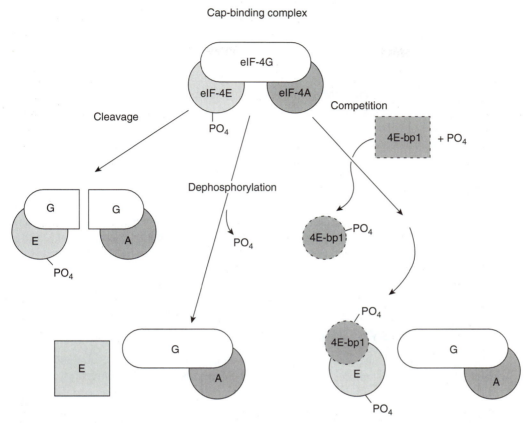

FIGURE 7.8 The cap-binding activity of eIF-4 can be eliminated in a variety of ways. The proteases of viruses like poliovirus may cleave eIF-GI/II so that it can no longer bind to the 5′ end of a mRNA. Poliovirus also inactivates the cap-binding complex by phosphorylating a small peptide 4E-bp1, allowing it to compete with eIF-4G for binding to eIF-4E. Removing the phosphate from eIF-4E to alter its binding to eIF-4G is also a strategy used by several viruses.

Although changes in membrane permeability are probably not the major means of altering cellular translation in most viruses, they do play a significant role in certain cases. Infection by the togavirus Sindbis, for example, produces an inhibition of the plasma membrane Na^+-K^+ATPase pump that is similar to that produced by the specific inhibitor ouabain. This inhibition requires viral protein synthesis and precedes any reduction in the efficiency of translation of host cell mRNAs, and may in fact be its cause.

Viral Inhibitory Factors

Another mechanism by which viruses inhibit host cell translation also involves an alteration in a step in the process that follows initiation. As we saw in the gel pictured in Figure 7.5, mengovirus very efficiently inhibits virtually all host protein synthesis. The host mRNAs are not degraded but instead shift from being associated with polysomes to being bound to only a single 80S ribosome. Ribosomes taken from infected cells are able to bind an mRNA and Met-tRNA^{met}, but the initiation complex does not then proceed to elongation. Two treatments of the ribosomes that would remove or inactivate nonribosomal proteins, rinsing with 0.5M KCl or heating to 70° for 5 minutes, restore their translational ability. One possible interpretation of these findings is that mengovirus inhibits host cell protein synthesis by means of a protein that binds to ribosomes and blocks the process after formation of the initiation complex.

You will recall that mengovirus was discussed previously as an example of a virus that used competition for initiation complexes between its RNAs and its host's as a means of inhibiting translation. In addition to competition and inhibition of elongation, mengovirus also uses a third mechanism, phosphorylation of the initiation factor eIF-2, as a means of inhibiting protein synthesis. This reaction does not occur until several hours after the elongation inhibitory protein has blocked most translation, however, producing the two-stage inhibition noted in Figure 7.5. It has been suggested that the viral RNAs are so very efficient at this competition that a complete termination of translation is necessary to liberate viral mRNAs, which are also the viral genomes, for incorporation into new virions.

HOW DO VIRUSES INHIBIT THE TRANSCRIPTION OF THEIR HOSTS' GENOMES?

Many cytocidal DNA and RNA viruses cause both transcription of host genes and synthesis of host DNA to slow or cease. In most cases the inhibition of these activities follows shutoff of host protein synthesis and can be seen as a secondary effect of that process. As factors required for transcription or replication of host genes are turned over at the normal rate, they are not replaced by newly synthesized factors, and so host RNA and DNA synthesis is gradually inhibited.

Various viruses do appear able to suppress the synthesis of host mRNA, generally by interfering with the action of RNA polymerase II, the eukaryotic enzyme specific for the transcription of protein-encoding genes. Inhibition often involves a specific viral protein. One of the poliovirus proteins, for example, appears to inhibit the action of some factor required by RNA polymerase II, but which protein is involved and its mechanism of action are still unknown. RNA polymerases I and III are also inhibited but to a lesser extent. During the replication of reoviruses, both RNA and protein synthesis are inhibited with the same kinetics. Genetic studies suggest that it is the same viral inner core protein σ3 that is responsible for both activities.

There is some evidence that viral RNAs themselves may inhibit transcription. A well-studied example of this occurs in vesicular stomatitis virus. Host cell transcription becomes reduced only after transcription of the VSV genome into a 48-base "leader" RNA and five mRNAs of varying lengths. Early studies found that host transcription inhibition occurs during replication of viral mutants that are incapable of transcribing anything except the leader sequence, so that small piece of RNA appears to be the responsible molecule. The leader RNA does not bind directly to host DNA sequences, which suggests that its action is on some transcription factor instead. Since initiation but not the elongation steps of transcription is blocked, an initiation factor is implicated.

Viral proteins may also inhibit transcription. Some genetic studies implicate the matrix protein of VSV in shutting down host cell transcription. For example, when a vector carrying only the matrix protein gene is introduced into host cells, transcription is quickly shut down. While the mechanism of this inhibition is unknown, the C-terminal end of the matrix protein appears to be required since deletion of the terminal 56 amino acids removes the inhibitory activity.

Inhibition of transcription is not the only means by which viral infections can affect host cell RNAs. As we noted in the preceding section, adenoviruses interfere with the transport of mRNAs from the nucleus to the cytoplasm, and influenza viruses cleave the capped ends from cellular RNAs to provide themselves with primers for synthesis of viral RNAs. Other viruses like the herpesviruses and bunyaviruses degrade host mRNAs.

HOW DO VIRUSES INHIBIT THE REPLICATION OF THEIR HOSTS' GENOMES?

As with host transcription, often host DNA synthesis is slowly inhibited as a secondary effect of reduced host-specific protein and/or RNA synthesis. In some cases specific viral proteins inhibit replication of host DNA. Reovirus blocks the initiation of new rounds of DNA synthesis in infected cells. Studies with recombinant viruses implicate the virion hemagglutinin σ1 protein as the inhibitory agent. Taken together with the protein and RNA synthesis-inhibiting activities of the core protein σ3, this means that reoviruses use virion proteins to inhibit all the steps in the expression and replication of host cell genetic information. Another RNA virus, the togavirus western equine encephalitis virus, also appears to induce an inhibitory agent that acts on DNA synthesis. The benefits, if any, that accrue to these RNA viruses by inhibiting host DNA synthesis are unknown.

DNA viruses in the herpesvirus and poxvirus groups interfere with DNA synthesis in direct fashions. The virion of the poxvirus vaccinia, for example, includes a protein that degrades single-stranded DNA, disrupting host cell DNA synthesis by destroying the DNA being used as template at replication forks. Herpesvirus infection leads to the displacement of host chromatin from its normal association with nuclear matrix proteins. This relocalization inhibits replication and transcription of host DNA and may also play a role in making host accessory factors for DNA synthesis available to the virus replication complex.

HOW DO BACTERIOPHAGES INHIBIT THE METABOLISM OF THEIR HOST CELLS?

As we discussed in Chapters 3 and 6, during productive infections nearly all types of bacteriophages lyse their host cells to release their newly formed virions and complete their life cycles. Like their eukaryotic counterparts, these lytic bacteriophages have evolved mechanisms that allow them to subvert the normal flow of precursor molecules in the steps of the Central Dogma, directing them away from host synthesis and toward viral synthesis. A striking difference between the two groups of viruses, however, is the "target of choice" in these mechanisms. As we saw earlier, animal viruses have a wide variety of strategies for inhibiting host cell protein synthesis, a lesser variety of ways of inhibiting transcription, and only a few direct means of inhibiting DNA replication. In the bacteriophages, the pattern is reversed. Host protein synthesis stops because host mRNAs are not made, often because the host genome itself has been destroyed. One possible reason for this difference may be that the normal rate of turnover of mRNA molecules in prokaryotes is very rapid (half-lives in the order of 3 to 5 minutes) compared to that of eukaryotic cells (half-lives in the order of 600 minutes). Protein synthesis will cease quickly when no new mRNA is made in a prokaryotic cell, but only slowly in a eukaryotic cell. Direct inhibition of translation is thus not required in order for a bacteriophage to take over a host cell rapidly but may be useful for a eukaryotic virus in the same circumstance.

The best studied strategies for inhibiting host cell DNA and RNA synthesis are those employed by the large DNA bacteriophages of the T-series coliphage group. Table 7.3 summarizes some of those strategies. DNA and RNA synthesis is stopped by the rather straightforward method of degrading the host genome in these and many other bacteriophages. A few viruses like T4 have evolved a mechanism of distinguishing their own DNA from that of their host cell, which allows them to target only the host DNA for destruction. T4 does this by "labelling" its DNA by using modified cytosine (hydroxymethylcytosine or hmCyt) bases during its synthesis. Viral nucleases that recognize only the natural base then will degrade only host DNAs. Virus-directed synthesis of the hmCyt nucleotide in place of normal Cyt is a corollary mechanism of inhibition of DNA synthesis as the cell's Cyt pool is depleted. Other bacteriophages like T3 and T5, which do not have modified bases in their DNAs, also degrade their host cells' genomes, but how their nucleases distinguish between host and viral DNA molecules is not known.

Although destruction of the genome is probably sufficient to prevent transcription and translation of host information, the T-series coliphages also inhibit transcription in several different ways. These mechanisms are not really directed specifically toward interfering with the host process, however, but that result occurs as a secondary effect of the mechanisms that these bacteriophages have evolved to regulate the sequential expression of their own genomes. The shift from expression of immediate-early genes to delayed-early genes in T7, for example, is induced by phosphorylating, and thereby inactivating, the host RNA polymerase so it can no longer recognize the early gene promoters. Host cell transcription is inhibited since the viral polymerase that recognizes the delayed-early and late promoters does not act on host promoters. ADP-ribosylation of the two α subunits of RNA polymerase has

Table 7.3 | Strategies for Inhibition of Bacterial Metabolism in T-Series Coliphages

DNA Synthesis
1. Degradation of host DNA
    Nonspecific nucleases: T5 and T7
2. Depletion of precursor pools: T4

RNA Synthesis
1. Covalent modifications of host RNA polymerase
    Phosphorylation: T7
    ADP-ribosylation: T4
2. Replacement of promoter recognition factor: T4

the same result in T4-infected cells. This covalent modification is coupled with a replacement of the host σ factor by viral promoter recognition factors.

HOW DO HOST CELLS FIGHT BACK AGAINST THE VIRUSES THAT INFECT THEM?

In all of the discussion thus far of viruses and host cells, the focus has been on the activities of the virus and the result of these activities on the host cell. That concentrated focus on the virus may have led to an impression that the host cell role in the virus–cell interaction is that of passive victim, destined to "go gently into that good night" after having graciously permitted the virus to produce more of its kind. This is most decidedly not the case! The virus–host interaction is better viewed as a continuous evolutionary thrust and counterthrust, as virus attempts to take over host cell, host cell responds with antiviral defenses, and virus counters with strategies to negate the antiviral defenses.

WHAT TYPES OF ANTIVIRAL DEFENSES DO PROKARYOTIC CELLS HAVE?

As we have seen before, the answers to questions asked by virologists frequently have importance far beyond the particular case under study. The answer to the question posed for this section is an example *par excellence* of this. Elucidation of the mechanism by which bacteria attempt to defend themselves against viral infections is one of the pivotal discoveries of modern science, leading directly to the development of recombinant DNA technology and the resulting revolution in biology and its related industries.

The first evidence for a prokaryotic antiviral defense came from studies on the efficiency of plating (EOP) of λ virus in different strains of *E. coli* by Werner Abner and Daisy Dussoix in the early 1960s. They found that the λ produced by a particular strain of *E. coli* would usually infect cells of that strain with a high EOP, but cells of other *E. coli* strains with only very low EOPs. The virus is said to be **restricted** against infection of those strains. Their data, summarized in Table 7.4, have several important features. First, in all cases the viruses produced from a given strain (λ·B from *E. coli* B, for example) can readily infect both that strain

Table 7.4 | The Plating Efficiencies of λ Virus Strains on Different *E. coli* Strains

λ Strain	Plating Efficiency on Indicated Host			
	C	**B**	**K12**	**K12(P1)**
λ·C	1	2×10^{-4}	4×10^{-4}	4×10^{-7}
λ·B	1	1	4×10^{-4}	7×10^{-7}
λ·K12	1	10^{-4}	1	2×10^{-5}
λ·K12(P1)	1	10^{-4}	1	1

The bacterial host cell type that produced a particular λ strain is indicated by the characters following the dot in each case. *E. coli* K12(P1) cells have the genome of the bacteriophage P1 incorporated into their genomes.

as well as *E. coli* C. λ C, in contrast, has low EOPs on all strains except its own. *E. coli* C would appear to be a "universal recipient" for λ variants but would appear to be unable to "donate" viruses that can infect any other cell types. Second, the effect of the host cell on the EOP of particular λ virus strains can be modified by the presence of another bacteriophage P1 whose genome has become incorporated into the *E. coli* genome. The effect of that added genetic information appears to be additive, since the EOP of λ·C on strain K12(P1) is one-thousandth that of the EOP on K12 alone. It is evident from these data that *E. coli* has a very effective mechanism for blocking infection by particular viruses, as long as those viruses originated in a different strain of bacterium. This phenomenon is termed **restriction**. Restriction is not universal, however, since strain C appears to lack protection.

A striking feature of this protective mechanism is that a virus assumes the pattern of whatever cell it does manage to infect. For example, very rarely a λ·C virus is able to infect an *E. coli* K12 cell and reproduce itself. The viruses subsequently released from that infected K12 cell are able to infect other K12 cells with the high EOP typical of λ·K viruses rather than the low EOP of λ·C viruses. In the reciprocal experiment, λ·K viruses lose their ability to infect *E. coli* K12 if they are passed through an *E. coli* C host cell; they become λ·C viruses. This phenomenon is called **host-induced modification.**

The basis of each restriction and host-induced modification pattern is a pair of enzymes with complementary activities. Modification is the result of enzymes that recognize particular DNA sequences, often palindromes, and covalently alter or modify one or more bases within those sequences. For example, the modifying enzymes in strains B and K recognize the sequences indicated in Table 7.5 and transfer a methyl group from S-adenosylmethionine to the adenosyl residue marked with the asterisk. These sequences, termed *Eco* K and *Eco* B recognition sites, are therefore a means of labelling "self" DNA, since DNAs from a cell that lacks that particular enzyme will not have methylated adenosyl residues in those sequences. These nonmodified sequences are recognized by the second enzyme in each pair, **restriction endonucleases,** which cleave the unmodified DNAs and thus destroy the "nonself" DNA. All the λ types are able to multiply in *E. coli* C because that strain has no restriction/modification system to protect it. In contrast, *E. coli* K12(P1) is doubly protected since it has not only the K12 system but also a separate system encoded by the virus P1 genome. λ·K(P1) can multiply in both K12 and K12(P1) host bacteria since its DNA has both modi-

Table 7.5 | Restriction/Modification Recognition Sequences in *E. coli* K12 and *E. coli* B

Site	Sequence
eco B	TGA̍NNNNNNNNTGCT
	ACTNNNNNNNNACGA
eco K	AAC̍NNNNNNNGTGC
	TTGNNNNNNNCACG

N represents any nucleotide. The location of methylation is indicated by asterisks.

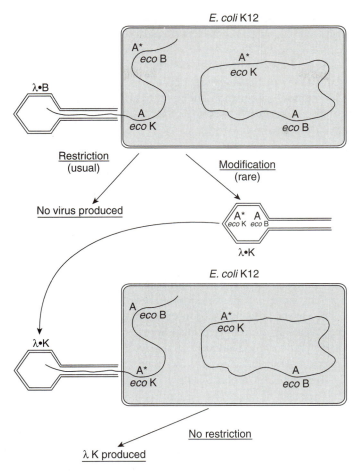

FIGURE 7.9 Host-induced modification of bacteriophage DNA. When a viral DNA without the proper methylation pattern enters a host bacterium, its DNA is degraded by host cell restriction endonucleases so no new viruses are produced. Occasionally a viral DNA without the proper methylation pattern escapes degradation and its newly replicated DNA becomes modified by the host's particular system. The new viruses produced in that cell have the restriction pattern of their host and can therefore infect other cells of that bacterial strain.

fication patterns; λ·K cannot multiply in K12(P1) cells since it is restricted for the P1 system. Since two separate, independent "escapes" are necessary for a λ·B or λ·C to infect a K12(P1) cell, the EOP for such attempts is extremely low.

The molecular events that underlie the host-induced modification pattern described previously are diagrammed in Figure 7.9. When λ·B viruses inject their DNAs into *E. coli* K12 cells, virtually all those molecules are destroyed by the restriction endonuclease *Eco* K because their K12 recognition sites are not methylated. The rare virus that escapes destruction multiplies in a K12 modification environment, however, so its newly synthesized DNA carries the characteristic K12 methylations and lacks the methylations of its parent λ·B molecule. They will therefore be λ·K viruses.

Hundreds of restriction/modification systems have now been identified. Unlike the two just described, many of these systems recognize short palindromic sequences and usually methylate cytosyl rather than adenosyl residues. These restriction endonucleases make possible the directed dissection of DNA that is fundamental to recombinant DNA technology.

WHAT TYPES OF ANTIVIRAL DEFENSES DO EUKARYOTIC CELLS HAVE?

Much of the research on antiviral host defense has been done in a medical context, so the best documented examples concern defenses against animal cell viruses, particularly those infecting mammalian cells. One of the most important cellular defense systems against viral invaders was discovered in 1957, when Alick Isaacs and J. Lindenmann found that after chicken eggs were exposed to inactivated influenza virus, the allantoic fluid from those eggs contained a substance that interfered with the infection of other chick cells by live influenza virus. They termed this inhibitor **interferon.** Over the next decade a great variety of other cell types were also found to produce interferon when exposed to viruses. RNA viruses were discovered to be good inducers of interferon, while DNA viruses, except for the poxviruses, were poorer inducers. Agents other than viruses were also found to be potent inducers: double-stranded RNA molecules (especially poly-inosine/cytosine), intracellular bacteria like Rickettsia, and the endotoxin component of the gram-negative bacterial cell wall.

At the same time, the nature of the interferon defense was also being probed. It soon became evident that interferon was a *cellular* defense that was very different from the immunological defenses occurring in the entire organism. The hallmark of immunological reactions is their specificity; an immunological response to a particular type of virus is directed specifically at that inducing virus and no other type. A hallmark of interferon activity is a total lack of this kind of specificity. Regardless of the inducing agent, be it a virus, polyIC, or endotoxin, the interferon produced will inhibit infection by numerous different types of viruses, RNA and DNA alike. The immune response and interferon do, however, share an important characteristic: The cells that produce the immune response or form interferon are acting in a somewhat altruistic manner to protect other cells in the organism rather than themselves. Nevertheless, this altruistic activity has several important differences in the two systems. First, the immune response is mounted only by a special set of cells (lymphocytes and their accessory cells), while interferon can be produced by virtually any cell, whether in an intact organism or in tissue culture. Second, the protection conferred by an immune reaction can be species independent, as when antibodies from cattle are used to provide passive immunization in humans, while the protection conferred by interferon has an extraordinary degree of species specificity. The interferon produced by cells of one species protects only other cells of that same species and confers virtually no protection on cells of a different species. Finally, the major proteins produced by the humoral arm of the immune system, the antibodies, provide direct antiviral activity, while interferon acts indirectly to enable other cells to defend themselves.

The timing of the activities of the immune system and interferon indicates that they serve separate functions in the overall protection of an organism. Clinical studies have demonstrated that interferon represents the first line of defense against viral infection. Its effects are seen within hours after the introduction of the virus and persist for several days, permitting the slower-developing primary immune response to be mounted. This time-buying protective role of interferon has been demonstrated by experiments showing that far more virulent infections occur when the action of interferon is blocked with anti-interferon antibodies.

All these observations led to intensive investigations into what interferon is and how it acts as a cellular defense against viral infection. As has so often been the case, questions first asked in a virological context have generated answers with significance throughout the biological world.

Interferon Proteins and Interferon Genes

During the 1960s three classes of human interferons were described: the "type I" interferons IFN-α and IFN-β produced primarily by leukocytes and fibroblasts respectively in response to viral infections and other inducing agents, and "type II" or IFN-γ produced by lymphoid cells in response to mitogens or specific antigen challenges. The α, β, and γ interferon proteins are of similar size (143, 145, and 146 amino acids long, respectively), but have surprisingly different sequences considering their common biological activities. Cloning studies have identified at least 15 INF-α genes, as well as a number of related pseudogenes, all located on chromosome 9. The INF-α genes show only about 30% homology with the INF-β gene also found on chromosome 9. The INF-γ gene is strikingly different from either of the other two. Its locus is on chromosome 12 rather than 9, and its sequence shows no homology to those of INF-α or β. Despite these DNA sequence differences, there do appear to be two sets of amino acids whose positions have been conserved in the interferon proteins. Their conservation suggests that they may be related to the functioning of the interferon proteins, but site-directed mutagenesis of cloned DNAs has not given much support to this hypothesis.

Synthesis of Interferon

Interferon is synthesized *de novo* in response to a variety of inducers, especially RNA viruses, double-stranded RNA, and many types of metabolic inhibitors. The most important feature of those inducers is that they are *inhibitors of protein synthesis*. Analysis of the regulatory sequences of the INF-β gene suggests how inducers may function to bring about transcription of that gene (Figure 7.10). There appear to be three functional domains within the regulatory region. Two sites permit binding of positive regulatory molecules that facilitate transcription, the usual mechanism for control of RNA synthesis in eukaryotic cells. The third site, nearest to the promoter, appears to bind a repressor molecule that would block transcription. In the current model for induction of the interferon gene, the inducing molecule begins the process by inhibiting protein synthesis. The labile repressor molecule on the interferon repressor regulatory site is therefore not replenished, quickly leaving the regulatory site empty. At the same time two positive regulatory proteins are phosphorylated and become able to bind to their sites upstream from the empty repressor site, activating transcription on the unblocked sequence. Interferon is then synthesized and eventually secreted from the induced cell. Even in the continuing presence of inducing molecules, synthesis stops after only a few hours.

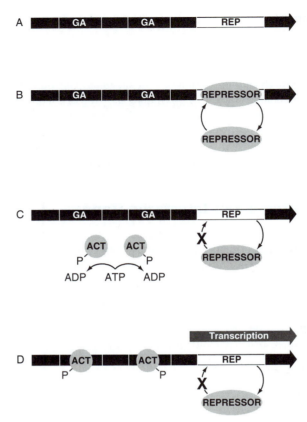

FIGURE 7.10 Induction of the interferon β gene requires interactions at three regulatory sites. **A,** Two gene activator sites (GA) and a single repressor binding site (REP), which normally is covered by a labile repressor protein **(B).** Induction begins **(C)** when host cell protein synthesis is inhibited so that the repressor protein is not replaced on the REP site. At the same time, two gene activator proteins become phosphorylated. Binding of these proteins to the gene activator sites initiates transcription of the interferon β gene **(D).**

The Action of Interferon

Interferon does not interact directly with viruses themselves. Instead, interferon is itself an inducer that causes the transcription of several other genes, creating an **antiviral state** in those cells that protects them from infection. Several lines of evidence demonstrate that mechanism of interferon induction is similar to that found with peptide hormones. A number of experiments, summarized in Figure 7.11, indicate that interferon acts by binding to specific receptors on the plasma membrane and thereby regulating transcription in the target cell. The studies show, for example, that interferon has no effect on the cell that synthesized it

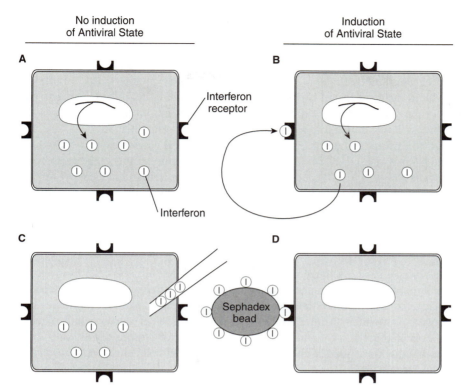

FIGURE 7.11 Evidence that induction of the antiviral state by interferon requires it to bind to a cell surface receptor. **A,** Newly made interferon within the cytoplasm of the cell that synthesized it has no effect. **B,** The synthesizing cell can be induced by interferon that is secreted and then binds to the plasma membrane. **C,** Microinjected interferon has no effect, but **(D)** interferon coupled to Sephadex beads can interact with plasma membrane receptors to induce the antiviral state.

(Figure 7.11, *A*) unless it is first secreted and then becomes bound to the secreting cell's surface (Figure 7.11, *B*). Likewise, interferon has no inductive effect when it is injected into a nonsecreting cell (Figure 7.11, *C*). Further evidence for a cell-surface interaction are experiments in which interferon bound to beads too large to be taken up by a cell was found to be fully active in inducing an antiviral state (Figure 7.11, *D*).

There appear to be two different interferon receptors. A common receptor for INF-α and INF-β is encoded on chromosome 21, while the gene for a separate INF-γ receptor is on chromosome 6. These are very high affinity receptors, permitting interferon to be effective at concentrations estimated at only 50 molecules per target cell. The strict species specificity of interferon action is probably the result of the specificity of the interaction of interferons with their cell-surface receptors.

When interferon binds to its cell-surface receptor, it appears to trigger a signalling cascade involving a number of protein kinases and signal transduction and transcription factors. These events result in inducing the transcription of several genes whose products create the antiviral state in the target cell.

The Antiviral State

The antiviral state induced by interferon is analogous to calling up the reserves when a country is threatened by a foreign invader. The threatened country is prepared to defend itself, but no shots are actually fired unless and until the invasion actually occurs. Similarly, interferon induces the synthesis of two different proteins that are in the form of inactive proenzymes. These proteins prepare the cell to defend itself, but until virus infection triggers their conversion into active forms, the metabolism of the cell is not altered.

What form does that defense take? Since the antiviral state protected cells from both RNA and DNA viruses, W.K. Joklik reasoned that a metabolic activity common to both types of viruses must be the target. That common activity must be translation of early viral mRNAs, or in the case of positive-strand RNA viruses, of the genome itself. Numerous experiments with many different types of viruses support this reasoning, since in all cases the viral life cycle is aborted or at least slowed as a result of inhibition of viral translation. This inhibition is the result of degradation of mRNAs (the 2,5-oligoA system), inactivation of the initiation factor eIF-2 (the protein kinase system), or both.

Although we have discussed interferon as an agent that protects cells from viral infection, the actual final result of interferon-induced inhibition depends on several factors. One is the relative concentrations of interferon and infecting virions. High multiplicities of infection appear to overwhelm the interferon system. Another is the particular cell involved. Some interferon-treated cells may actually suffer cytotoxic effects more rapidly once infected, and therefore die more quickly, than untreated cells. Other cells are spared, however, since new virions are not synthesized in the infected cells.

The 2,5-OligoA System

One of the proenzymes synthesized in response to interferon is *oligoisoadenylate (oligoA) synthetase,* which catalyzes the formation of adenylic acid oligomers from ATP. The phosphodiester linkage in these chains is not the conventional 3′, 5′ form found in other normal RNAs (like the 3′ tail on mRNAs, for example), but is a 2′,5′ linkage. The product is therefore called **2,5-oligoA.**

Figure 7.12 illustrates how the 2,5-oligoA system is thought to protect cells from viral infection. The 2,5-oligoA synthetase proenzymes induced by interferon are only part of the system. The other component is a ribonuclease (RNase L) proenzyme that is continuously present in the cytoplasm. When a virus attempts to infect, 2,5-oligoA synthetase is activated by the presence of double-stranded RNA in a manner not yet understood. The short strands of 2,5-oligoA then produced are effectors that activate RNase L. RNase L cleaves both viral and host RNAs at numerous locations. Translation is thus inhibited as both viral mRNAs and host mRNAs and rRNAs are degraded.

In addition to its direct role in preventing viral infection by degrading mRNAs, RNase L is also a "death effector" that induces apoptosis, or programmed cell death. The molecular mechanism of this induction is not yet understood, but clearly apoptosis of infected cells is an effective way to prevent further spread of a virus.

Since 2,5-oligoA synthetase requires double-stranded RNA for activation, it is not surprising that the 2,5-oligoA system is more effective against infections by RNA viruses than DNA viruses. All RNA viruses produce double-stranded RNAs as they replicate and/or

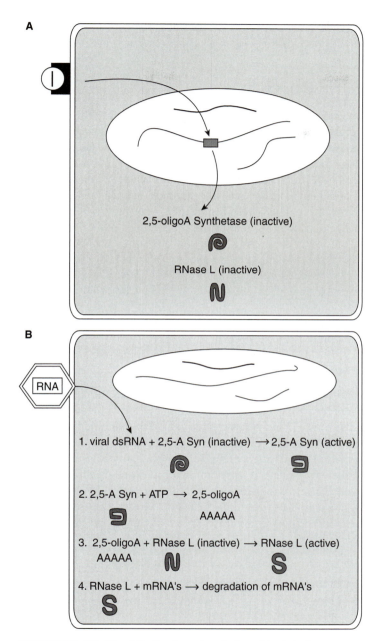

FIGURE 7.12 The 2,5-oligoA system induced by interferon inhibits protein synthesis when a virus attempts to infect a cell in the antiviral state. **A,** The antiviral state. Interferon binds to a cell surface receptor to induce the synthesis of a proenzyme 2,5-oligoA synthetase. Another proenzyme, RNase L, is always present in the cytoplasm. **B,** Double-stranded RNA activates the synthase to produce 2,5-oligoA molecules. These activate RNase L, which degrades both viral and cellular RNAs to inhibit translation.

express their genomes. Only DNA viruses that transcribe concurrently from both strands at a given location will produce RNAs that are complementary and thus able to form double-stranded molecules.

The Protein Kinase System

The second interferon-induced system employs a mechanism we have earlier seen used by several viruses to inhibit their hosts' protein synthesis. In the interferon system, the proenzyme synthesized in response to interferon is the 67 kd protein kinase, RNA-dependent (PKR) (Figure 7.13). This protein kinase, sometimes also called P1, is activated by the presence of double-stranded RNA during an attempted viral infection. When activated, P1 first phosphorylates itself and then phosphorylates the small (α) subunits of the translation initiation factor eIF-2. As noted previously (see Figure 7.6), phosphorylation of eIF-2α blocks the GDP/GTP exchange necessary for that factor to participate in the formation of a 40S initiation complex. Translation then ceases since new polysomes cannot be formed.

Obviously the interferon-induced antiviral activities are also antihost since they do not readily distinguish between viral and host translation. There is evidence to suggest, however, that there may be some discrimination between invader and host in both the interferon antiviral defenses. This is done by localizing the effects of both systems to the immediate cytoplasmic vicinity of the double-stranded RNA that serves as the effector activating both the 2,5-oligoA and protein kinase systems.

Other Interferon-Related Activities

Two other antiviral effects of interferon have been described, but their causes have not been identified. One effect that has been described in numerous situations is inhibition of transcription of viral genomes in interferon-treated cells. This inhibition may be related in some manner to the second effect, which is to produce changes in the plasma membranes of induced cells. These changes in the plasma membrane may play protective roles at both ends of the virus life cycle. They may interfere with viral adsorption and penetration, and possibly with transcription of the earliest viral genes, or with proper assembly of enveloped viruses that bud from the plasma membrane. Vesicular stomatitis virus virions released from interferon-treated cells have reduced amounts of the membrane protein M and spike protein G, for example. Multiplication of retroviruses may also be inhibited at a similar stage.

WHAT COUNTERMEASURES DO VIRUSES HAVE TO COMBAT THE HOST INTERFERON SYSTEM DEFENSES?

Several viruses appear to have anti-interferon defenses, especially against the eIF-2 kinase system, but these defenses do not always result in the viruses becoming able to multiply freely in interferon-induced cells. For example, the reovirus core protein σ3 is capable of binding to double-stranded RNA and preventing its activation of the protein kinase *in vitro*. Nevertheless, reovirus infection is blocked by interferon. Likewise, the poxvirus vaccinia encodes a protein that may be able to bind double-stranded RNA and thus prevent activation of both the kinase and 2,5-oligoA systems, but vaccinia multiplication is

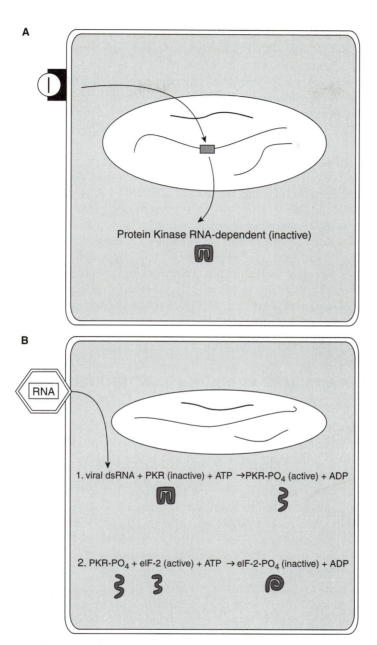

FIGURE 7.13 The protein kinase system induced by interferon inhibits the formation of translation initiation complexes. **A,** The antiviral state. Interferon induces that synthesis of an inactive 67 kd protein kinase called P1. **B,** Double-stranded RNA induces the phosphorylation of P1. P1 then phosphorylates the α subunit of eIF-2, inactivating it and preventing formation of the 40S initiation complex (see Figure 7.6).

inhibited in many cells by interferon. These observations highlight that the antiviral state induced by interferon probably takes a number of forms, and that the role of each individual system may be quite different in infections by different viruses.

Some anti-interferon defenses are effective, however. Both the DNA adenoviruses and the RNA influenza viruses encode a function that blocks the eIF-2 kinase systems and allows the virus to multiply in interferon-induced cells. The adenovirus system is presently the better described of the two. Adenoviruses are resistant to the action of interferon due to the action of a small RNA molecule called VA1. VA1 is transcribed along with the viral late genes, but by the eukaryotic RNA polymerase III responsible for synthesis of tRNA rather than the RNA polymerase II that generates mRNAs. The VA1 RNA molecule does not encode a polypeptide but instead assumes a folded configuration similar to that of a tRNA. Nevertheless, it appears to be a regulatory molecule responsible for controlling the rate of translation of viral proteins as well as protecting the virus from the antiviral state defenses. VA1 RNA prevents the binding of double-stranded RNA to the eIF-2 kinase, blocking its activation and consequently phosphorylation of that initiation factor. This seems to be a bizarre and even unnecessary action for adenovirus to take, however, since *adenovirus itself inactivates eIF-2 by phosphorylating it.* This has been taken to indicate that host and viral translation must take place in separate compartments within the cell, so the action of the VA1 RNA permits viral translation to continue while host translation is inhibited by earlier viral activity.

WHAT CLINICAL USE IS MADE OF INTERFERON AND OTHER ANTIVIRAL AGENTS?

The previous discussion focused on the role of interferons as a natural first line of defense against viral infections. For the past 25 years, particularly as cloning of the interferon genes into bacterial host made large amounts of interferons available, interferons have been studied as therapeutic agents for the treatment of a variety of different viral infections. One widely heralded study, for example, found that intranasal administration of α-interferon or β-interferon helped reduce the incidence of colds produced by a variety of respiratory viruses. Attractive as the prospect of eliminating the universal bane of the common cold may be, it is likely that interferon treatments will be more useful in cases of life-threatening infections such as viral encephalitis, or of persistent infections by herpesvirus, hepatitis B virus, hepatitis C virus, or papillomavirus.

One of the most exciting and provocative recent discoveries in molecular cell biology has been that interferons play major roles in the regulation of numerous activities related to cell growth and differentiation, as well as in regulation of the immune system. These activities sparked enormous interest in interferons, not just as a means to combat viral infections, but also as possible therapeutic agents against cancers of various types. α-interferon, for example, has been found to produce at least partial remissions in 90% of patients with hairy cell or chronic myelocytic leukemias and 50% of those with T-cell lymphoma. Interferons have not been the "magic bullets" they were first hoped to be, however, since the large doses required in therapy also produce a number of serious toxic side effects. Combination therapies using interferon as one component appear promising.

Antiviral Drugs

The development of antibiotics to treat bacterial infections was one of the most important medical advances of the twentieth century. These agents are based on *selective toxicity,* that is, the antibiotic must block or inhibit some crucial function in the infecting bacteria but not cause unacceptable toxicity in the patient. Since prokaryotes use fundamentally different enzymes, ribosomes, and cofactors than eukaryotes for the processes of the Central Dogma, many different antibiotics have been developed whose selective toxicity exploits those differences. Streptomycin, for example, specifically inhibits the activity of the prokaryotic 70S ribosome during translation but has no effect on the eukaryotic 80S ribosome at the same concentration.

Like antibiotics, antiviral drugs must also possess selective toxicity in order to be effective clinical agents. Since the viruses rely on their host cells for most of the components used in the expression and replication of their genetic material, finding compounds that will interfere with viral infection without significantly harming the patient is very challenging. That difficulty may be appreciated by observing that after more than 30 years of research only a handful of antiviral drugs are available for clinical use; in contrast, during the same period many hundreds of antibiotics were discovered or synthesized.

Nucleoside Analogues   In theory, any of the steps in the viral life cycle (attachment, penetration, uncoating, synthesis of viral proteins and nucleic acids, assembly, and release) might be a target for an antiviral drug. Most antiviral drugs in current use or under development are nucleoside analogues targeted against the viral enzymes that are involved in viral synthesis of DNA. These include thymidine kinase (TK) and DNA polymerase in DNA viruses and reverse transcriptase (RT) in the retroviruses. Figure 7.14 illustrates the chemical structures of some of the nucleoside analogues that are major antiviral drugs in current use.

A number of antiviral drugs are guanosine analogues. Acyclovir is a guanine analogue that exhibits a high degree of selective toxicity for herpesviruses, particularly herpes simplex. Selective toxicity in this case is based on differences in the ability of acyclovir to bind to enzymes involved in DNA synthesis. Both viral TK and viral DNA polymerase are able to bind the compound much more efficiently than their host cell counterparts. In addition, only the herpesvirus TK can phosphorylate acyclovir to convert it to the form required for it to be incorporated into a DNA molecule. Since incorporation of acyclovir into the growing point of a DNA strand blocks any further additions to that strand, this and analogues with similar activities are *chain terminators* that selectively inhibit viral DNA synthesis. Newer derivatives of acyclovir, such as valaciclovir (1-valyl ester of acyclovir) and famciclovir, are converted to acyclovir in the liver and thus have the same therapeutic effect, but can be taken by mouth on a less frequent basis than can acyclovir. Another acyclovir derivative that is especially effective against cytomegalovirus is ganciclovir. It, too, has a valyl ester form, valganciclovir, that can be administered by mouth rather than intravenously. Vidarabine (adenine arabinoside or Ara A) is another nucleoside analogue that interacts far more readily with herpesvirus DNA polymerase than with the host enzyme. It is also a chain terminator.

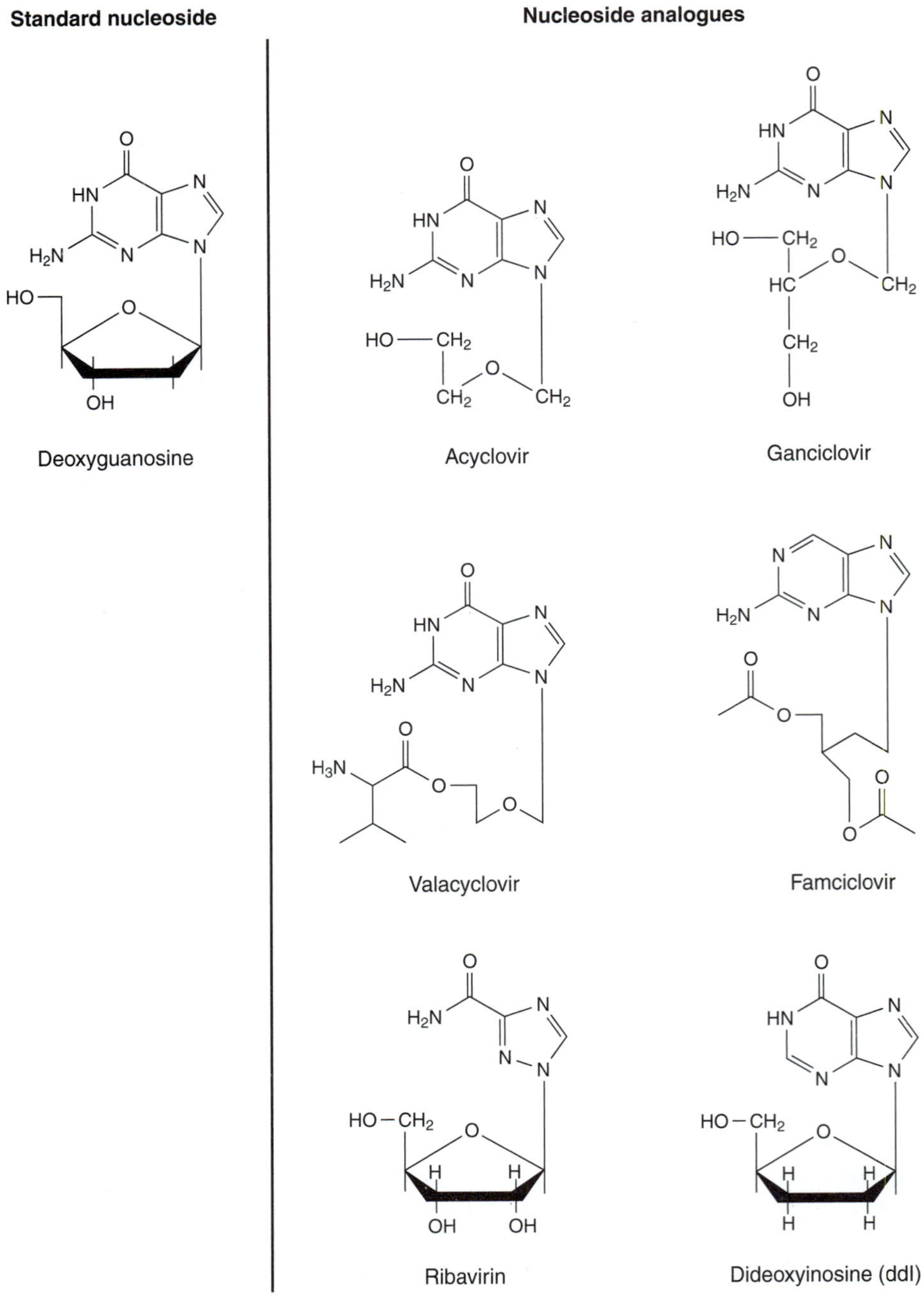

Standard nucleoside

Nucleoside analogues

Deoxyguanosine

Acyclovir

Ganciclovir

Valacyclovir

Famciclovir

Ribavirin

Dideoxyinosine (ddI)

FIGURE 7.14 Many antiviral drugs are nucleoside analogues. They usually function as chain terminators during DNA synthesis by DNA polymerase or reverse transcriptase. When the analogue becomes attached to the growing points of a nucleotide chain, further elongation is blocked because the analogue is missing the required 3′ hydroxyl group.

Standard nucleoside

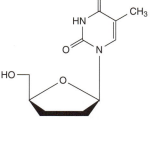

Deoxyadenosine

Nucleoside analogues

Vidarabine

Thymidine

Zidovudine (AZT)

Stavudine (d4T)

Cytidine

Zalcitabine (ddC)

Lamivudine (3TC)

Another guanine analogue, ribavirin, has shown promise against a great variety of both RNA and DNA viruses. The drug has been extensively studied in influenza virus infections, where it has been shown to inhibit several steps in virus replication. These include both initiation and elongation by the viral RNA replicase and the 5′ capping of mRNAs.

Several nucleoside analogues are effective in blocking the reverse transcription step in the life cycle of retroviruses like HIV. Zidovudine (azidothymidine, or AZT) is a thymine analogue with an azido group N_3 replacing the 3′ hydroxyl group. The selective toxicity of zidovudine (after it has been phosphorylated by host enzymes) is based on its having about a 100-fold greater affinity for RT than host DNA polymerases. Unfortunately, serious adverse side effects and the development of viral mutants resistant to the drug both occur with zidovudine. Other, possibly less toxic agents for treatment of HIV infections are currently available or under study. These include dideoxynucleoside analogues of cytidine (ddC), adenosine (ddA), and inosine (ddI), all of which act as chain terminators since they lack 3′ hydroxyl groups.

After HIV's reverse transcriptase makes a DNA copy of the viral genome, that molecule must be transported into the nucleus and inserted into a host cell chromosome. The HIV integrase responsible for these activities has no equivalent in human cells, so it is also being investigated as a target for anti-HIV drugs. A wide variety of drugs have been shown to be effective in cell cultures, but none is presently available as a therapeutic agent.

Other Antiviral Agents Not all antiviral drugs are nucleoside analogues. Figure 7.15 shows two drugs that block viral uncoating. Amantadine and rimantadine are tricylic cage molecules with activity against influenza A virus. They appear to block the uncoating step of the viral life cycle by interacting with the matrix protein M2 associated with the viral genomic RNA molecules. Amantadine also appears to decrease the pH of the *trans*-Golgi compartment, which in turn causes the acid-induced conformational change in the influenza A hemagglutinin molecules to occur during maturation of the virion rather than when the virion is penetrating a new host cell.

The experimental drug, Pleconaril, blocks the release of viral RNA from the capsid of rhinoviruses and enteroviruses (members of the picornavirus group). Pleconaril is the result of a process called "rational drug design." The agent was originally discovered nearly 20 years ago in a large-scale screening assay to seek drugs that might be useful against picornaviruses. While it showed some antiviral activity, its effectiveness was very low. Pleconaril appears

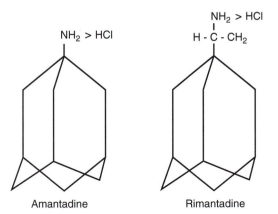

Amantadine Rimantadine

FIGURE 7.15 Several antiviral drugs that are active against influenza A have cage-like structures. These agents prevent uncoating of the virion.

to act by fitting into the canyon that lies between the viral capsomers (see Figure 2.11) and preventing movement of the viral genome out of the capsid. When the structure of the rhinovirus capsid was described in detail in X-ray crystallographic studies, however, it became possible to alter the drug in specific ways to fit better into the canyon, thus making the drug more potent.

A number of antiviral agents target cleavage events that are required for maturation of new virions. A variety of *protease inhibitors* now in use to treat HIV infections have produced dramatic improvements in the quality of life and longevity of persons with AIDS. These agents mimic the structure of the eight locations where the Gag-Pol polyprotein of HIV is cut by viral protease to produce individual capsid proteins and three enzymes (see Figures 4.31 and 7.16). By filling its active site of the enzyme in preference to the polyprotein, protease inhibitors inhibit the maturation process. Similar protease inhibitors are being sought to block maturation of

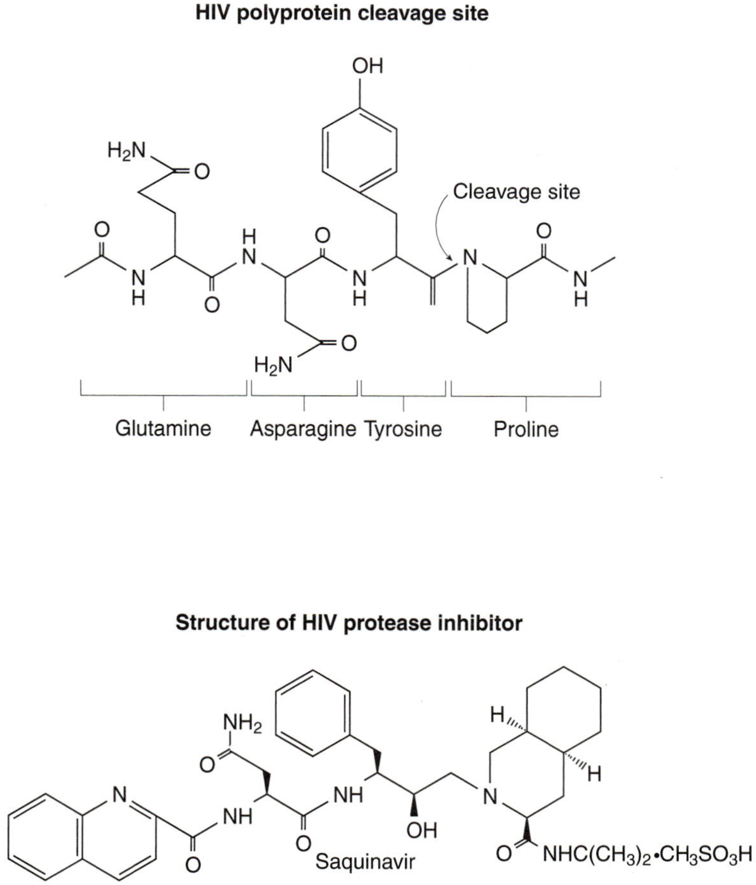

HIV polyprotein cleavage site

Structure of HIV protease inhibitor

FIGURE 7.16 Protease inhibitors are designed to compete for the viral proteases that cleave polyproteins into individual proteins during maturation. The inhibitors mimic the structure of the amino acids near the protein cleavage site.

herpesviruses like cytomegalovirus and flaviviruses like hepatitis C virus. The experimental drug Bay-38-4755 blocks a different cleavage event in cytomegalovirus. The agent specifically inhibits cleavage of DNA concatemers into individual viral genomic molecules and thus blocks the packaging step of maturation.

An intriguing new approach to designing antiviral agents is illustrated by the development of Zanamivir, an inhibitor of influenza A virus neuraminidase. Neuraminidase is one of two spike proteins on the viral envelope. It promotes viral spread by removing N-acetyl-neuraminic acid (sialic acid) from glycoproteins and glycolipids on cell surfaces and thereby releasing viruses that are stuck to those cell surfaces. As Figure 7.17 illustrates, Zanamivir has been designed as an analog of the transition state formed during the cleavage reaction. As such, it binds to the viral neuraminidase with very high affinity and prevents the enzyme from recognizing its normal substrate.

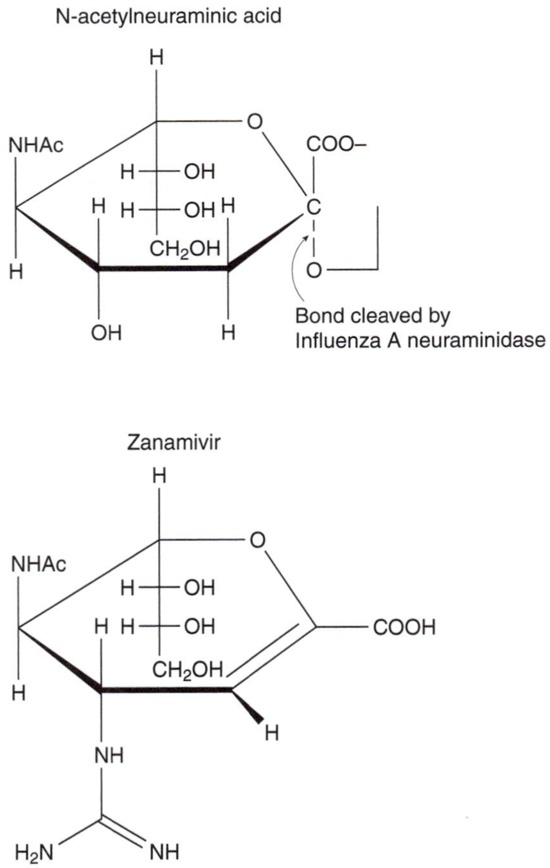

FIGURE 7.17 Zanamivir mimics the structure of the transition state created during cleavage of N-acetylneuraminic acid (sialic acid) from a glycoprotein or glycolipid by influenza A neuraminidase.

SUMMARY

Viruses have evolved a variety of strategies to enhance synthesis of their own proteins and nucleic acids at the expense of their host cells' macromolecular synthesis. Most of these strategies involve inhibition of host cell translation. Eukaryotic host cell translation may be inhibited by (1) inhibition of transport of host cell mRNAs from the nucleus into the cytoplasm; (2) differential degradation of host cell mRNAs; (3) blockage of initiation complex formation by host cell mRNAs; (4) covalent modification of translation-related components to convert them to viral use; (5) increases in intracellular cation concentrations to levels more favorable to viral enzymes; and (6) production of viral factors that directly inhibit host cell translation. The processes of host cell transcription and replication are also often inhibited, but this usually occurs as a secondary effect of inhibition of host cell translation. Bacteriophages inhibit their hosts by degrading the bacterial DNA or by altering the specificity of host enzymes involved in transcription.

Both prokaryotic and eukaryotic cells have, in turn, evolved methods of combating viral infections. Prokaryotic cells have restriction/modification systems that monitor DNA within the cell for the proper covalent modifications and destroy DNA molecules without them. Animal cells have a two-step defense: the interferon system. In the first step a virus infects a cell causing de-repression of the genes encoding a class of proteins called interferons. Interferons are then secreted from the infected cell and interact with uninfected cells (the second step) to prepare them to combat viral infection. Interferon induces the synthesis of two sets of inactive proenzymes, the 2,5-oligoA system and the protein kinase system, to create an "antiviral state." When a virus attempts to infect a cell in the antiviral state, the proenzymes become activated and alter the translation and transcription processes of the host cell to make it more difficult for viruses to express and replicate their genomes. Interferons are being used clinically to treat some types of viral infections as well as certain types of cancers. Other antiviral drugs are being developed. Many of these interfere with viral DNA synthesis while others block viral enzymes needed for expression of the viral genome, construction of new virions, or release of virions.

SUGGESTED READINGS

1. Three review articles discussing virus-host cell interactions are: "Inhibition of cell functions by RNA-virus infections" by L. Kaariainen, and M. Ranki, *Ann Rev Microbiol* 38:91–109, (1984), "Impact of virus infection on host cell protein synthesis" by R.J. Schneider and T. Shenk, *Ann Rev Biochem* 56:317–332 (1987), and "Reciprocity in the Interactions between the poxviruses and their host cells" by S. Dales, *Ann Rev Microbiol* 44:173–192 (1990).

2. "Inhibition of HeLa cell protein synthesis during adenovirus infection—restriction of cellular messenger RNA sequences to the nucleus" by G.A. Beltz and S.J. Flint, *J Molecular Biol* 131:353–373 (1979), and "Effect of adenovirus on metabolism of specific host mRNAs: transport control and specific translation discrimination" by A. Babich, L.T. Feldman, J.R. Nevins, J.E. Darnell Jr., and C. Weinberger, *Mol Cell Biol* 3:1212–1221 (1983), present evidence that adenoviruses inhibit host protein synthesis by blocking transport of mRNA from the nucleus into the cytoplasm.

3. "Herpes simplex virus-infected cells contain a function(s) that destabilizes both host and viral mRNAs" by A.D. Kwong and N. Frenkel, *Proc Natl Acad Sci USA* 84:1926–1930 (1987), describes degradation of mRNAs in HSV-infected cells as a mechanism of inhibiting host translation.

4. "Inhibition of 40S-met-tRNA^{met} ribosomal initiation complex formation by vaccinia virus" by A. Person, F. Ben-Hamida, and G. Beaud, *Nature* 287:356–357 (1980), presents evidence that poxviruses have a virion protein that prevents formation of the translation initiation complex.

5. "Inhibition of HeLa cell protein synthesis following poliovirus infection correlates with the proteolysis of a 220,000-dalton polypeptide associated with eukaryotic initiation factor 3 and a cap binding protein complex" by D. Etchison, S.C. Milburn, I. Edery, N. Sonenberg, and J.W.B. Hershey, *J Biol Chem* 257:14806–14810 (1982), describes the central element of the poliovirus mechanism for inhibiting host cell translation.

6. "Reovirus inhibition of cellular RNA and protein synthesis: Role of the S4 gene" by A.H. Sharpe and B.N. Fields, *Virology* 122:381–391 (1982), presents studies using recombinant viruses to determine the viral protein that inhibits host transcription and translation.

7. Phosphorylation of the translation initiation factor eIF-2 is described in several papers on mengovirus: "A kinase able to phosphorylate exogenous protein synthesis initiation factor eIF-2α is present in lysates of mengovirus-infected L cells" by A Pani, M. Julian, and J. Lucas-Lenard, *J Virol* 60:1012–1017 (1986), and "The α subunit of eukaryotic initiation factor 2 is phosphorylated in mengovirus-infected mouse L cells" by J. DeStefano, E. Olmsted, R. Panniers and J. Lucas-Lenard, *J Virol* 64:4445–4453 (1990).

8. Phosphorylation of ribosomal proteins as a means of inhibiting host cell translation is discussed in "Ribosomal protein phosphorylation in vivo and in vitro by vaccinia virus" by B. Buendia, A. Person-Fernandez, G. Beaud, and J-J. Madjar, *Eur J Biochem* 162:95–103 (1987).

9. Many papers claim that host cell translation is inhibited by virally-induced changes in intracellular cation concentrations. A sample of those dealing with picornaviruses includes "Selective blockage of initiation of host protein synthesis in RNA-virus-infected Cells" by D.L. Nuss, H. Oppermann and G. Koch, *Proc Natl Acad Sci USA* 72:1258–1262 (1975), "Sodium ions and the shutoff of host cell protein synthesis by Picornaviruses" by L. Carrasco and A.E. Smith, *Nature* 264:807–809 (1976), and "Reversion by hypotonic medium of the shutoff of protein synthesis induced by encephalomyocarditis virus" by M.A. Alonso and L. Carrasco, *J Virol* 37:535–540 (1981). Two papers on sindbis virus are "Na$^+$ and K$^+$ concentrations and the regulation of protein synthesis in sindbis virus-infected chick cells" by R.F. Garry, J.M. Bishop, S. Parker, K. Westbrook, G. Lewis, and M.R.F. Waite, *Virology* 96:108–120 (1979), and "Differential effects of ouabain on host- and sindbis virus-specified protein synthesis" by R.F. Garry, K. Westbrook, and M.R.F. Waite, *Virology* 99:179–182 (1979).

10. Others rebut the claims made for the role of cation concentration in inhibition of translation: "Guanidine-sensitive Na$^+$ accumulation by poliovirus-infected HeLa cells" by C.N. Nair, J.W. Stowers, and B. Singfield, *J Virol* 31:184–189 (1979), and "Protein synthesis in cells infected with Semliki forest virus is not controlled by intracellular cation changes," by M.A. Gray, K.J. Micklem, and C.A. Pasternak, *Eur J Biochem* 135:299–302 (1983).

11. "Evidence for the presence of an inhibitor on ribosomes in mouse L cells infected with mengovirus" by M.N. Pensiero and J.M. Lucas-Lenard, *J Virol* 56:161–171 (1985), describes modification of ribosome function by means of a viral protein.

12. "Inhibition of DNA-dependent transcription by the leader RNA of vesicular stomatitis virus: Role of specific nucleotide sequences and cell protein binding" by B.W. Grinnell and R.R. Wagner, *Mol Cell Biol* 5:2502–2513 (1985), and "LaCross virus gene expression in mammalian and mosquito cells" by C. Rossier, R. Raju and D. Kolakofsky, *Virology* 165:539–548 (1988), describe how these negative-strand viruses inhibit host-cell transcription. "Inhibition of transcription factor activity by poliovirus" by N. Crawford, A. Fire, M. Samuels, P.A. Sharp, and D. Baltimore, *Cell* 27:555–561 (1981), gives the same for a positive-strand virus.

13. "Inhibition of adenovirus DNA replication by vesicular stomatitis virus leader RNA" by J. Remenick, M.K. Kenny, and J.J. McGowan, *J Virol*, 62:1286–1292 (1988), and "Reovirus inhibition of cellular DNA synthesis: role of the S1 gene" by A.H. Sharpe and B.N. Fields, *J Virol* 38:389–392 (1981), discuss cases in which inhibition of DNA synthesis appears to be due to viral action rather than to be a secondary effect of inhibition of protein synthesis.

14. "Host specificity of DNA produced by *Escherichia coli*. I. Host controlled modification of bacteriophage λ" by W. Arber and D. Dussiox, *J Mol Biol* 5:18–36 (1962), and "Host specificity of DNA produced by *Escherichia coli*. II. Control over acceptance of DNA from infecting phage λ" by D. Dussoix and W. Arber, *J Mol Biol* 5:37–49 (1962), are the classic papers describing host-induced modification.

15. "Concerning the mechanism of action of interferon" by W.K. Joklik and T.C. Merigan, *Proc Natl Acad Sci USA* 55:558–565 (1966), describing the action of interferon on vaccinia virus infection suggests what the role of interferon might be in preventing infections.

16. "pppA2' p5' A2' p5': An inhibitor of protein synthesis synthesized with an enzyme fraction from interferon-treated cells" by I.M. Kerr and R.E. Brown, *Proc Natl Acad Sci USA* 75:256–260 (1978), and "Interferon, double-stranded RNA, and RNA degradation: activation of an endonuclease by (2'-5') A_n," by E. Slattery, N. Ghosh, H. Samanta, and P. Lengyel, *Proc Natl Acad Sci USA* 76:4778–4782; describe the 2,5-oligoA system induced by interferon.

17. Interferon, double-stranded RNA, and protein phosphorylation" by B. Lebleu, G.C. Sen, S. Shaila, B. Cabrer, and P. Lengyel, *Proc Natl Acad Sci USA* 73:3107–3111 (1978), describes the protein kinase system induced by interferon. "Adenovirus VA1 RNA antagonizes the antiviral action of interferon by preventing activation of the interferon-induced eIF-2α Kinase" by J. Kitajewski, R.J. Schneider, B. Safer, S.M. Munemitsu, C.E. Samuel, B. Thimmappaya, and T. Shenk, *Cell* 45:195–200 (1986), and "Modification of protein synthesis initiation factors and the shutoff of host protein synthesis in adenovirus-infected cells" by R.P. O'Malley, R.F. Duncan, J.W.B. Hershey, and M.B. Mathews, *Virology* 168:112–118 (1989), describe how adenovirus subverts that defense. "Translational control by influenza virus: suppression of the kinase that phosphorylates the alpha subunit of initiation factor eIF-2 and selective translation of influenza viral mRNAs" by M.G. Katze, B.M. Detjen, B. Safer, and R.M. Krug, *Mol Cell Biol* 6:1741–1750 (1986), and "Influenza virus regulates protein synthesis during infection by repressing autophosphorylation and activity of the cellular 68,000-M_r protein kinase" by M.G. Katze, J. Tomita, T. Black, R.M. Krug, B. Safer, and A. Hovanessian, *J Virol* 62:3710–3717 (1988), do the same for influenza virus.

18. "Disarming flu viruses" by W.G. Laver, N. Bischofberger, and R.G. Webster, *Scientific American* 280:78–87 (1999), "'Flu' and structure-based drug design" by R.C. Wade, *Structure* 5:1139–1145 (1997), and "Rational design of potent sialidase-based inhibitors of influenza virus replication" by M. Von Itzstein, W.Y. Wu, G.B. Kok, M.S. Pegg, J.C. Dyson, B. Jin, T. Van Phan, M.L. Smythe, H.F. White, S.W. Oliver, P.M. Colman, J.N. Varghese, D.M. Ryan, J.M. Woods, R.C. Bethell, V.J. Hotham, J.M. Cameron, and C.R. Penn, *Nature* 363:418–423 (1993), discuss the development of drugs that are structured to interact with specific viral targets.

19. "Viruses and apoptosis" by A. Roulston, R.C. Marcellus, and P.E. Branton, *Ann Rev Microbiol* 53:577–628 (1999), reviews the role of programmed cell death as a host defense and how viruses both induce and suppress it.

20. "Defeating AIDS: What will it take?" *Scientific American* 179: 81–107 (1998), is a special report on developing strategies for prevention and treatment of AIDS.

Chapter 8

Effects of Viral Infection on Host Cells: Integrated Viruses and Persistent Infections

- Besides integration into the host cell genome, how may a virus infection persist?
- What effect does a persistent infection have on the host organism?
- How do the herpesviruses establish their latent infections?
- Summary

It has been known for more than 70 years that a **productive infection** leading to the release of newly synthesized virions is not the only possible outcome of a virus–host cell interaction. In this chapter we shall consider three other possible outcomes of that interaction. The first of these involves integration of all or part of the viral genome into the genome of the host cell, leading to a variety of phenotypic changes in that cell. The second and third involve the establishment of a long-term infection, either by reducing the lytic potential of the virus in order to maintain a low overall level of infection in a culture or an organism, or by entering into an infection in which the viral genome is present within some cells but no virus is produced.

These three types of virus–host cell interactions can have profound effects on the nature of the host cell. The most significant of these effects is the conversion of an animal cell from its normal condition into a cancer cell. We shall examine how different types of viruses cause such conversions, and the relationship between the cancer genes found in some viruses and genes that are parts of the host cell genome.

WHY IS EVERY VIRAL INFECTION NOT PRODUCTIVE?

A set of simple experiments illustrates that some viruses can interact with host cells in a manner that does not cause every infected cell to be destroyed as new viruses are produced. If *E. coli* is infected in a plaque assay, with two different viruses, T4 and λ, the plaques produced are about the same size, but their appearance is different; T4 plaques are clear in the center, while λ plaques are hazy or *turbid*. Examination of these plaques under low magnification reveals the cause for this difference: T4 plaques are devoid of cells in their centers while the λ plaques contain many intact cells.

If plaques are the result of many rounds of lytic viral reproduction, why are there still intact bacteria in the λ plaque? The simplest explanation is that they are mutant cells that λ is unable to infect because of some change in their cell walls or metabolism. Such a change should be comparatively rare, however, so one would expect to find only a few (if any) resistant cells in any particular plaque. This is indeed the case with T4 plaques, but *every* λ plaque is filled with intact cells. Nevertheless a second experiment supports the hypothesis that the cells within the λ plaques are resistance mutants; when these cells are exposed to λ again, the virus is indeed unable to infect them.

Despite this result, the fact is that these cells are not resistant to a λ virus infection at all. A third experiment suggests a reason for the inability of λ to infect cells taken from the interior of a λ plaque. When a culture derived from one of the cells taken from the middle of a λ plaque is cross-streaked against a cell culture that can be infected by λ virus, a hazy

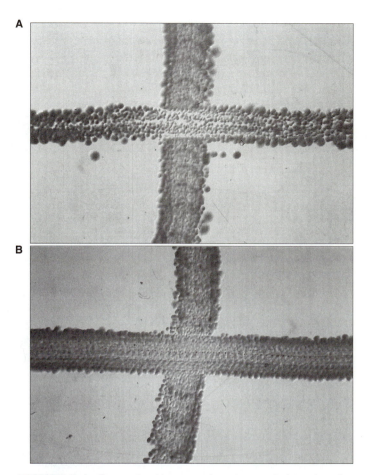

FIGURE 8.1 Cross-streaking experiments permit detection of very low levels of viral release from certain cultures. **A,** *E. coli* K12(λ) was streaked across *E. coli* W1485, a strain capable of infection by λ virus. Virus released from the K12(λ) cells lyses W1485 cells at the crossing point. **B,** *E. coli* B cells taken from a colony that arose in a T4 plaque streaked across *E. coli* W1485. No lysis occurs at the crossing point.

band of lysis occurs where the original culture crosses the plaque-derived culture (Figure 8.1, *A*). A similar experiment performed with one of the very rare cells that appears in a T4 plaque does not show this feature (Figure 8.1, *B*). These experiments indicate that the cells from the T4 plaque are indeed T4-resistant mutants, but that cells from the λ plaque were not. The fact that the λ plaque-derived culture seems to produce a low level of λ virus that can infect and lyse at least some of the original culture cells indicates that those cells were not resistant mutants at all, but were infected along with all the other cells in the plaque area. The outcome of that infection was not lysis, however, but some relatively

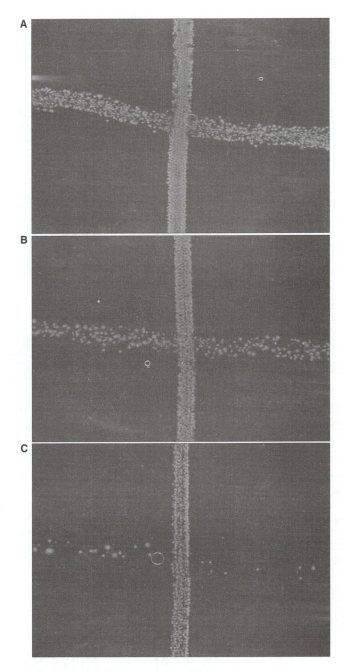

FIGURE 8.2 The effect of ultraviolet irradiation of *E. coli* K12(λ) and *E. coli* W1485. *E. coli* K12(λ) was streaked across *E. coli* W1485, the plates exposed to 1, 3, or 5 second pulses of ultraviolet light, and then incubated overnight. The *E. coli* W1485 cells are virtually unaffected, while the *E. coli* K12(λ) cells show a dose-dependent decrease in cell number.

stable association of the virus with its host such that only a few virions are produced and released at any one time.

A fourth experiment supports this hypothesis. Figure 8.2 shows the result of exposing cross-streaked cultures to a brief ultraviolet light treatment. The uninfected host culture is apparently unaffected by the UV treatment, while the λ plaque-derived culture has been largely destroyed. Furthermore, UV irradiation appears to have caused λ virus within infected cells to multiply and lyse their host cells since infectious virus can be recovered from the region of the plate where the cells were lysed. From these experiments, it is clear that the cells in the original plaque were indeed infected by λ, but that the virus was carried from generation to generation in its cell in a special, noninfectious form: a **provirus** (*pro* = Latin for "before," meaning that this form leads to an actual virus). Viruses that can become proviruses are also termed *temperate* viruses, a term that was introduced in Chapter 3 when the lytic life cycle of λ was discussed. Although λ virus is the archetype for temperate viruses and will be discussed at length here, a number of temperate viruses enter the lysogenic state in a different fashion than does λ. The *E. coli* P1 virus is carried as a "plasmid" or independently replicating circular DNA molecule, for example.

WHAT INTERACTION BETWEEN A BACTERIOPHAGE AND HOST CELL GIVES RISE TO THE LYSOGENIC STATE?

Early studies found that the lysogenic state behaved like any other physical or biochemical marker in *E. coli* genetic crosses. These experiments suggested that in lysogenized cells, λ becomes incorporated into the bacterial DNA to become, in effect, a set of *E. coli* genes. This hypothesis was first confirmed by conjugation mapping experiments using lysogenic Hfr's, cells in which the F plasmid has become inserted into the *E. coli* genome. These early studies indicated that λ becomes integrated into the *E. coli* genome at a specific site between the loci for genes involved in tryptophan biosynthesis (*trp* genes) at 28 minutes and galactose utilization (*gal* genes) at 17 minutes. Deletion mapping experiments on the λ virus itself suggested that a specific viral location called the *b2 region* was required for a virus to become lysogenic when it infected a new host cell. These locations are now termed *att* B and *att* P for bacterial and phage "attachment" loci, respectively. Finer scale mapping experiments now place *att* B at 17.4 minutes on the *E. coli* map, between the *gal* genes (17.0 minutes) and the *bio* genes (17.5 minutes) involved in biotin synthesis.

But how can a linear λ DNA molecule become incorporated into a specific site in the circular *E. coli* genome without any loss of *E. coli* genes? When a linear fragment of DNA enters during conjugation, for example, it *replaces* a segment of the recipient cell's own DNA as it becomes incorporated by a pair of recombinational events (Figure 8.3, *A*). No host genes are lost when the linear λ genome becomes incorporated, however, so λ must not become lysogenic by this form of replacement insertion. Instead, the linear λ DNA molecule must *circularize* before insertion occurs, so that only a single recombinational event is necessary to combine the λ and *E. coli* genomes, as diagrammed in Figure 8.3, *B*. Note that insertion by this mechanism appears to reorder some of the phage genes, but has no effect on the

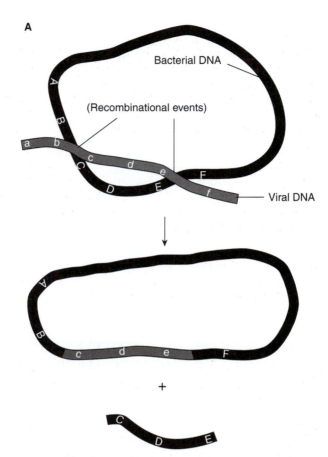

A

FIGURE 8.3 Insertion of a linear DNA molecule into another DNA molecule. **A,** Two recombinational events result in the replacement of a portion of recipient (bacterial host) DNA by donor (phage) DNA. *(Continued.)*

order of the host genes flanking the insertion, although the distance between them is obviously increased.

Genetic and biochemical experiments clearly demonstrate that this is exactly how lysogenic insertion occurs. The first requirement for insertion by a single recombinational event is circularization of the λ genome. As we saw in Chapter 3 (see Figure 3.12), analysis of λ virus's linear genome reveals that it has unusual termini called "cohesive" or "sticky" ends." The 5′ end of each DNA strand extends 12 bases beyond the 3′ end of the opposite strand. Since these 5′ extensions are complementary to each other, they can base pair to create a circular molecule from the linear genome. Comparison of the order of several genes in λ genomes taken from virions with those same genes as mapped by conjugation in lysogenic cells (Figure 8.4) provides genetic evidence for circularization followed by a sin-

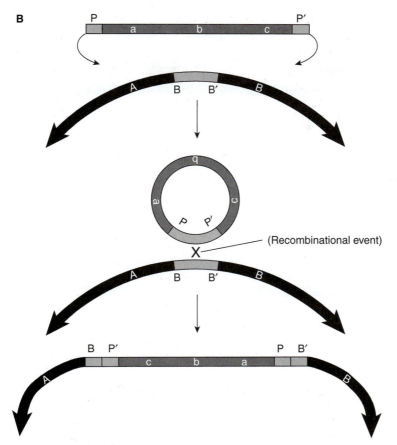

FIGURE 8.3 CONT'D B, Circularization of the donor DNA and insertion by a single recombinational event does not cause any loss of recipient DNA.

gle recombinational event. λ markers called *m6* and *mi* are near the ends of the genomic molecules found in virions, but are near each other in the interior of the genome when it becomes incorporated into *E. coli.* Conversely, two other markers, *h* and *c,* which were near each other on either side of the *att* P site in the virion molecule, are found at opposite ends of the inserted molecule. The order of the *bio* and *gal* genes in the host cell has not been changed, however.

The fact that insertion occurs only at one location in the *E. coli* genome, the *att* B site, is evidence that insertion requires *site-specific* recombination. Figure 8.5, *A,* gives the common sequence of 15 base pairs, which is found at both the *att* B and *att* P sites. Notice that 12 of the 15 base pairs are A/Ts. This sequence, held together by base pairs with only two hydrogen bonds, can be expected to open to permit recombination to occur.

The mechanism for this recombination has been worked out in *in vitro* systems employing plasmids bearing the *att* P site, fragments of the *E. coli* genome containing the *att* B

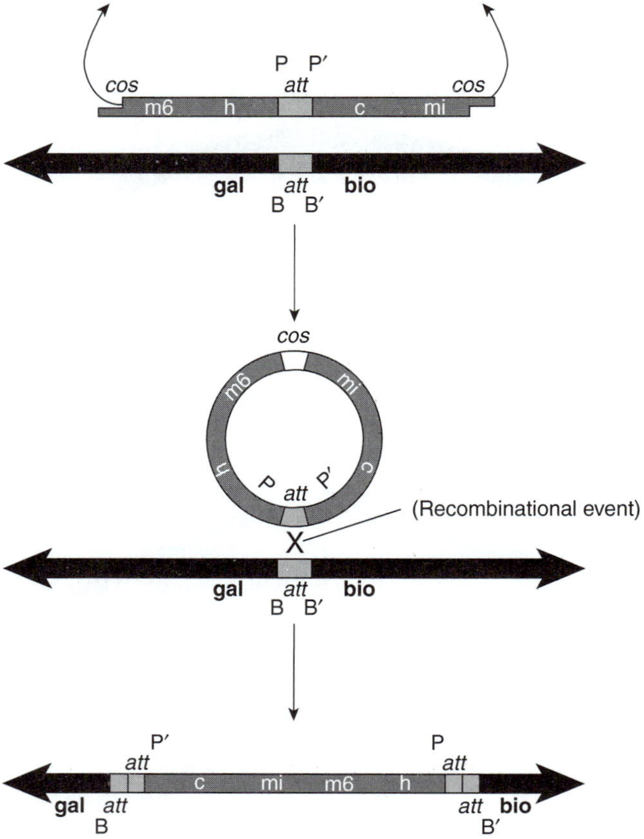

FIGURE 8.4 λ virus integration involves circularization of the viral DNA followed by site-specific recombination between the viral and host cell genomes. Note that the order of the λ markers *(m6, h, c, mi)* appears changed *(c, mi, m6, h)* as a result of this process. P*att*P′ = phage attachment site; B*att*B′ = bacterial attachment site.

site, a λ protein called an *integrase,* and a bacterial protein called the *integration host factor,* or IHF. Figure 8.5 illustrates that integration involves the same reactions as normal genetic recombination: the formation and resolution of a Holliday intermediate. When the *att* P and *att* B sites are lined up, the viral integrase makes staggered cuts at the indicated locations (Figure 8.5, *B*) and facilitates a strand transfer (Figure 8.5, *C*). Branch migration for three bases then occurs as a form of "checking" to ensure that the proper sequences are being exchanged (Figure 8.5, *D*). If the "check" is satisfactory, integrase repeats the exchange with the other two DNA strands (Figure 8.5, *E*). Resolution or cutting of the crossed strands completes the insertion process (Figure 8.5, *F*). The result is equivalent to staggered recombination since a series of base pairs within the *att* sequence is heteroduplex DNA.

A   Bacterial and phage attachment sites

D   Branch migration moves the cross-over point

B   Viral integrase nicks two strands

E   Nicking and strand exchange occur again

C   Single strand exchange occurs

F   Resolution creates staggered recombination product

FIGURE 8.5 Site-specific recombination between the λ *att*P sequence and the *E. coli att*B sites. After alignment of the *att*P and *att*B sequences **(A)**, the viral integrase makes cuts in two of the four strands **(B)** and facilitates single-strand exchange **(C)**, followed by branch migration **(D)**. A second exchange **(E)** follows. Cutting and ligation complete the process to create staggered recombination **(F)** at each join between the viral and host DNAs.

WHAT GENETIC EVENTS PERMIT A LAMBDA VIRUS TO BECOME LYSOGENIC?

Two things must happen in order for a virus to become lysogenic. First, the events leading to the lytic cycle must be prevented, and second, the viral genome usually becomes incorporated into its host cell's genome. In Chapter 3 we saw that shifts in gene expression during lytic infections by λ are controlled by a series of antiterminator proteins, which permit transcription past weak stop signals. The lytic cycle is thus under a form of positive control in which certain proteins facilitate the transcription of other genes that lead to the formation of new virions and cell lysis. Lysogeny is the result of both positive transcriptional regulation in the form of gene activator proteins and its opposite, repression, operating in competition with the events leading to production of new virions.

The events leading to establishment of lysogeny by λ virus are illustrated in Figure 8.6; the proteins involved in the lytic cycle have been omitted in this figure since they are given in Figure 3.15. Immediate-early gene expression is the same, regardless of the ultimate outcome of the infection. When the antiterminator protein N binds to the weak termination sequences T_{R1} and T_{L1} (Figure 8.6, *B*), transcription continues past those sites to permit

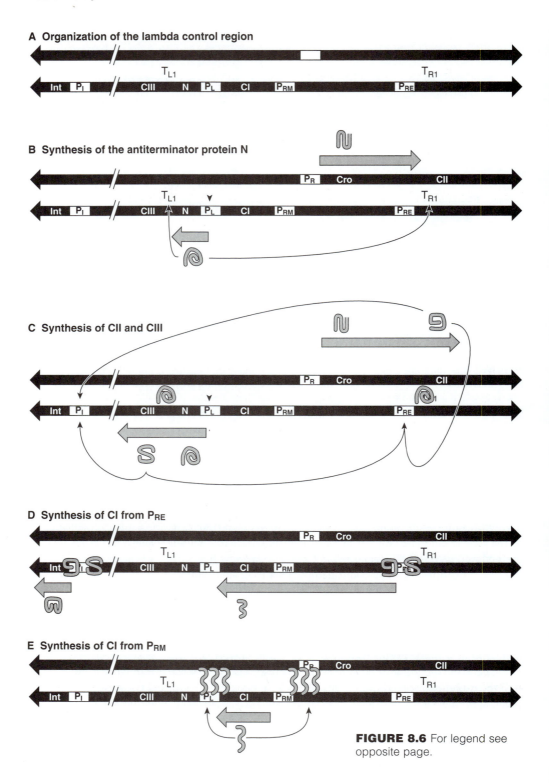

A Organization of the lambda control region

T_{L1} T_{R1}

Int P_I CIII N P_L CI P_{RM} P_{RE}

B Synthesis of the antiterminator protein N

P_R Cro CII

T_{L1} T_{R1}

Int P_I CIII N P_L CI P_{RM} P_{RE}

C Synthesis of CII and CIII

P_R Cro CII

Int P_I CIII N P_L CI P_{RM} P_{RE}

D Synthesis of CI from P_{RE}

P_R Cro CII

T_{L1} T_{R1}

Int P_I CIII N P_L CI P_{RM} P_{RE}

E Synthesis of CI from P_{RM}

P_R Cro CII

T_{L1} T_{R1}

Int P_I CIII N P_L CI P_{RM} P_{RE}

FIGURE 8.6 For legend see opposite page.

synthesis of two delayed-early gene products called CII (from *cII* on the right) and CIII (from *cIII* on the left) (Figure 8.6, *C*). CII is a positive transcriptional regulator involved in establishing lysogeny, while CIII protects CII from too rapid degradation by a host protease called HflA (for *high frequency of lysogenation*). CII binds to a site called P_{RE} (*promoter for repressor establishment*) located between the *cro* and *CII* genes, which are transcribed from P_R. When CII binds at P_{RE} it acts as a gene activator facilitating transcription *to the left* from that location. The RNA molecule then synthesized is complementary to *cro* mRNA (that is, "anti-*cro*" RNA) at its 5′ beginning and then contains the information for the repressor protein CI at its 3′ end (Figure 8.6, *D*).

How Are the Events Leading to the Lytic Cycle Prevented?

CI protein acts in a remarkable fashion. It functions, at the same time, both as a gene activator for transcription of its own gene and as a repressor of transcription of other genes. CI acts as a repressor by binding to both P_R and P_L to block further transcription from those sites (Figure 8.6, *E*). This prevents synthesis of the gene products necessary for progression to the late genes whose products are necessary for the formation of new virions. When CI binds to P_R, however, it also functions as a gene activator for a formerly cryptic promoter called P_{RM} (*promoter for repressor maintenance*). P_{RM} is transcribed to the left to yield an mRNA for the CI protein that does not contain the "anti-*cro*" portion. CI is thus synthesized by means of a feedback loop in which CI is a gene activator for its own promoter! The rate of CI synthesis can be quite finely tuned since there are actually three CI binding sites in a tandem array between P_R and P_{RM}. The binding of CI to the two sites farthest to the right blocks P_R, but permits RNA polymerase access to P_{RM}. When levels of CI become high, all three sites become filled, inhibiting transcription from the *cI* gene until the concentration of CI protein falls back to within the appropriate range.

How Is λ DNA Integration Accomplished?

The synthesis of CI carries out the first requirement for establishment of lysogeny: repression of the lytic cycle. The second requirement, integration of the viral genome into its host's genome, is mediated by a specific viral integrase that is encoded by a gene located on the left side of the regulatory gene loci. Transcription of the integrase gene (*int*) requires the presence of the gene activator protein CII bound to its promoter (called P_I for

FIGURE 8.6 The molecular events of establishing λ virus lysogeny. **A,** The pertinent sites in the control region of the λ genome. **B,** Immediate-early gene expression of the cro and N proteins. The cro protein attempts to bind to P_R and P_L, while the antiterminator protein N attempts to bind to T_{R1} and T_{L1}. **C,** Antitermination by N permits synthesis of the CII and CIII gene activator proteins as transcription proceeds past T_{R1} and T_{L1}. CII and CIII gene activator proteins then bind to P_{RE} and P_I. **D,** Synthesis of the CI repressor and Int proteins begins at P_{RE} and P_I, respectively. **E,** CI repressor protein blocks further transcription from P_R and P_L. The CI proteins bound to P_R also act as gene activators permitting transcription of the CI gene from P_{RM}.

promoter for integration) (Figure 8.6, *D*). Other genes transcribed from P_I encode peptides involved with site-specific recombination, as well as an enzyme called the **excisionase,** which reverses the reactions involved in integration.

WHAT DETERMINES IF A PARTICULAR LAMBDA VIRUS INFECTION RESULTS IN LYSOGENY OR THE COMPLETE LYTIC CYCLE?

The initial synthesis of proteins is the same in every λ-infected cell, regardless of the eventual outcome of that infection. Thereafter, it is a game of molecular musical chairs between the CI repressor and cro proteins as each attempts to occupy the binding sites of P_R. If the CI repressor proteins win, the virus will be committed to become lysogenic, while if the cro proteins win, it will complete the lytic cycle.

Since cro is an immediate-early gene product, while CI repressor is made only after the delayed-early gene products CII and CIII are synthesized, why does cro not always win the race for P_R? Two factors appear to even the odds. First, the affinity of cro for binding to all three sites on P_R is quite low, so unless the concentration of cro is high, it is likely that P_R will not be blocked. The second factor allowing CI repressor to compete for P_R is the order in which the three binding sites become occupied by the different proteins. Cro proteins bind to the three sites from left (near P_{RM}) to right, while the CI repressor proteins bind from right (near P_R) to left. In theory, each is therefore in a position to inhibit its competitor's synthesis immediately, but cro tends to fall off readily while the CI repressor binds relatively tightly.

Many physiological factors participate in tilting the race one way or the other. It is likely that CII actually plays a pivotal role in the outcome as it regulates the rate of synthesis of the CI repressor protein. The synthesis of CII is determined by the rates of a series of processes: transcription of the *CII* gene, translation of the CII mRNA, cleavage of the first two amino acids from the resulting protein, and the assembly of CII peptides into a tetramer. The activities of host cell proteins like the protease HflA that degrades the CII protein are also involved in the decision.

Synthesis of HflA is regulated by the *E. coli* cAMP signaling system. Low levels of intracellular glucose or other energy sources activate adenylate cyclase, leading to an increase in cAMP concentration in the cell. This results in inhibition of transcription of the *hfl*A gene so that the amount of the protease in the cytoplasm is reduced, which in turn reduces the degradation of CII. This means that starving cells whose overall biosynthetic rates are low are more likely to become lysogenized than actively growing cells. Lysogeny is more advantageous than the lytic cycle to the virus because starving host cells would not be able to produce numerous virions quickly.

HOW IS LYSOGENY REVERSED?

As we saw at the beginning of this chapter, a few lysogenized cells in any population will spontaneously shift into the lytic cycle to produce new virions. Since the lysogenic state is a dynamic one that must be constantly maintained by replacement of CI proteins on P_R, chance alone dictates that occasionally P_R will become open and the lytic cycle will begin. This inherent instability of the lysogenic state also allows a provirus to "bail out" when its host cell becomes injured, particularly by damage to its DNA. When a bacterial cell suffers DNA damage, an emergency system with the name "SOS repair" is induced. The SOS genes are under the control of a repressor protein called LexA. Damage to DNA induces the synthesis of a coprotease called RecA that stimulates latent proteolytic activity in its substrate, the LexA repressor. Self cleavage of LexA permits the synthesis of SOS DNA repair enzymes. The λ virus has evolved a mechanism of responding to this molecular signal when the host cell is in trouble. The repressor CI is also activated and self-cleaved by RecA, tipping the provirus into the lytic cycle at the same time that the SOS repair response is initiated. The host cell, which was already fighting for its life, is then finished off as its former passenger synthesizes new virions; gratitude for services rendered is apparently not a feature of the viral lifestyle.

Since the excision process involves a reversal of the single recombinational event that led to insertion (recall Figure 8.3, B), it generally results in the formation of a circular DNA molecule that contains the entire viral genome, but no more and no less. Occasionally, however, the excision is a bit sloppy, so that the separated DNA molecule contains not only viral genomic sequences, but also a small segment of *E. coli* DNA as well. Since the λ capsid is capable of holding slightly more than a genome's length of DNA, a host gene like *bio,* which lies very near the *att* B site, may be carried along with an entire λ genome. Since all of the progeny virions produced by copying that excised genomic molecule contain the *E. coli bio* gene, each new host cell infected by those viruses will gain an extra copy of that particular gene. This type of transfer of a bacterial gene incorporated into a viral genome to a new bacterial cell is called **specialized transduction.** λ *bio* transducing phages retain all of their viral genes. Other transducing λ phages may "trade" viral genes for host ones during the incorrect excision process and thus become defective viruses that can only reproduce in cells coinfected with normal λ viruses. They are still able to transfer the incorporated bacterial gene to new bacterial cells since that process requires only normal attachment and penetration.

WHAT EFFECTS MAY LYSOGENY HAVE ON THE HOST BACTERIUM?

Lysogeny appears responsible for a great variety of changes in the structure or physiology of the bacterium harboring the prophage. Table 8.1 gives some examples of *lysogenic conversion,* as these changes are sometimes called. A cardinal feature of lysogeny is that the lysogenic cell cannot be infected by additional members of the same virus type. **Superinfection immunity** is the term applied to this phenomenon. Reflection on the mechanism by which lysogeny of a phage like λ is maintained makes clear the mechanism for superinfection immunity. When a λ virus introduces its genome into a lysogenic cell, CI repressor proteins are already present in the cytoplasm of that cell. Consequently, these proteins can immediately bind to P_R and P_L in the newly introduced virus to block any expression of that virus's genome, thus preventing it from entering the lytic cycle. Note that this mechanism is quite different from that which prevents a virulent phage like T4 from infecting a cell already infected by the same virus. In this case, called **superinfection exclusion,** the incoming viral DNA is degraded rather than having its expression inhibited.

In addition to preventing further infections by the same type of virus, some viruses alter their hosts' ability to support infection by different viral species. For example, the

Table 8.1 | Changes Produced by Temperate Phages in the Phenotypes of Their Host Bacteria

Change	Example
1. Superinfection immunity	
- To the same type of virus	λ by λ prophage
- To other viruses	T4 rII mutants by λ prophage
2. DNA methylation patterns	*dam* methylase of P1 virus
3. Insertional mutagenesis	Inactivation of lipase and β-hemolysin genes in *S. aureus*
4. Specialized transduction	*gal* and *bio* by λ virus
5. Changes in cell wall	Changes in O antigen carbohydrates of *Salmonella* by epsilon virus
6. Production of exotoxins	See Table 8.2

λ *rex* gene product is synthesized from the same mRNA as the CI repressor protein in *E. coli* K12(λ). Since wild-type T4 phage can replicate in the lysogenic host but rII mutants of T4 cannot (hence "rex" from *rII ex*clusion), this system permits the detection of even extremely rare wild-type recombinant phage from crosses between different rII mutants. Despite nearly 40 years of study, the mechanism by which the *rex* gene product inhibits T4 rII mutants is still unclear. Overcoming *rex* exclusion is clearly an active process in viruses like T4 since mutations can cause the virus to lose that ability. The rex protein appears to act during DNA replication, while the rII proteins are associated with the host cell cytoplasmic membrane. The functions of all of these proteins are still uncertain.

A wide variety of host cell enzymes can be altered by the presence of a lysogenic phage. In some cases enzymes normally synthesized are inactivated when viral prophage insertion interrupts the genes for those enzymes. In other cases, information for new or different enzymes is carried into the genome by the temperate phage. These can be host cell genes, as occurs during specialized transduction, or viral genes (probably of host origin at some point, however) not normally found in the host. Table 8.2 presents a few examples of exotoxins that play significant roles in the diseases caused by the host organisms

Table 8.2 | Exotoxins Encoded by the Prophage Genome in Lysogenic Bacteria

Bacterium	Toxin
Clostridium botulinum	Type C and D toxins
Corynebacterium diphtheriae	Diphtheria toxin
E. coli	Shiga toxin
Streptococcus, group A	Erythrogenic toxin
Staphylococcus aureus	Enterotoxin A
	Staphylokinase
	Fibrinolysin
Vibrio cholerae	Cholera toxin

that are either encoded by temperate phages or appear to have been introduced by transduction. The classic example of this phenomenon is that of the diphtheria toxin produced by *Corynebacterium diphtheriae.* The gene for this toxin, *tox,* is carried in the genome of the temperate β phage, so only bacteria infected by or lysogenic with this virus produce diphtheria toxin and cause disease. Expression of the *tox* gene is controlled by a host cell transcription regulator that is responsive to iron levels within the cell. A similar situation occurs with the shiga-like toxin 1 produced by certain strains of *E. coli.* The gene for this toxin is transduced by several different bacteriophages (H-19B and 933) that are relatives of bacteriophage λ. Production of this toxin is also increased under conditions of iron starvation.

Pathogenic strains of the gram-positive bacterium *Staphylococcus aureus* can produce a variety of different toxins in response to infection by one or more temperate phages. In this bacterium, phage integration can cause opposite effects on toxin production. Integration of a serotype B phage leads to formation of toxins such as staphylokinase or enterotoxin A. Integration of the serotype A phage φ42E, however, causes insertional inactivation of a cellular gene for the toxin β-lysin. Another phage called φ13 causes a double conversion since it both produces insertional inactivation of the β-lysin gene and introduces the staphylokinase gene. Another gram-positive pathogen, *Streptococcus pyogenes,* also produces a wide array of exotoxins, a number of which are encoded by different bacteriophages.

The situation in the gram-negative organism *Vibrio cholerae* is especially interesting since it requires infection by two different temperate viruses. The disease cholera is caused by an exotoxin (cholera toxin, or CT) that the bacterium releases when it is attached to mucosal cells lining the intestine by means of thread-like structures called TCP pili (for *t*oxin *c*oregulated *p*ili). Both the toxin and the pili are necessary for bacterial virulence. The two subunits of CT are encoded by the *ctxAB* operon that is part of the genome of a temperate bacteriophage called CTXΦ, which can either become integrated as a prophage or replicate as a plasmid. This filamentous phage uses the TCP pili as the cellular attachment site for infection. The genes for the TCP pili are encoded in a gene cluster in the host cell called the *Vibrio* pathogenicity island (VPI). VPI turns out to be the genome of another filamentous phage now called VPIΦ. The TcpA subunit of the TCP pili is actually the coat protein of the VPIΦ. Thus, the coat protein of one phage generates the cellular receptor for another phage, with both phages required for pathogenicity.

In a number of cases it appears that pathogenicity islands have been introduced to cells through transduction but the transducing phage itself is not present as a complete prophage. For example, the *tst* gene in *Staphylococcus aureus* encodes the toxin responsible for toxic shock syndrome. The *tst* gene is excised and circularized by the lytic φ13 and 80α bacteriophages, both of which transduce it at very high frequency to previously nontoxin-producing strains of the bacterium. In other cases, the bacterial pathogenicity island contains phage sequences that indicate that transduction was probably the process that introduced those genes. The three pathogenic species of the gram-negative genus *Yersinia,* including *Y. pestis,* which causes plague, share a very large pathogenicity island that includes a cluster of genes involved in uptake of iron. The island is flanked by a putative integrase gene that is highly homologous to that of the temperate phage P4 and by two putative DNA-binding proteins homologous to those in the phages P4 and P2.

WHAT IS THE EVOLUTIONARY SIGNIFICANCE OF LYSOGENY?

Studies with agents that cause damage to DNA, such as ultraviolet light and the antibiotic mitomycin C, have revealed that numerous different types of bacteria appear to support lysogeny. Treatment with these agents causes the lysogenized cells to lyse, in some cases releasing infective virions. In other cases the lysed cells produce only virion components such as tail assemblies or empty capsids, suggesting that the prophages they harbored were defective in some essential function. These studies suggest that lysogeny may be very widespread among the bacteria.

Lysogeny has an obvious advantage over an endless series of productive infections since it exchanges a chancy existence for one in which reproduction of progeny is assured. Rather than sending progeny virions out into the environment to attempt to encounter new host cells, an event that in natural circumstances would be anything but certain, a prophage is reproduced each time its host bacterium divides. Moreover, a prophage cannot exhaust its supply of host cells within a given environment as viruses reproducing lytically can.

UNDER WHAT CIRCUMSTANCES DO ANIMAL VIRUSES BECOME INSERTED INTO THEIR HOST CELLS' GENOMES?

Lysogeny of the type seen with temperate bacteriophages, in which a DNA virus with a circular genome becomes integrated into its normal host's genome by a single recombinational event rather than reproducing progeny viruses, does not occur in the animal viruses. This is not to say, however, that the genomes of some types of animal viruses do not become inserted into their hosts' genomes. Ironically, the group of viruses for which insertion of the viral genome into the host cell's genome is obligatory are not DNA viruses at all, but the *RNA* retroviruses. As we saw in Chapter 4, retroviruses use reverse transcription to create DNA copies of their RNA genomes and insert that DNA into their hosts' chromosomes. The proviral DNA is then transcribed like the cellular genes to produce the mRNAs for viral polyproteins, as well as new viral genomic molecules. This mode of reproduction means that retroviruses are spread to new host cells by two different mechanisms. The first, typical of productive infections in general, is **horizontal transmission;** that is, virions are released from one cell and infect another. If the infected cells are part of the germ line, however, retroviruses may also show **vertical transmission** to new generations of organisms as the integrated proviruses pass through the gametes. These routes of transmission are illustrated in Figure 8.7. The BALB/c strain of laboratory mouse has a vertically transmitted **endogenous** mouse mammary tumor virus (MMTV) called MMTV-O carried in its genome, and passed via its germ line cells. The mammary epithelial cells of newborn BALB/c mice can also be infected by the horizontally transmitted **exogenous** strain MMTV-S if the newborns suckle a mouse shedding that virus in her milk.

The situation with DNA viruses is less straightforward since the result of an infection depends, not only on the virus itself, but on the cell it has entered. If the virus is not defective and if the host cell is permissive, or able to support all the activities necessary to synthesize new virions, the infection is productive and leads to the formation of progeny virions. In most cases the genome of a DNA virus does not become integrated into a permissive host cell's chromosomes, but rather functions as an independent genetic element (episome) within the nucleus or cytoplasm. If the cell is **nonpermissive,** or not able to support a productive infection for some reason, one result may be the insertion of all or part of the viral genome into its host's genome. For example, polyomavirus (a papovavirus) reproduces productively in permissive cells like

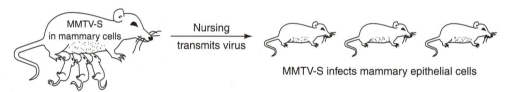

Vertical transmission: all cells, including gametes,
have viral sequences in their genomes

MMTV-O
in all cells

Breeding
transmits virus

MMTV-O in all cells in all offspring

Horizontal transmission: mammary epithelial cells produce
virions that are passed to offspring in milk

MMTV-S
in mammary cells

Nursing
transmits virus

MMTV-S infects mammary epithelial cells

FIGURE 8.7 Vertical and horizontal transmission of different strains of mouse mammary
tumor virus (MMTV). MMTV-O is present in all cells of a BALB/c mouse and is therefore passed
to the next generation through the gametes (vertical transmission). Newborn BALB/c mice can
be infected with MMTV-S if they suckle on a mouse shedding that virus in her milk (horizontal
transmission).

mouse kidney, but becomes integrated into the genome of nonpermissive baby hamster kidney
cells. This pattern is typical; cells from the species that the virus normally infects (mouse for
polyomavirus) are permissive, while cells from other species (hamster) are nonpermissive.

We saw earlier in this chapter that insertion of temperate bacteriophage like λ into
its host's DNA was a carefully orchestrated process resulting in a single, complete copy
of the virus being incorporated into the host. Specific sites in both the viral and host DNA
molecules are required for this process. Insertion of the DNA copy of a retroviral genome
into its target cell's DNA is a similarly precise process, since a faithful copy of the viral
genome is obviously required for synthesis of new virions. Insertion in this case involves
a specific site in the proviral DNA but not in the host cell DNA. The situation when a DNA
virus attempts to infect a nonpermissive host is quite different, however, since it involves
random sites in both the viral and host cell DNAs. In some instances insertion of the
entire genome of small viruses like those in the papova groups may occur, but rather than
the careful process seen with λ phage or the retroviruses, duplications, deletions, or

rearrangements are common. With larger viruses like the adenoviruses and herpesviruses, only small portions of the genome may become incorporated into the host cell's chromosomes, while the rest of the viral genetic information is lost. Insertion in the DNA viruses thus represents a manifestation of a mismatch between a virus and its nonpermissive host cell, rather than a useful or necessary step in a normal reproductive cycle.

WHAT EFFECT DOES INTEGRATION OF AN ANIMAL VIRUS HAVE ON A HOST CELL?

We have seen that the presence of a prophage can alter the phenotype of its host cells in a variety of ways, ranging from conferring superinfection immunity to altering the architecture of the cell wall. The changes that can occur in animal cells as a result of incorporation of viral genomes or sequences can be even more dramatic. The cells may, in fact, take on the phenotype of a neoplastic (from Greek, *neo* = "new" + *plasma* = "formation") or cancer cell in a process called **neoplastic transformation** (or just "transformation" for short). Since most members of the retroviruses and papovaviruses can cause transformation of their host cells, they are sometimes referred to as RNA and DNA **tumor viruses,** respectively.

The most rigorous demonstration that neoplastic transformation has occurred is made by transplanting the cells in question into genetically suitable animals and awaiting the development of neoplasms or tumors. This procedure is difficult, time-consuming, and expensive at best, and in the case of human cells requires the use of "nude" mice, which have impaired immune systems. Fortunately a number of morphological and biochemical changes in cultured fibroblastic cells have been found to correlate with neoplastic transformation. Some of these changes, called "transformation markers," are listed in Table 8.3. Normal

Table 8.3 | Changes Occuring in Fibroblasts as a Result of Transformation by Tumor Viruses

1. *Morphology*
   - From spindle-shaped to rounded
2. *Growth in Culture*
   - Loss of contact inhibition
     - From ordered to disordered cell layer pattern on surface of culture vessel
     - From single layer of cells to piled-up colonies
     - Higher cell densities achieved
   - Loss of anchorage dependence
     - From growth only on solid surface to growth also in suspension
   - Decrease in amount of serum supplement required
3. *Plasma Membrane Architecture*
   - Change in pattern of gangliosides
   - Change in pattern of sugars on glycoproteins
   - Change in agglutinability by plant lectins
4. *Metabolic Processes*
   - Increased rate of glycolysis (the Warburg effect)
   - Increased rate of transport of sugars
   - Increased amounts of proteases released

fibroblast cells have the shape of elongated diamonds that align with one another in a layer without nuclear overlaps (monolayer) on the surface of a plastic or glass culture vessel. Transformation causes changes in cytoskeletal components, causing the cells to assume a more rounded morphology and to become less well attached to the culture vessel. They also lose their ability to recognize cell surface signals, which regulate their growth patterns. As a result, they do not align with one another, but instead pile up to form disorganized layers of cells (Figure 8.8). At the biochemical level, transformation causes changes in the composition of numerous glycoproteins and glycolipids on the cell surface, as well as an increase in the rate of glycolysis (called the *Warburg effect*) and the transport of sugars. These transformation markers are not universal, however, since not all transformed fibroblasts show all of these changes. Moreover, other types of cells, such as mammary or kidney epithelial cells, do not exhibit the same types of morphological changes as fibroblasts when they become transformed.

This panoply of markers for neoplastic conversion in fibroblasts has made it possible to study the mechanisms by which tumor viruses transform animal cells under controlled *in vitro* conditions. A striking finding of these studies is that transformation induced by RNA or DNA viruses requires the continuous expression of only one or a few viral genes. Temperature-sensitive mutants have been isolated in which the transformed phenotype can

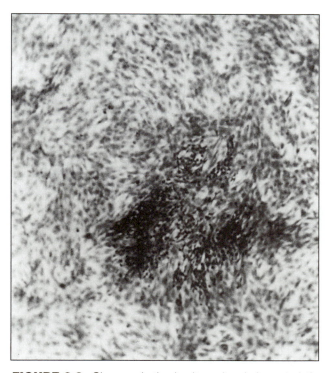

FIGURE 8.8 Changes in the *in vitro* cultural characteristics of fibroblasts as a result of transformation. The elongated normal cells form swirling patterns of nonoverlapping cells, while the shortened transformed cells form disordered piles or foci.

be "turned on" or "turned off" at the experimenter's will, indicating that transformation is a dynamic state that must be maintained by the continued expression of viral information. Since in many cases only a single viral gene is involved in inducing transformation, it is evident that viral genes that produce those changes must be very pleiotropic in their actions within the cell.

HOW DO ANIMAL CELLS REGULATE THEIR CELL CYCLES?

Transformation always involves an alteration of the regulation of the **cell cycle.** As Figure 8.9 illustrates, the cell cycle has four phases: M, during which the cell divides; G_1, during which the cell grows larger; S, during which DNA synthesis occurs; and G_2, during which the cell continues to grow and prepare for mitosis. The cycle is regulated at several points. The *restriction point* in late G_1 phase is the time when a "decision" is made whether to continue the cycle or to exit the cycle into a nondividing state called G_0. Cells in G_0 may differentiate and assume specialized functions. A cell can remain in G_0 indefinitely or may re-enter the cell cycle in response to signals from a variety of growth factors. Once the cell passes the restriction point in G_1, the cycle continues unless it is arrested at one of several *checkpoints* in response to some problem that needs to be corrected. Progression is halted in late G_1 and in late G_2 if DNA damage has occurred. These checkpoints allow time for the cell to repair the damaged DNA before the cycle resumes. The checkpoint in G_2 also responds to the presence of unreplicated DNA and prevents mitosis from occurring until all of the DNA has been copied. A checkpoint in late M phase halts the cell cycle until all chromosomes are properly aligned

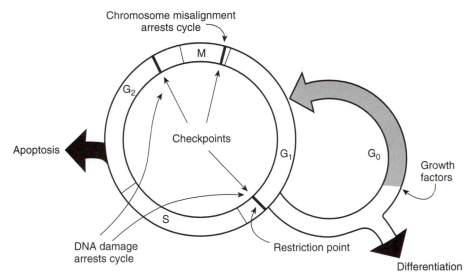

FIGURE 8.9 The eukaryotic cell cycle. The cell divides (M), then grows and prepares (G_1) for DNA synthesis (S), followed by further growth (G_2) in preparation for division again. At the restriction point in G_1, the cell may exit the cycle into G_0. Growth factors may direct the cell back into the cycle. DNA damage or chromosome misalignment induces cell cycle arrest at three different checkpoints. DNA damage may also induce apoptosis.

on the mitotic spindle so that daughter cells receive a full complement of chromosomes. In the event that a cell enters S phase with damaged DNA, the process of *apoptosis,* or programmed cell death, may be triggered to prevent the mutant cell from reproducing itself.

In eukaryotes, proteins in several different families interact to regulate progression through the cell cycle. *Cyclins, cyclin-dependent protein kinases* (Cdks), and *Cdk inhibitors* (CKIs) all interact either to block or unblock phases of the cycle. Cyclins and Cdks act together as a dimer, functioning as the regulatory and catalytic subunits, respectively, since binding of a cyclin to Cdk alters its conformation to induce its phosphorylating activity. Cyclins are degraded at the end of their functional period, which in turn inactivates the Cdk partner in the dimer. The assembly of cyclin/Cdk dimers is, in turn, regulated by a variety of different proteins.

An example of cell cycle regulation is illustrated in Figure 8.10, which shows the pathway by which a serum growth factor signals a cell to progress from G_0 through the G_1 restriction point and reinitiate the cell cycle. The transduction of an extracellular signal by means of sequential phosphorylation of a series of kinases leads to the transcription of the cyclin D gene. The cyclin D protein then associates with either a Cdk4 or a Cdk6 protein. The

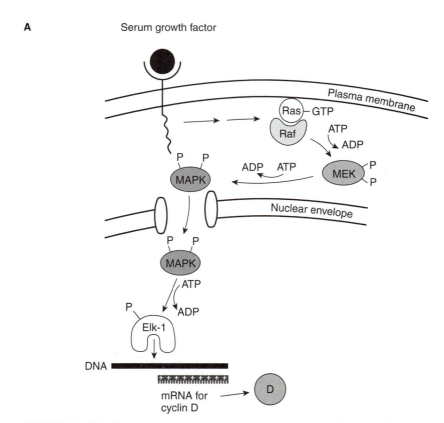

FIGURE 8.10 Regulation of the cell cycle by a serum growth factor. **A,** Binding of serum to a receptor in the plasma membrane generates a cascade of phosphorylations leading to activation of the gene for cyclin D. *(Continued.)*

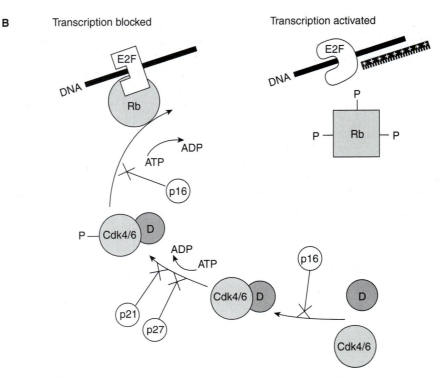

FIGURE 8.10 CONT'D B, Transcription is blocked by the binding of the retinoblastoma (Rb) to the transcription factor E2F. When cyclin D binds to Cdk4/6, the Cdk kinase phosphorylates itself and then the Rb protein, causing Rb to release from E2F and permitting transcription of genes needed for progression to S phase.

Cdk protein becomes phosphorylated. The target of the active dimer is the **Rb** (*retinoblastoma*) protein that is bound to a transcription factor called E2F. The Rb protein/E2F dimer blocks transcription of genes needed for the cell to enter S phase, so its presence creates the cell cycle restriction point in G_0 cells. When the Rb protein is phosphorylated by Cdk4/6-cyclin D, it is no longer able to bind to E2F, so the S phase genes are activated.

Figure 8.10 also shows that a variety of other proteins (p16, p21, p27) may act to block the assembly, activation, and phosphorylating activity of the Cdk/cyclin D dimer. These proteins, in turn, are regulated by a protein called **p53** that functions to block the cell cycle in G_1 if there has been damage to the cell's DNA that needs to be repaired before it is replicated. Synthesis of p53 is induced by DNA damage, especially that caused by irradiation.

P53 and Rb, as well as p16, p21, and p27 are examples of a group of proteins called **tumor suppressors** since their normal function is to prevent the growth of cells with damaged DNA or cells that should have exited the cell cycle into G_0. Tumor suppressors take their name from the observation that mutations in these proteins allow cancer cells to grow. In addition to its role at the G_1 checkpoint, p53 also responds to unrepaired DNA damage by triggering apoptosis of the injured cell. Although the details of this triggering reaction are still unclear, the outline of the pathway is shown in Figure 8.11. P53 probably interacts with a

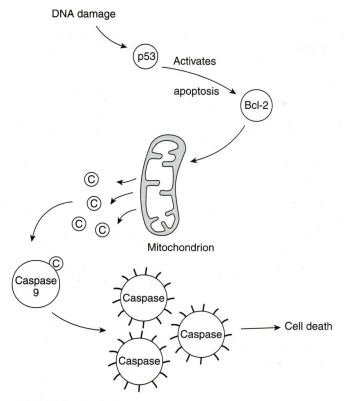

FIGURE 8.11 DNA damage may induce apoptosis. In one possible scenario, DNA damage causes p53 to interact with a protein of the Bcl-2 family. The Bcl-2 protein the causes mitochondria to release cytochrome C, an activator of caspase 9. Caspase 9, in turn, activates other caspases, leading to cell death.

member of the Bcl-2 family of proteins, which contains both positive and negative regulators of apoptosis. Bcl-2 proteins, in turn, activate special enzymes called *caspases* (cysteine proteins that cleave at *asp*aragine residues) to initiate a protease cascade that results in digestion of DNA into nucleosome-sized fragments and formation of apoptotic vesicles, leading ultimately to the destruction of the cell.

HOW DO RETROVIRUSES TRANSFORM ANIMAL CELLS?

The discovery that viruses may contain specific, cancer-causing oncogenes has been of great significance to biology, not just for what we have learned about the viruses *per se,* but also for what we have learned about the ways in which animal cells control their own growth and interactions with other cells. This is particularly true for the retroviruses since the genetic information that leads to transformation of infected cells has no role in the viral reproductive cycle and in some cases, is not even contained in the viral genome at all!

As we saw in Chapter 4 (Figure 4.27), the basic pattern of the retrovirus genome contains three genes, 5'-*gag-pol-env*-3', encoding polyproteins containing structural components of the nucleocapsid, the reverse transcriptase that permits formation of a DNA copy of the viral RNA genome, and glycoproteins that are part of the viral envelope.

As we also noted in Chapter 4, the *oncovirinae* subfamily of retroviruses contains many members that cause leukemias, sarcomas, and mammary carcinomas in a variety of different host species. Rous sarcoma virus (RSV), a chicken virus, holds "pride of place" in this subfamily since it was the first tumor virus to be described in the literature (by Peyton Rous in 1911). Sixty years later, intensive research in a number of labs revealed that RSV contains a fourth gene called *src* (for sarcoma) that is responsible for its ability to cause tumors in chickens. Such tumor-causing genes are termed **oncogenes.** The *src* gene lies at the 3' end of the genome and is transcribed either as part of an *env-src* unit, or by itself to yield a single protein of about 60 kd size. This protein is a protein kinase that phosphorylates tyrosine residues in proteins rather than the serines or threonines that are the targets of other kinases. Genetic studies employing both deletion mutants and temperature-sensitive mutants have demonstrated that the *src* protein is both necessary and sufficient to cause transformation in cultured cells.

Since phosphorylation of proteins had long been known to be a regulatory device in a variety of biochemical pathways, the discovery that RSV had a gene encoding a tyrosine kinase suggested that transformation might be the result of a cascade of regulatory changes begun by phosphorylation of particular proteins in the infected cell. The exact mechanism of transformation by the *src* protein is still not known and may in fact require more than just the activity of the *src* protein. In order to cause transformation, the *src* protein must become associated with the inner surface of the plasma membrane by means of a myristic acid bound to the amino-terminal end of the protein. In that location the *src* protein must then phosphorylate one or more cellular target proteins. Although numerous candidate proteins have been advanced for that "honor," including the Fak (focal adhesion) kinase that phosphorylates cytoskeletal proteins and the Shc (SH-2 containing) protein that is a component of several growth factor signalling pathways, the actual target is still unidentified.

Exciting as the discovery that a retrovirus could contain an oncogene like *src* was, a more significant discovery came when *src* sequences were used as hybridization probes with the DNAs of chicken and other cells. Not only did uninfected chicken cells contain the *src* gene sequences, but all other vertebrate cells, including human cells, did also. Thus, it appears that the viral *src* gene (now termed *v-src*) is not actually a viral gene at all, but a *cellular* gene (called *c-src*) that has become incorporated into the viral genome. RSV is therefore a *specialized transducing virus,* even though it has an RNA genome that has acquired an RNA copy of a cellular **proto-oncogene** (from Greek *proto* = "first" + oncogene). Antibodies against the *v-src* tyrosine kinase also react with a tyrosine kinase protein in normal, uninfected chicken cells, so expression of the *c-src* proto-oncogene appears to be a normal cellular event. This makes the term *proto-oncogene* somewhat unfortunate since it suggests that the role of the proto-oncogene product is also related to formation of tumors, but this is clearly not the case.

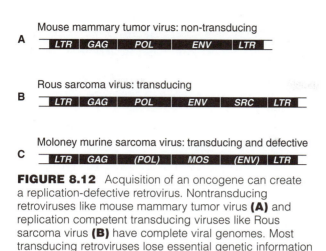

FIGURE 8.12 Acquisition of an oncogene can create a replication-defective retrovirus. Nontransducing retroviruses like mouse mammary tumor virus **(A)** and replication competent transducing viruses like Rous sarcoma virus **(B)** have complete viral genomes. Most transducing retroviruses lose essential genetic information when they acquire their oncogenes. Moloney murine sarcoma virus **(C)**, for example, has lost parts of its *pol* and *env* genes when it acquired the *mos* oncogene.

Transducing retroviruses, and the cellular proto-oncogenes they have acquired, have now been identified in a broad range of cells from a wide variety of species. A striking feature of these viruses is that they are nearly all *replication-defective*. The recombinational event by which they acquired a proto-oncogene also caused the loss or disruption of one or more essential viral genes, so they are unable to carry out all of the steps necessary to produce new virions. For example, the *mos* oncogene in Moloney murine sarcoma virus has replaced parts of the *pol* and *env* genes (Figure 8.12), rendering the virus replication-defective. These viruses can only replicate if the cell they infect is also infected by a replication-competent virus that supplies the missing retroviral function.

WHY DO VIRAL ONCOGENES CAUSE TRANSFORMATION WHEN THEIR EQUIVALENT CELLULAR PROTO-ONCOGENES DO NOT?

The activities associated with a number of viral oncogenes and their corresponding cellular proto-oncogenes have been determined for an array of viruses. As might be expected, many of the gene products are kinases, with either tyrosine or serine-specific activities, but other products are growth factors and their receptors, nuclear proteins, or GTP-binding proteins. All of these classes of proteins have known roles in the regulation of growth and gene expression in animal cells, so it is not unexpected that incorporation of cellular genes for these functions into viruses might lead to disturbances in growth and even to transformation. Table 8.4 presents a few examples of the different types discovered thus far.

If proto-oncogenes are normal cellular genes, why are their viral counterparts capable of transforming cells that the viruses infect? The answer to this question appears to involve either

Table 8.4 | Retroviral Oncogenes and Their Associated Cellular Proto-oncogenes

Oncogene	Retrovirus	Oncogene Protein	Proto-oncogene Protein	Function of Proto-oncogene Protein
PLASMA MEMBRANE-ASSOCIATED				
Tyrosine Kinases				
src	Rous Sarcoma Virus	pp60$^{v\text{-}src}$	pp60$^{c\text{-}src}$	Signal transduction
abl	Abelson Murine Leukemia Virus	p160$^{gag\text{-}abl}$	p160$^{c\text{-}abl}$	Signal transduction
Tyrosine Kinase Growth-factor Receptors				
erbB	Avian Erythroblastosis (AEB) Virus	gp74^{erbB}	gp180$^{c\text{-}erbB}$	Epidermal growth factor receptor
fms	Feline Sarcoma Virus	gp180$^{gag\text{-}fms}$	gp150$^{c\text{-}fms}$	Colony stimulating factor-1 receptor
G-Proteins				
H-*ras*	Harvey Murine Sarcoma Virus	p21^{ras}	p21$^{c\text{-}ras}$	GTP-binding/GTPase
CYTOPLASMIC				
Hormone Receptors				
erbA	AEB Virus	p75$^{gag\text{-}erbA}$	p46$^{c\text{-}erbA}$	Thyroid hormone (T3) receptor
Serine Kinases				
mos	Moloney Murine Sarcoma Virus	p37$^{env\text{-}mos}$	Unknown	Germ cell maturation factor
raf	3611-MSV	p79$^{gag\text{-}raf}$	p74$^{c\text{-}raf}$	Signal transduction
NUCLEOPLASMIC				
Nuclear Proteins				
fos	FBJ Murine Sarcoma Virus	p55^{fos}	p62$^{c\text{-}fos}$	Transcription activator
jun	Avian Sarcoma Virus 17	p55$^{gag\text{-}jun}$	p47$^{c\text{-}jun}$	AP-1 transcription factor
SECRETED (extracellular)				
Growth Factors				
sis	Simian Sarcoma Virus	p28$^{env\text{-}sis}$	p28–35$^{c\text{-}sis}$	Platelet-derived growth factor

The protein products of viral oncogenes and their corresponding cellular proto-oncogenes are designated as follows: p = protein, gp = glycoprotein, or pp = polyprotein of given molecular weight. The superscripts of viral oncogene products indicate the fusion polyprotein encoded.

quantitative or qualitative differences in expression of the two genes. When a proto-oncogene is recombined with a retroviral genome, that gene becomes subject to transcriptional regulation by viral rather than host control elements. This can result in overexpression of the new viral oncogene compared to its cellular counterpart. The *v-mos* oncogene, for example, has an identical amino acid sequence to the *c-mos* proto-oncogene, but is fused to the 5′ end of the viral *env* gene. In that location the *v-mos* gene is overexpressed compared to the *c-mos* proto-oncogene. The amount of the *c-mos* proto-oncogene's product detected in normal cells averages less than one peptide per cell, so the overexpression of the *v-mos* oncogene appears to be sufficient for transformation. Further evidence for this mechanism comes from experiments showing that normal cellular genes deliberately inserted into retroviruses, and therefore under viral transcriptional control, can cause transformation.

The qualitative differences between viral oncogenes and their cellular proto-oncogenes frequently involve either deletions or substitutions in the viral gene. Deletions of portions of the proto-oncogene may occur during insertion into the viral genome. These deletions appear to produce structural changes in the oncogene protein that alter its function or regulation within the cell. For example, a cellular gene for the epidermal growth-factor receptor (EGFR) appears to have served as the proto-oncogene for the *erb*B viral oncogene (Figure 8.13). EGFR spans the cytoplasmic membrane, binding epidermal growth factor on the outer surface of the membrane and transducing that signal into a mitotic stimulus by means of kinase activity on the inner surface of the membrane. The *erb*B protein has these same activities but is missing the extracellular binding site on the N-terminus of the EGFR molecule, as well as a segment at the C-terminus that appears to regulate the activity of the

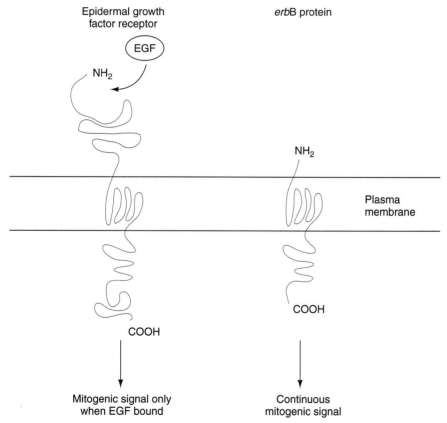

FIGURE 8.13 Comparison of the epidermal growth factor (EGF) receptor and *erb*B oncogene protein. The EGF receptor gives a mitogenic signal only when the growth factor is bound to the N-terminal binding site. Because the *erb*B protein is missing the EGF binding site, as well as regulatory sequences at the C-terminus, it gives a continuous mitogenic signal.

protein. These deletions mean that the *erb*B protein does not require the binding of an external effector molecule to give the mitogenic signal, so that the cell constantly receives a signal to proliferate.

Single amino acid changes in the sequence of a viral gene product compared to its cellular progenitor may also occur. The first example of this type of change to be described was the substitution of a valine for a glycine at position 12 of the *ras* oncogene. Later, other strains of murine sarcoma virus were found to have point mutations altering other amino acids. The point mutations that caused the proto-oncogene to acquire transforming ability all involved amino acids that played a role in the interaction of the *ras* protein with GTP. The mutations either decrease the GTPase activity of the *ras* protein or increase the rate that bound GDP is exchanged for free GTP. Both of these changes cause the *ras* protein to function abnormally as a plasma membrane signal-transducing protein.

WHAT MECHANISMS BESIDES VIRAL ONCOGENES DO RETROVIRUSES USE TO TRANSFORM CELLS?

The transducing retroviruses are able to transform cultured cells or to produce tumors in animals within a relatively short period after infection. Other retroviruses, like mouse mammary tumor virus (MMTV), the avian leukosis viruses (ALV), and a variety of leukemia viruses, do not cause the changes characteristic of transformation in cultured cells and produce tumors in animals only months or years after infection. These slow-acting retroviruses have two distinctive characteristics: They are fully replication competent, and *they have no viral oncogenes.* How can they cause the production of tumors in that case?

These "non-transducing" retroviruses appear to cause transformation by activating cellular proto-oncogenes in the location where the proviral DNA is inserted into the host cell's chromosome. MMTV proviral sequences, for example, can be inserted next to one of four cellular genes called *int*-1, *int*-2, *int*-3, and *int*-4/*int* 5. As illustrated in Figure 8.14, after integration of MMTV at those locations, expression of the *int* genes becomes governed by enhancer sequences occurring in the long terminal repeats (LTR) that flank the inserted proviral genome (see Figure 4.31). An especially interesting feature of the MMTV-LTRs is that they contain hormone receptor-binding sequences, so that expression of the *int* gene becomes hormone-dependent after proviral insertion. For this reason the development of tumors in MMTV-infected mice is stimulated by the presence of mammotropic hormones.

Despite the fact that MMTV activates them both (and their similar names), the *int*-1 and *int*-2 proto-oncogenes have no structural relationship to one another and are located on different chromosomes. *Int*-1 appears to have some homology to the developmental gene *wingless* in *Drosophila*, so it may also have a developmental role in vertebrate cells. *Int*-2 is structurally related to the fibroblast growth factor genes. *Int*-3 appears to encode a protein of the *notch* family involved in development, while *int*-4 encodes a protein involved in estrogen synthesis. The mechanism of transformation by these genes has not yet been established.

The avian leukosis viruses provide another example of insertional activation. These viruses can integrate near the *c-myc* proto-oncogene (*v-myc* is the viral oncogene of the transducing avian myelocytoma viruses). Depending on whether the integration is upstream or downstream for the *c-myc* gene, the ALV-LTR sequences act as promoters or enhancers to increase transcription of the proto-oncogene.

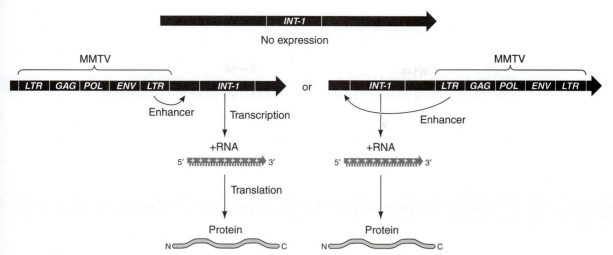

FIGURE 8.14 Insertional activation of a cellular proto-oncogene. When an MMTV provirus integrates near the proto-oncogene *int*-1, enhancer sequences in the long terminal repeats flanking the provirus cause expression of the *int*-1 gene.

An unusual variation on insertional regulation of cellular oncogenes is the situation seen in transformation by Friend murine leukemia virus. When this virus becomes inserted into a particular site in the host chromosome, it *inactivates* a gene encoding p53, the tumor-suppressor protein. Although the details of the process are complicated (involving both replication-competent and replication-defective viruses), it is clear that transformation is the result of two steps: The first is the creation of a premalignant state, and the second is the removal or inactivation of a cellular tumor-suppressor protein that prevents expression of that state. Inactivation of the p53 tumor-suppressor protein also plays a major role in transformation by DNA viruses, as we shall discuss shortly.

The previous examples are of retroviruses that act in the *cis* configuration; that is, on a DNA sequence linked directly to the proviral sequence. There is a suggestion that *trans*-activation leading to transformation may also occur. *Trans*activation involves the production of peptides that influence the activities of other genes that may or may not be on the same DNA molecule as that encoding the *trans*activator. Human T-cell lymphotropic virus (HTLV-1) may play a role in human adult T-cell leukemia since HTLV-1 provirus is associated with all cases of that syndrome. The sequence of the HTLV-1 genome indicates that it is not a transducing virus since there is no oncogene present, nor has HTLV-1 been found to integrate in the same genomic location in each case. These observations rule out transformation by viral oncogenes, as well as *cis* activation of cellular oncogenes. As we saw in Chapter 4, however, oncoviruses like HTLV-1 have several nonstructural genes in addition to the three large structural genes common to all retroviruses. The proteins encoded by these regulatory genes play roles in the efficient replication of the virus. It is possible, but not yet demonstrated, that these proteins may also be *trans*activators for other cellular genes leading to transformation. When one of these genes called *tax* is introduced into transgenic mice under the control of the viral long-terminal repeat sequence, the tax protein is found in muscle cells and the animals develop sarcomas. This suggests that the tax protein may have transforming activity.

HOW IS TRANSFORMATION OF ANIMAL CELLS BY DNA VIRUSES DIFFERENT FROM THAT CAUSED BY RNA VIRUSES?

The mechanism of transformation by DNA viruses differs from that of the RNA viruses in several fundamental ways. The first difference to note is that among all the groups of RNA viruses, only the retroviruses are capable of transforming their host cells. In contrast, all the families of double-stranded DNA viruses appear to have species that either can transform cultured cells under laboratory conditions or are associated with naturally occurring tumors in whole organisms.

Another important difference between RNA and DNA viruses is their *oncogenic efficiency,* as measured by the number of infected cells that become transformed or are necessary to produce a single tumor. The oncogenic efficiency of the retroviruses is very high; in some cell culture systems, each infected cell becomes transformed. The oncogenic efficiency of the DNA viruses, in contrast, is very low, with transformation occurring only rarely even under the most propitious conditions. Moreover, when a population of cells that appears to have some of the *in vitro* characteristics of transformation is observed over time, most of these cells revert to a normal phenotype in a process termed **abortive transformation.**

The reason for these phenomena appears to lie not so much with the viruses themselves, but with the cells they infect. Retroviruses transform the cells in which they normally produce virions (their permissive hosts), and the transformed cells (with the exception of HTLV-1 and HTLV-2) continue to produce new viruses. On the other hand, small DNA viruses like the papovaviruses and adenoviruses, which can be shown to transform *in vitro,* do so only in cells from species that the virus does not naturally infect (nonpermissive cells). For example, the DNA virus SV40 grows productively in cultures of monkey cells but transforms nonpermissive hamster cells that are unable to support the synthesis of new virions. As a consequence, cells transformed by small DNA viruses do not produce new virions. The situation is less clear-cut with larger DNA viruses like the herpesviruses and poxviruses that may be involved with the production of certain kinds of tumors in their usual host organism.

WHAT IS THE MECHANISM OF TRANSFORMATION BY DNA VIRUSES?

The differences outlined previously are manifestations of underlying differences in the mechanisms by which transformation is produced by RNA and DNA viruses. The normal reproductive cycle of the retroviruses requires that they incorporate a DNA copy of their entire genomes into their host cells. Transformation is the result of the action either of activated cellular proto-oncogenes or their transduced viral counterparts. Transformation by DNA viruses, in contrast, results from continued expression of a limited number of normal viral early genes that encode proteins usually involved with replication of the viral genome in a productive infection. Since DNA synthesis in eukaryotic cells occurs only during S phase of the cell cycle, an important role of early gene products is to force infected cells from G_0/G_1 into S phase. In nonpermissive cells, some essential host cell function is missing so that late viral genes are never expressed and the virus reproductive cycle is aborted. Instead, expression of these same viral early genes causes the cell to receive a continuous signal for DNA synthesis and cell division.

If DNA viruses do not possess special oncogenes like those of the retroviruses, how do they transform their host cells? Studies with a variety of DNA viruses suggest that these viruses directly target one or more of the normal cellular proteins that function as *anti-oncogenes* or tumor suppressors. One of these proteins, p53, was mentioned previously in its role in transformation by the Friend murine leukemia retrovirus. A second is the Rb protein (pRb), an anti-oncogene whose expression blocks the formation of a type of cancer in the eye called retinoblastoma. As we saw earlier, both p53 and pRb may have roles in regulating the cell cycle, particularly G_1 arrest. They also appear to function as repressors or transactivators of transcription of genes that are essential for cell proliferation. The loss or impairment of those functions results in escape from normal cell-cycle controls.

If the virus does not become associated with the host cell's genome, the viral genome may be degraded or simply diluted out over several cell generations, resulting in abortive transformation. In order for stable transformation to occur, at least the portion of the viral genome containing the normal viral genes whose expression creates the transformed phenotype usually must become physically incorporated into the host chromosomes. Such insertion is mediated solely by the cellular enzymes responsible for genetic recombination, so viral integration is both rare and random in the site of occurrence. In some cases, however, the viral genome is maintained as an independently replicating episomal DNA. Epstein-Barr virus (a herpesvirus) and some papillomaviruses are found in this form in transformed cells.

The Papovaviruses

SV40 and polyoma are small papovaviruses that transform nonpermissive cells in tissue cultures. The lytic cycles for these viruses were discussed in Chapter 5. These viruses express two sets of genes: the early genes encoding proteins called collectively the **T antigens,** and the late structural protein genes. As Figure 5.3 depicts, the T antigens are generated by alternative splicings of the same transcript, so the N-terminal amino acids of each set are the same, but the C-terminal ones are different. Polyomavirus has three T antigens called, not too imaginatively, large, middle, and small, while SV40 has only two (large and small). In the lytic cycle the proteins encoded by these genes are required for forcing the host cell into S phase. Since this complex cellular process must be induced by only two or three viral T antigens, it is not surprising that they are multifunctional proteins as described next.

The Polyoma T Antigens   The three T antigens of polyomavirus appear to divide up the various tasks necessary to create a transformed phenotype in a nonpermissive host cell. Genetic studies indicate that the primary role of the large T antigen in transformation is during initiation of the process, and that it is not required for maintenance of the transformed state. Temperature-sensitive mutants of the large T gene cannot initiate transformation at the nonpermissive temperature, but if cells are first transformed at the permissive temperature, they retain that phenotype even when shifted to nonpermissive conditions. The function of the large T protein during productive infections is to initiate viral DNA synthesis by binding to the Rb protein and preventing it from combining with the E2F transcription factor. E2F is then free to activate genes leading to entry into S phase. It is likely that the large T protein operates similarly at the beginning of the transformation process. Concurrent viral and host cell DNA synthesis would facilitate the recombinational events necessary to integrate the viral genome into some random location in the host chromosomal complement.

A second role of the large T antigen is to "immortalize" cultured cells. After being placed in culture, cells normally divide only a limited number of times and then become senescent and eventually die. Studies with temperature-sensitive mutants indicate that the large T protein is able to overcome this tendency toward senescence, so that the cells divide indefinitely.

The middle T antigen of polyomavirus appears to be responsible for induction of numerous phenotypic changes associated with transformation and with maintenance of the transformed state. Figure 8.15 diagrams the action of middle T antigen. The middle T protein becomes associated with the plasma membrane, where after being phosphorylated, it activates two different protein kinases. The first of these is the *c-src* protein, which, as we saw earlier in this chapter, is a tyrosine kinase that phosphorylates a number of cellular proteins involved with control of growth. Interestingly, one of the *c-src* protein's targets is the middle T antigen bound to the *c-src* protein itself. This phosphorylation creates a binding site for the second enzyme, phosphatidylinositol kinase (Pi3K), which adds phosphates to the 3 rather than the usual 4 position in the inositol ring of the phospholipid. The exact fate of the resulting phosphoinositides is not known, but they may interact with growth factor receptors and thus play roles in mitogenesis.

Although it is a membrane protein, middle T antigen also activates a number of cellular genes, including the *c-raf* gene, and induces the phosphorylation of a ribosomal protein. These pleiotropic effects within the cell appear to follow some sort of dose-response curve, since the growth and morphological characteristics of transformation become more pronounced as the amount of middle T antigen in the cell increases.

Small T proteins are found both in the nucleus and in the cytoplasm. The activities of the small T antigen are not as well understood as those of its larger relatives, in part because the expression of the small T antigen appears to be different in different cell types. In the nucleus, the small T antigen protein seems to act cooperatively with the large T antigen in induction of transformation. It may also play a role in maintenance of the transformed state by binding to and inactivating the enzyme Pp2A (protein phosphatase), which removes phosphates from proteins like the MAP kinases. The MAP kinases are therefore held in their active (that is, phosphorylated) state and continuously signal the cell to divide.

The SV40 T Antigens

As we saw in Chapter 4, the large T antigen of SV40 is truly multifunctional. It is a DNA binding protein that interacts with both cellular and viral DNAs and is able to bind to a number of cellular enzymes and regulatory proteins as well. In addition, it has two enzymatic activities (ATPase and helicase). Evidence has been found for six different posttranslational covalent modifications to the large T protein that may be related to its different functionalities.

In transformed cells SV40 large T antigen is found primarily in the nucleus. Since SV40 does not have a middle T antigen, the large T antigen carries out the functions associated with both the large and small T antigens of polyomavirus. Like the polyoma large T antigen, it is responsible for initiation of transformation and immortalization of the trans-

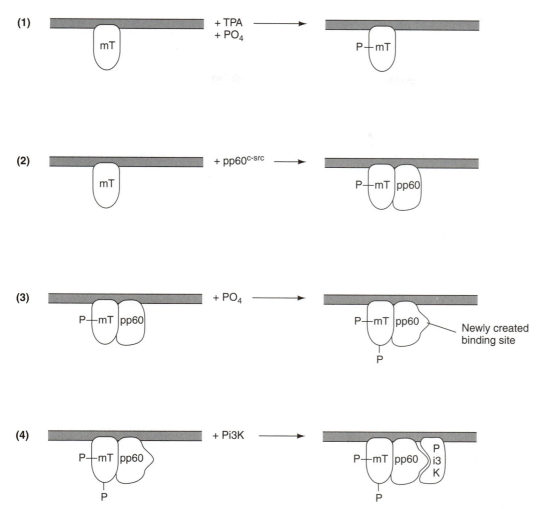

FIGURE 8.15 A scheme for the interaction of the polyomavirus middle T antigen (mT) with *c-src* protein (pp60^{c-src}) and phosphatidylinositol 3-kinase (Pi3K). (1) The initial phosphorylation of mT is mediated by a phorbol ester (TPA). (2) After phosphorylation, mT associates with pp60^{c-src}. (3) pp60^{c-src} then phosphorylates mT again, creating a binding site for phosphatidylinositol kinase (Pi3K). (4) Pi3K is activated by binding to mT-pp60^{c-src}.

formed cells, probably by facilitating integration of the viral genome. It appears to act in concert with the SV40 small T antigen to maintain the transformed state.

Only two of the many functions associated with large T antigen are required for its transforming abilities. Mutations that inactivate the ATPase and helicase enzymatic activities, as well as those functions involved in DNA binding and viral replication, do not affect transformation. The ability of large T antigen to bind to two cellular tumor suppressor proteins, however, is necessary. When the large T antigen protein binds to p53 and pRb, inactivating their growth-regulating activity, the cells become immortalized and transformed.

Table 8.5 | Comparison of Transforming Characteristics of RNA and DNA Viruses

RNA Viruses	DNA Viruses
Only retroviruses can transform.	All families of double-stranded viruses have transforming species.
Oncogenic efficiency is very high.	Oncogenic efficiency is very low.
Permissive host cells are transformed.	Nonpermissive host cells are transformed.
Transformed cells usually produce virions.	Transformed cells usually do not produce virions.
Entire viral genome becomes incorporated into transformed cell.	All or only a portion of the viral genome may become incorporated into host cell.
Oncogenes are normal or modified cellular genes.	Oncogenes are normal viral genes.

The Adenoviruses

Adenoviruses from a number of host species, including humans, will transform nonpermissive cells. Human adenoviruses, for example, can transform rodent and hamster cells in culture or produce tumors in these animals. Although the adenoviruses are much larger than the papovaviruses, the mechanisms by which they transform appear to be very similar. The transforming genes are immediate-early genes in the E1A and E1B transcriptional groups (see Figure 5.4). Like the T antigens, the E1A and E1B transcripts are spliced to yield several different proteins, two large and several small in each case. The activities of the E1A proteins show several similarities to those of the papovavirus T antigens: They confer immortalization on cultured cells, they bind to the *c-ras* and Rb proteins, and they *trans*activate cellular genes involved in regulation of cell growth. The smaller E1B protein is responsible for some of the cell-surface changes that characterize transformed cells, like increased lectin binding, while the larger protein of some viral strains has been shown to bind to p53.

The differences between the ways in which RNA and DNA viruses transform are summarized in Table 8.5.

WHAT HUMAN TUMORS MAY BE CAUSED BY DNA VIRUSES?

Two groups of DNA viruses have been clearly implicated in production of particular human cancers, while epidemiological evidence hints at viral roles in other tumors. The best candidates for human cancer viruses are the *papillomaviruses,* which cause benign keratinized skin tumors (warts) and are strongly implicated in cervical and other carcinomas, and *Epstein-Barr virus,* a herpesvirus associated with Burkitt's lymphoma and nasopharyngeal carcinomas. Human herpesvirus 8 is thought to have a role in Kaposi's sarcoma. Hepatitis B virus and hepatitis C virus may also have roles in production of hepatomas, although the evidence is more circumstantial.

The Papillomaviruses

The papillomaviruses, members of the papovavirus group, take their name from their ability to produce warts or papillomas in their natural hosts. These viruses have been little stud-

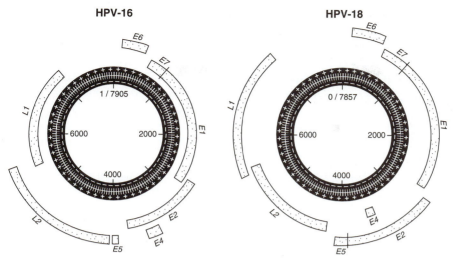

FIGURE 8.16 Genomic maps for human papillomaviruses 16 and 18 based on DNA sequences. E = ORFs for possible early genes; L = ORFs for possible late genes.

ied in culture since they are highly species-specific and require squamous epithelial cells, a differentiated tissue, as host cells. Sequence data suggest that their 8000 base-pair genomes contain about 10 open reading frames **(ORFs),** all of which are on the same strand of the double helix. Gene expression involves formation of a number of mRNAs by a combination of multiple promoters and alternative splicings. Deduced maps for two human viruses are depicted in Figure 8.16.

For over 100 years, epidemiological evidence has been accumulated that suggests that cervical carcinoma is a form of venereal disease. For example, women who have had few, if any, sexual partners develop cervical carcinomas at a far lower rate than female prostitutes who are known to have had numerous partners. These observations were first correlated with the absence or presence of antibodies against human herpesvirus type 2, leading to the proposal that HSV-2 was responsible for causing cervical carcinoma. Analysis of viral sequences in cervical cancer cells, however, indicates that papillomaviruses are more likely to be the transforming agent.

More than 85 strains of human papillomaviruses (HPV) have been described, but only a few of these are strongly associated with human cancer. HPV16, HPV18, and HPV33 are linked with cervical squamous cell carcinomas with high malignant progression, while HPV6 and HPV11 are linked to tumors with lower risk of malignant progression. The genomes of these viruses are found integrated into the DNA of tumor cells in such a fashion that the E2 ORF is disrupted; evidently this ORF contains the viral *att* sites where the circular genome is cut during integration. Disruption of the E2 ORF obviously prevents expression of its protein, which is thought to regulate the expression of two other ORFs, E6 and E7. The products of those two ORFs have been linked to the natural transformation of human cells. The E7 region encodes a multifunctional protein with sequence homologies to regions of the adenovirus E1A and SV40 large T proteins. These regions are significant because they are the locations where the *c-ras* protein or the Rb protein is bound. Additional E7 protein homologies to the adenovirus E1A protein indicate that HPV16 may be able to regulate host cell transcription. The E7 protein alone is not sufficient for transformation of human cells, however. The E6 protein also plays a role in the process. Strains of HPV that are associated with high cancer progression risks appear to encode E6 proteins that very rapidly degrade p53 *in vitro* and in infected cells. These similarities strongly suggest that the human papillomaviruses transform by the same mechanisms employed by other DNA viruses: inactivation of tumor suppressor proteins.

The Herpesviruses

Another herpesvirus is clearly implicated in human cancers, however. Epstein-Barr virus (EBV) was isolated from cultures of lymphoid cells taken from a patient with Burkitt's lymphoma. EBV has now been shown to cause a variety of lymphoproliferative diseases, ranging from infectious mononucleosis to lymphomas, that involve immortalization of the infected lymphocyte host cells. The exact role played by the virus in Burkitt's lymphoma is still uncertain since multiple transformation steps appear to be required. For example, a characteristic of the tumor cells of Burkitt's lymphoma is a translocation of distal portion of the q arm of chromosome 8 bearing the *c-myc* gene to locations adjacent to the immunoglobulin heavy chain gene on chromosome 14 or the light chain genes on chromosomes 22 or 2 (Figure 8.17). In these locations the *c-myc* gene becomes deregulated, a condition that may play a role in production of a completely transformed phenotype. The relationship between EBV infection and this translocation event is unclear—does immortalization by EBV precede or follow the translocation? The picture is further complicated by genetic and immunological factors that are poorly understood.

In Burkitt's lymphoma, the EBV genome may not be present in the tumor cells. In another type of cancer caused by EBV, nasopharyngeal carcinoma, however, copies of the viral genome are integrated into the transformed cells. Unfortunately, cause and effect have not been demonstrated, so it is possible that EBV simply preferentially integrates into these tumor cells.

Kaposi's sarcoma-associated herpesvirus (KSAH, but also called human herpesvirus 8) is strongly implicated as the cause of Kaposi's sarcoma and several other AIDS-associated B-lymphocyte proliferative diseases. Although the exact mechanism of transformation is not yet known, KSAH may play several roles in producing Kaposi's sarcoma. Several viral proteins are angiogenic and may stimulate the formation of the new blood vessels that infiltrate the sarcoma to give it its characteristic coloration. More recently, the protein encoded by ORF K8 has been shown to be a protein of the *bZIP* (*b*asic leucine*zip*per) family of transcription factors. This protein interacts with p53 and suppresses the transcriptional activity of that tumor suppressor that would lead to apoptosis. KSAH also encodes several homologues of cell cycle regulatory proteins, such as cyclin D, so the virus may alter cell cycle regulation as part of its transforming activities.

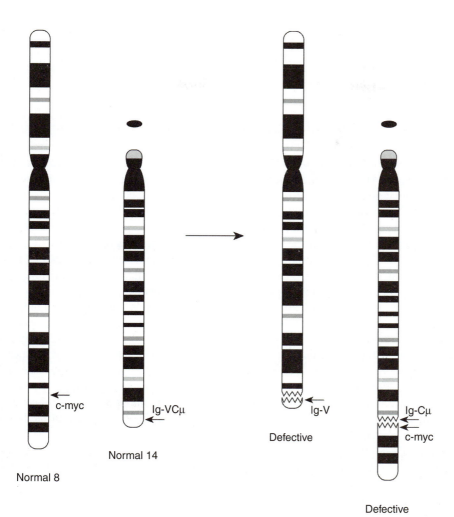

c-myc

Ig-VCμ

Normal 14

Normal 8

Ig-V

Defective

Ig-Cμ

c-myc

Defective

FIGURE 8.17 A reciprocal translocation found in Burkett's lymphoma tumor cells occurs between the distal portions of the q arms of chromosomes 8 and 14. The translocation places the *c-myc* proto-oncogene under the transcriptional regulation of the immunoglobulin Cμ gene.

BESIDES INTEGRATION INTO THE HOST CELL GENOME, HOW MAY A VIRUS INFECTION PERSIST?

Our discussions of the viruses have concentrated primarily on the interactions between a virus and its own host cell as the virus attempts to reproduce itself. In most cases such reproduction results in the lysis of the infected cells as they release newly synthesized virions. If a number of cells in a multicellular organism are infected and then destroyed, a disease results. Although these diseases may range from the merely inconvenient to the life-threatening or fatal, they have tended to focus the attention of scientists on the acute aspects of viral infections. It is becoming evident, however, that viruses and organisms interact in more complex ways than focusing on a disease state might suggest, and that many viruses establish more or less permanent associations with their host cells and organisms. In other words, many viruses establish **persistent infections.**

Persistent infections have been divided into two types, but it should be kept in mind that these categories reflect our convenience rather than clearly distinct entities in the real world. **Chronic infections** are those in which the infected cells continue to produce virions over a period of time. Hepatitis B virus, for example, may establish a chronic infection of human hepatocytes. In **latent infections** the viral genome can be detected in infected cells, but infectious virions are not produced except under certain conditions. The lysogenic state produced by temperate bacteriophages is a prototypic example of a latent infection. Among the animal viruses, the herpesviruses establish latent infections that are characterized by long periods without expression of the virus alternating with episodes of viral production. The course that a particular infection takes depends on the interaction of a number of factors, including the nature of the virus, the type of cell that it has infected, the metabolic state of that cell, and the overall immunological status of the organism.

The first requirement for establishing a persistent infection is that a virus must not destroy its host cells. Enveloped viruses like the arenaviruses, which do not greatly disrupt their host cells' overall metabolism and are released by budding rather than lysis, easily establish persistent infections. Other viruses are lytic in one type of host cell, but in another type of cell express only a few viral genes and therefore establish the latent form of persistent infection. Herpes simplex virus, for example, normally destroys its epithelial host cell. When it infects neurons, however, only a single region of the genome may be transcribed, and no viral proteins synthesized, so the infection becomes latent. The importance of the host cell in the outcome of an infection is dramatically illustrated when cytomegalovirus (CMV) infects teratoma cells *in vitro*. Teratomas are tumors formed from very early embryonic cells that maintain the ability to differentiate into a full range of cell types. If the teratoma cells are undifferentiated, CMV expresses only a few of its genes and establishes a persistent infection. When the infected teratoma cells are forced to differentiate, however, CMV shifts to full expression of its genome in a productive infection.

Another type of persistent infection, in which such a small number of cells is infected at any one time that few progeny viruses are released, may be created in cell cultures by fully lytic viruses like human adenovirus. This condition, given the poetic name of "smoldering infection," may be due to the presence of compounds like interferon in the medium or to the condition of the cells themselves. It is possible that similar "smoldering infections" occur naturally, and may be responsible in part for such banes of humanity as the ever-present common cold by maintaining a pool of virus in the population.

The second requirement for establishment of a persistent infection is evasion of the host organism's immunological defenses. Specificity is the hallmark of immunological responses. All immunological defenses are directed at particular small subsections (the antigenic determinants) of larger molecules like proteins and glycoproteins (the antigens). These host defenses take two forms. The first defense is the production of antibodies, which are specific either for the virions themselves or for virally infected cells. Antibodies directed against virions (neutralizing antibodies) generally bind to capsid proteins or glycoproteins and interfere with the interactions necessary for viral attachment and penetration. Antibodies directed at viral proteins/glycoproteins associated with the infected cell's surface usually induce complement-mediated cytolysis. The second type of specific host immunological defense is the activation of T lymphocytes, which act against infected cells after recognizing viral proteins on the cell surface. T lymphocytes either destroy infected cells directly or release proteins that stimulate a variety of other host defenses.

Viruses can escape destruction by the immune system through a variety of mechanisms involving both the virus itself and the immune system. For example, when the number of infected cells is low, the organism may mount an immune response that produces some antibodies with low affinities (low avidity) for the viral antigenic determinants. Such antibodies may actually interfere with the effects of other neutralizing antibodies by forming non-neutralizing complexes with virions as they are released. In other cases, the binding of antibodies to viral proteins embedded in the infected cell's plasma membranes may lead to shedding of those portions of the cell surface rather than to complement-mediated cytolysis.

T-lymphocyte-mediated reactions may be blocked by viruses that alter the expression of normal cell surface molecules (the class I histocompatibility antigens) that the lymphocyte requires as part of its antigen recognition system. Human adenoviruses, for example, encode proteins that prevent proper glycosylation of nascent histocompatibility molecules, preventing their normal transport to the cell surface. Poxviruses produce soluble cytokine receptors that bind host cytokines before they can interact with cell-surface receptors.

Since the immune response is specific for a particular antigenic determinant, even a slight change in that determinant's composition may have a dramatic effect on the effectiveness of a response. Viruses may therefore reduce the efficiency of the immune response by altering the structure of the antigenic determinants on their virions or the proteins that are expressed on the surface of infected cells. The architecture of each antigenic determinant is genetically determined, so these viruses have a high frequency of spontaneous mutation at various sites in the genes coding for structural proteins. Such antigenic variation occurs in the lentiviruses like HIV, for example, to hinder both the activities of neutralizing antibodies and T-lymphocyte-mediated responses.

The ultimate means of evading a host's immune response is to destroy the ability of that host to mount any immune response. This, of course, is the route taken by viruses like HIV, which infect the helper class of T lymphocytes required for both the production of antibodies by B lymphocytes and the antiviral activities of cytotoxic T lymphocytes. HIV evokes a strong immune response during the initial acute phase of its infection, but this response does not completely clear the virus from the individual, probably due in part to viral antigenic variation. The person then enters a long period called "clinical latency" in which a large number of cells harbor proviruses but only a low level of virus is present in the circulation. The HIV persistent infection is very different from that produced during herpesvirus latency as described earlier. Clinical latency in HIV infection is steady-state in which there is a balance between ongoing viral replication that produces 1 billion to 10 billion HIV virions per day and the rapid removal of those viruses from the circulation. Similarly, millions of T lymphocytes are infected and die but are replaced by new cells in the circulation. Eventually, however, the number of CD4$^+$ T lymphocytes in the body decreases to a point at which the balance is tipped in favor of the virus and another viremia ensues as the person develops clinical AIDS.

WHAT EFFECT DOES A PERSISTENT INFECTION HAVE ON THE HOST ORGANISM?

An interesting feature of many persistent infections is that the viruses manifest themselves very differently in those circumstances than they did in the original acute infection of the host. Table 8.6 compares the acute and persistent infections established by some human viruses. It is obvious from this table that some viruses become persistent in different tissues from those that were initially infected. The acute infections of the three herpesviruses indicated, as well as measles and rubella virus, involve the skin, while the persistent infections all involve infection of the nervous system.

The herpesviruses become latent in cells of the nervous system, so there is no manifestation of viral infection until some change in the physiological and/or immune status of the host causes a reactivation of the virus. The symptoms associated with that reactivation can be quite different from those of the original acute infection. Varicella-zoster, for example, causes the characteristic generalized skin eruptions of chicken pox in its first infection, but produces shingles with its agonizingly painful lesions localized to the surface processes of infected sensory neurons during its reactivations.

Measles virus appears to induce latent infections that are not subject to reactivation. As a result, the symptoms associated with a persistent measles infection, subacute scle-

Table 8.6 | Comparison of Acute and Persistent Infections by Some Human Viruses

Virus	Tissues and Symptoms of Acute Infection	Tissues and Symptoms of Persistent Infection
Herpes simplex type 1	Oral skin and mucous membranes; gingivostomatitis	Sensory neurons of oral skin and mucous membranes; "cold sores"
Herpes simplex type 2	Genital skin and mucous membranes; vesicular lesions	Sensory neurons of genital skin and mucous membranes; vesicular lesions
Varicella-zoster (herpesvirus)	Capillary endothelial cells in skin; chicken pox	Sensory ganglia; shingles (zoster)
Cytomegalovirus (herpesvirus)	Leukocytes, neurons, salivary glands; mononucleosis	Unknown
Epstein-Barr (herpesvirus)	B lymphocytes; infectious mononucleosis	B lymphocytes; transformation
Hepatitis B	Hepatocytes; acute hepatitis	Hepatocytes; chronic hepatitis
Measles	Epithelial cells of respiratory tract $\longrightarrow$ skin; measles	Brain and neurons; subacute sclerosing panencephalitis (SSPE)
Rubella	Epithelial cells of respiratory tract $\longrightarrow$ skin; rubella; birth defects	Brain and neurons
HIV	T lymphocytes; generalized malaise	T lymphocytes; AIDS

rosing panencephalitis, develop very gradually and progressively many years after the initial infection.

HOW DO THE HERPESVIRUSES ESTABLISH THEIR LATENT INFECTIONS?

The most common latent infections of humans are those caused by the human oral and genital herpesviruses. Although the manifestations of latent herpesvirus infections and their reactivations have been discussed in the medical and scientific literature for nearly 90 years, the mechanism by which these viruses establish and maintain their latent infections is still only imperfectly understood. The authors of the review article cited in the Suggested Readings for this chapter, in fact, note that they gave their paper exactly the same title as a review written 20 years previously in order to emphasize the intractability of the problem. A number of factors contribute to the difficulties encountered in studying the phenomenon of latency. Latency clearly involves a complex series of interactions between the virus, its host cells, and the host's immune system. Latency is therefore presumably far more complicated than the establishment and maintenance of bacteriophage lysogeny within a single, independent, host cell.

Tantalizing discoveries continue to be made, however. For example, for many years latent herpesviruses were thought to become completely inactive, but now a set of RNA species, called LAT for "latency associated transcript," have been found in latently infected cells. The large LAT RNA molecule appears to be transcribed in part from the "anti-sense" strand opposite the 3′ end of the earliest expressed herpesvirus gene, the α0 gene (see Chapter 5 for a discussion of the organization and expression of the herpesvirus genome). This 8.3 kilobase RNA is the precursor for two small RNAs. A 2.0 kilobase LAT RNA, which appears to have a lariat structure like that of introns after they have been excised during RNA transcript splicing, is detected in both productive infections, where it is produced as a late gene product, and in latent infections. A smaller LAT RNA (1.5 kilobases) that also has a lariat structure is found only in neurons during latent infections. It has been suggested that latency could result from some form of RNA-RNA regulatory interaction that blocks further gene expression. Support for this idea comes from experiments indicating that the LAT RNA reduces expression of several early genes, especially *ICP0*. LAT RNA also plays a role in efficient reactivation of the herpesvirus from latency. It also appears to promote survival of infected neurons by reducing apoptosis, which in turn would provide a larger pool of cells in which the virus might become latent.

SUMMARY

In addition to the production of new virions, many viruses can enter into more permanent interactions with their host cells that may involve either insertion of the viral genome into the host cell genome (lysogenic bacteriophage, the animal retroviruses) or the establishment of persistent or latent infections (herpesviruses and others). The effects of integration of a viral genome into its host's DNA can be profound. Retroviruses can cause neoplastic conversion (transformation) either by introducing viral oncogenes (derived from cellular proto-oncogenes) or by altering the expression of host cell oncogenes at the site of their insertion. DNA viruses can also transform their host cells, but transformation results when normal viral genes usually involved with replication of the viral genome in a productive infection inactivate the p53 or pRb tumor suppressor proteins.

SUGGESTED READINGS

Three articles discussing the lysogeny of λ virus are:

1. "A genetic switch in a bacterial virus" by M. Ptashne, A.D. Johnson, and C.O. Pabo, *Scientific American* 247; #5: 128–140 (1982).

2. "λ repressor and *cro*—components of an efficient molecular switch" by A.D. Johnson, A.R. Poteete, G. Lauer, R.T. Sauer, G.K. Ackers, and M. Ptashne, *Nature* 294:217–223 (1981).

3. "Viral integration and excision: Structure of the lambda *att* sites" by A. Landy and W. Ross, *Science* 197:1147–1160 (1977).

These articles discuss phage conversion in three different groups of bacteria:

1. "Infectious CTXΦ and the Vibrio pathogenicity island prophage in *Vibrio mimicus:* Evidence for recent horizontal transfer between *V. mimicus* and *V. cholerae*" by E.F. Boyd, K.E. Moyer, L. Shi, and M.K. Waldor, *Infection and Immunity* 68:1507–1513 (2000).

2. "The gene for toxic shock toxin is carried by a family of mobile pathogenicity islands in *Staphylococcus aureus*" by J.A. Lindsay, A. Ruzin, H.F. Ross, N. Kurepina, and R.P. Novick, *Mol Micro* 29:527–543 (1998).

3. "The 102-kilobase *pgm* locus of *Yersinia pestis:* Sequence analysis and comparison of selected regions among different *Yersinia pestis* and *Yersinia pseudotuberculosis* strains" by C. Buchrieser, C. Rusniok, L. Frangeul, E. Couve, A. Billault, F. Kunst, E. Carniel, and P. Glaser, *Infection and Immunity* 67:4851–4861 (1999).

Descriptions of several transformation markers are found in:

1. "Observations on the social behavior of cells in tissue culture II. 'Monolayering' of Fibroblasts" by M. Abercrombie and J.E.M. Heaysman, *Exptl Cell Res* 6:293–306 (1954).

2. "'Contact inhibition' of cell division in 3T3 cells" by R.W. Holley and J.A. Kiernan, *Proc Natl Acad Sci USA* 60:300–394 (1968).

3. "Topoinhibition and serum requirements of transformed and untransformed cells" by R. Dulbecco, *Nature* 227:802–806 (1970).

4. "An enzymatic function associated with transformation of fibroblasts by oncogenic viruses, I. Chick embryo fibroblast cultures transformed by avian RNA tumor viruses" by J.C. Unkeless, A. Tobia, L. Ossowski, J.P. Quigley, D.B. Rifkin, and E. Reich, *J Exp Med* 137:85–111 (1973).

The discovery of retroviral oncogenes is discussed in the following papers:

1. "Differences between the ribonucleic acids of transforming and nontransforming avian tumor viruses" by P.H. Duesberg and P.K. Vogt, *Proc Natl Acad Sci USA* 67:1673–1680 (1970).

2. "Rous sarcoma virus: a function required for the maintenance of the transformed state" by S. Martin, *Nature* 227:1021–1023 (1970).

3. "The size and genetic composition of virus-specific RNAs in the cytoplasm of cells producing Avian sarcoma-leukosis viruses" by S.R. Weiss, H.E. Varmus, and J.M. Bishop, *Cell* 12:983–992 (1977).

4. "Nucleotide sequence of Rous sarcoma virus" by D.E. Schwartz, R. Tizard, and W. Gilbert, *Cell* 32:853–869 (1983).

These papers discuss the discovery of cellular proto-oncogenes:

1. "DNA related to the transforming gene(s) of Avian sarcoma viruses is present in normal avian DNA" by D. Stehelin, H.E. Varmus, and J.M. Bishop, *Nature* 260:170–173 (1976).

2. "Nucleotide sequences related to the transforming gene of avian sarcoma virus are present in DNA of uninfected vertebrates" by D.H. Spector, H.E. Varmus, and J.M. Bishop, *Proc Natl Acad Sci USA* 75:4102–4106 (1978).

3. "Characterization of a normal avian cell protein related to the avian sarcoma virus transforming gene product" by M.S. Collett, J.S. Brugge, and R.L. Erickson, *Cell* 15:1363–1369 (1978).

4. "Comparison between the viral transforming gene (src) of recovered avian sarcoma virus and its cellular homolog" by T. Takeya and H. Hanafusa, *Mol Cell Biol* 1:1024–1037 (1981).

The formation of transducing retroviruses is presented in the following papers:

1. "Generation of novel, biologically active Harvey sarcoma viruses via apparent illegitimate recombination" by M.P. Goldfarb and R.A. Weinberg, *J Virol* 38:136–150 (1981).

2. "Form and function of retroviral proviruses" by H.E. Varmus, *Science* 216:812–820 (1982).

3. "Transduction of a cellular oncogene: the genesis of Rous sarcoma virus" by R. Swanstrom, R.C. Parker, H.E. Varmus, and J.M. Bishop, *Proc Natl Acad Sci USA* 80:2519–2523 (1983).

4. "High-frequency transduction of c-erbB in avian leukosis virus-induced erythroblastosis" by B.D. Miles, and H.L. Robinson, *J Virol* 54:295–303 (1985).

The relationship between viral oncogenes and their cellular proto-oncogenes is discussed in the following papers:

1. "Rous sarcoma virus variants that carry the cellular *src* gene instead of the viral *src* gene cannot transform chicken embryo fibroblasts" by H. Iba, T. Takeya, F.R. Cross, T. Hanafusa, and H. Hanafusa, *Proc Natl Acad Sci USA* 81:4424–4428 (1984).

2. "Amino acid substitutions sufficient to convert the nontransforming p60^{c-src} protein to a transforming protein" by J. Kato, T. Takeya, C. Grandori, H. Iba, J.B. Levy, and H. Hanafusa, *Mol Cell Biol* 6:4155–4160 (1986).

3. "Activation of the transforming potential of P60^{c-src} by a single amino acid change" by J.B. Levy, H. Iba, H. Hanafusa, *Proc Natl Acad Sci USA* 83:4228–4232 (1986).

4. "Cell transformation by pp60^{c-src} mutated in the carboxy-terminal regulatory domain" by C.A. Cartwright, W. Eckhart, S. Simon, and P.L. Kaplan, *Cell* 49:83–91 (1987).

Insertional activation of cellular proto-oncogenes is discussed in the following papers:

1. "Dexamethasone-mediated induction of mouse mammary tumor virus RNA: A system for studying glucocorticoid action" by G.M. Ringold, K.J. Yamamoto, G.M. Tomkins, J.M. Bishop, and H. Varmus, *Cell* 6:299–305 (1975).

2. "Many tumors induced by the mouse mammary tumor virus contain a provirus integrated in the same region of the host genome" by R. Nusse and H.E. Varmus, *Cell* 31:99–109 (1982).

3. "Tumorigenesis by mouse mammary tumor virus: Evidence for a common region for provirus integration in mammary tumors" by G. Peters, S. Brookes, R. Smith, and C. Dickson, *Cell* 33:369–377 (1983).

4. "A retrovirus vector expressing the putative mammary oncogene *int-1* causes partial transformation of a mammary epithelial cell line" by A.M.C. Brown, R.S. Wildin, T.J. Prendergast, and H.E. Varmus, *Cell* 46:1001–1009 (1986).

The mechanisms of action of several oncogenes is discussed in the following papers:

1. Transforming gene product of Rous sarcoma virus phosphorylates tyrosine" by T. Hunter and B.M. Sefton, *Proc Natl Acad Sci USA* 77:1311–1315 (1980).

2. "Expression of a new tyrosine protein kinase is stimulated by retrovirus promoter insertion" by A.F. Voronova and B.M. Sefton, *Nature* 319:682–685 (1986).

3. "Activation of a cellular *onc* gene by promoter insertion in ALV-induced lymphoid leukosis" by W.S. Hayward, B.G. Neel, and S.M. Astrin, *Nature* 290:475–480 (1981).

4. "Translocation of the *c-myc* gene into the immunoglobulin heavy chain locus in human Burkitt lymphoma and murine plasmacytoma cells" by R. Taub, I. Kirsch, C. Morton, G. Lenoir, D. Swan, S. Tronick, S. Aaronson, and P. Leder, *Proc Natl Acad Sci USA* 79:7837–7841 (1982).

5. "Specific chromosomal translocations and the genesis of B-cell-derived tumors in mice and men" by G. Klein, *Cell* 32:311–315 (1983).

6. "Close similarity of epidermal growth factor receptor and v-erb-B oncogene protein sequences" by J. Downward, Y. Yarden, E. Mayes, G. Scarce, N. Totty, P. Stockwell, A. Ullrich, J. Schlessinger, and M.D. Waterfield, *Nature* 307:521–527 (1984).

7. "Different structural alterations upregulate in vitro tyrosine kinase activity and transforming potency of the *erbB-2* gene" by O. Segatto, C.R. King, J.H. Pierce, P.P. Di Fiore, and S.A. Aaronson, *Mol Cell Biol* 8:5570–5574 (1988).

8. "Disease tropism of *c-erbB:* Effects of carboxyl-terminal tyrosine and internal mutations on tissue-specific transformation" by R.J. Pelley, N.J. Maihle, C. Boerkoel, H. Shu, T.H. Carter, C. Moscovici, and H. Kung, *Proc Natl Acad Sci USA* 86:7164–7168 (1989).

Transformation by polyoma and SV40 viruses is discussed in the following papers:

1. "The roles of individual polyoma virus early proteins in oncogenic transformation" by M. Rassoulzadegan, A. Cowie, A. Carr, N. Glaichenhaus, R. Kamen, and F. Cuzin, *Nature* 300:713–718 (1982).

2. "Polyoma virus transforming protein associates with the product of the *c-src* cellular gene" by S.A. Courtneidge and A.E. Smith, *Nature* 303:435–439 (1983).

3. "Enhancement of cellular *src* gene product associated tyrosyl kinase activity following polyoma virus infection and transformation" by J.B. Bolen, C.J. Thiele, M.A. Israel, W. Yonemoto, L.A. Lipsich, and J.S. Brugge, *Cell* 38:767–777 (1984).

4. "SV40 large tumor antigen forms a specific complex with the product of the retinoblastoma susceptiblity gene" by J.A. DeCaprio, J.W. Ludlow, J. Figge, J. Shew, C. Huang, W. Lee, E. Marsillo, E. Paucha, and D.M. Livingston, *Cell* 54:275–283 (1988).

Transformation by adenoviruses is discussed in the following papers:

1. "Size and location of the transforming region in human adenovirus type 5 DNA" by F.L. Graham and A.J. Vander Eb, *Nature* 251:687–691 (1974).

2. "Adenovirus E1b-58kd tumor antigen and SV40 large tumor antigen are physically associated with the same 54 kd cellular protein in transformed cells" by P. Sarnow, Y.S. Ho, J. Williams, and A.J. Levine, *Cell* 28:387–394 (1982).

3. "Association between an oncogene and an anti-oncogene: The adenovirus E1A proteins bind to the retinoblastoma gene product" by P. Whyte, K.J. Buchkovich, J.M. Horowitz, S.H. Friend, M. Raybuck, R.A. Weinberg, and E. Harlow, *Nature* 334:124–129 (1988).

Transformation by human papilloma viruses is discussed in the following papers:

1. "The human papillomavirus type 16 E7 gene encodes transactivation and transformation functions similar to those of adenovirus E1A" by W.C. Phelps, C.L. Yee, K. Munger, and P.M. Howley, *Cell* 53:539–547 (1988).

2. "The human papilloma virus-16 E7 oncoprotein is able to bind to the retinoblastoma gene product" by N. Dyson, P.M. Howley, K. Muner, and E. Harlow, *Science* 243:934–937 (1989).

3. "The E6 oncoprotein encoded by human papillomavirus types 16 and 18 promotes the degradation of p53" by M. Scheffner, B.A. Werness, J.M. Huibregetse, A.J. Levine, and P.M. Howley, *Cell* 63:1129–1136 (1990).

Virus and host cell interactions in persistent and latent infections are discussed in the following papers:

1. "Cytomegalovirus causes a latent infection in undifferentiated cells and is activated by induction of cell differentiation" by F.J. Dutko and M.C.A. Oldstone, *J Exp Med* 154:1636–1651 (1981).

2. "Cytomegalovirus replicates in differentiated but not in undifferentiated human embryonal carcinoma cells" by E. Gonczol, P.W. Andrews, S.A. Plotkin, *Science* 224:159–161 (1984).

3. "An inquiry into the mechanisms of herpes simplex virus latency" by B. Roizman and A.E. Sears, *Ann Rev Microbiol* 41:543–571 (1987).

4. "The latency-associated transcripts (LAT) of herpes simplex virus: Still no end in sight" by T.M. Block and J.M. Hill, *J Neurovirol* 3: 313–321 (1997).

5. "Herpes simplex virus type 1 latency-associated transcripts suppress viral replication and reduce immediate-early gene mRNA levels in a neuronal cell line" by N. Mador, D. Goldenberg, O. Cohen, A. Panet, and I. Steiner, *J Virol* 72: 5067–5075 (1998).

6. "Virus-induced neuronal apoptosis blocked by the herpes simplex virus latency-associated transcript" by G.-C. Perng, C. Jones, J. Ciacci-Zanella, M. Stone, G. Henderson, A. Yukht, S.M. Slanina, A.B. Nesburn, and S.L. Wechsler, *Science* 287: 1500–1506 (2000).

Subviral Entities, Viral Evolution, and Viral Emergence

As we have seen in the preceding chapters, viruses come in many sizes and shapes, and with a daunting array of molecular adaptations to permit their replication in and transmission to new host cells. Much of what we have learned about these processes has come from studies of diseased plants or animals since the replication of a virus can disturb the normal functioning of their host cells, resulting in cellular damage and possibly pathology to the entire multicellular plant or animal. In 1967 the infectious agent of a disease that causes potatoes to become too thin and spindly for sale was found to have some virus-like features, but it was emphatically not a virus. The term **viroid** was coined for this agent. Viroids and other subviral entities are the subject of the first part of this chapter. We shall then conclude our exploration of the viruses by examining theories on how viruses and the various subviral entities may have come into being and then how they may have evolved into their present myriad forms.

WHAT STRUCTURAL FEATURES DISTINGUISH VIROIDS FROM VIRUSES?

The first indication that potato spindle tuber viroid (PSTV) was not a conventional virus was that the infectious agent consisted solely of *naked RNA* at all stages of the infection. Virions characteristic of other RNA viruses were not found either intracelluarly or extracellularly in these infected plants. By 1971, further study of PSTV added other characteristics that distinguished this (and all other) viroids from the viruses. The most striking was the discovery that PSTV's RNA molecule was *extraordinarily small,* only about 1.2×10^5 daltons in size. This is fully an order of magnitude smaller than the smallest known infectious RNA viral genome. Nevertheless, this tiny molecule, only 359 bases long, is capable of *autonomous replication* and does not require the aid of any helper virus to provide some essential function.

In the 30 years since these first studies were done, other significant structural differences between viroids and viruses have been described. These are summarized in Table 9.1. Viroids are indeed small, ranging in size from 246 to 375 nucleotides long, while the RNA viruses are 10 to 100 times larger. The tiny size of the RNA molecules found in infected cells suggested that perhaps viroids divided their genetic information up into several different molecules, but this is not the case. Each type of viroid appears to have only a single molecule of RNA. A fourth difference is biochemical. Viruses frequently have chemically modified ribonucleotides in their genomes (especially methylated bases), whereas viroids contain only the standard four ribonucleotides.

Finally, although all viral RNAs are susceptible to degradation by exonucleases, viroid RNAs are not, indicating that viroid nucleic acid is circular rather than linear in form. This was

Table 9.1 | Comparison of Viroid and RNA Virus Structures

RNA Viruses	Viroids
RNA enclosed in capsid	RNA naked
RNA size 1.2×10^6 to 1.2×10^7 daltons	RNA size 1.1–1.3×10^5 daltons
May have modified nucleotides	Only the standard four nucleotides
Genome may be segmented	Only a single molecule of RNA
Linear RNA molecules that may be double stranded	Circular RNA molecule that is self-complementary
Exonuclease sensitive	Exonuclease resistant

confirmed by visualization of the genome. Denatured viroid RNA molecules appear in electronmicrographs as circles with about a 100 nm circumference. Native viroid molecules appear
to be rods about 50 nm long. This suggests that the genome is self-complementary such that
the sides of the circle opposite each other can form base pairs (see the poxviruses for a DNA
example of this type of arrangement). The two sides of the circle do not match perfectly,
however. Thermal denaturation experiments indicate that short helical runs (average of
5 base pairs) are interrupted by loops of one or two unpaired bases. Figure 9.1 shows the probable structure of PSTV based on its nucleotide sequence.

Complete nucleotide sequences of a number of viroids have confirmed this model, as
well as demonstrated a regular pattern within the viroid group. As shown in Figure 9.1, all
viroids appear to have five regions called *domains*. As its name indicates, the central conserved domain (C) contains about 30 bases that are highly similar in a number of viroids.

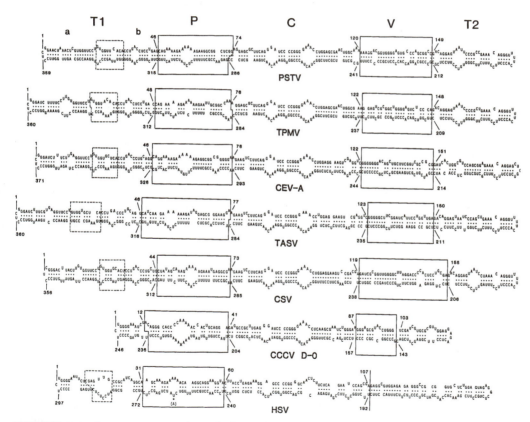

FIGURE 9.1 Sequence comparisons indicate that viroids share a common structural
organization. The five domains are C (conserved central), P (pathogenic), V (variable), and T1
and T2 (left and right termini). The viroids included in this figure are potato spindle tuber viroid
(PSTV), tomato planto macho viroid (TPMV), citrus exocortis viroid (CEV), tomato apical stunt
viroid (TASV), chrysanthemum stunt viroid (CSV), coconut cadang cadang viroid (CCCV), and
hop stunt viroid (HSV).

Flanking the center of the C domain are one or more inverted repeat sequences of 9 to 10 base length. The conservation of these structures throughout the viroid groups suggests that they play an important role in viroid function.

A classification scheme that divides viroids into three subgroups has been proposed based on the C domain sequences since three distinct patterns, which have essentially no overlap, have been identified. These groups are named for their "type" viroids. The potato spindle tuber viroid group and the apple scar skin viroid group show some structural homologies and appear to reproduce in similar fashions. The avocado sunblotch viroid group, if it can be called a "group" at all, contains only that one viroid. Avocado sunblotch viroid is structurally distinct from all other viroids thus far described, and appears to reproduce in a unique fashion as well.

The P and V domains flank the C domain. The P domain, characterized in part by a run of purines (usually adenines), appears to be involved with the pathogenic effects of the viroid. Site-directed mutagenesis experiments in PSTV have demonstrated that base changes in the P domain alter the severity of the disease produced in infected tomato plants. Although the V domain is named for the variable sequences found there, a regular feature of the region is a helix formed with a run of purines on one side and pyrimidines on the other. The left and right terminal domains (T1 and T2, respectively) end in small loops and are characterized by short, conserved sequences; the cytosine at the end of the T1 loop conserved sequence is the base at which numbering of the molecule begins.

WHAT PROTEINS DO VIROIDS ENCODE?

Since viroids average only about one-tenth the number of bases of even the smallest RNA viruses, they obviously do not contain much genetic information. We have seen, however, that the viruses can encode a surprising number of proteins in only short nucleotide sequences by employing a variety of strategies for "re-using" the same bases. For example, overlapping reading frames, different start or stop codons within the same reading frame, or a combination of both, are all options used by different viruses. An average viroid of about 315 bases might theoretically be able to produce a protein about 100 amino acids long if every base were used in a single reading frame, or a longer protein if frame-shifting occurred during translation.

Many lines of evidence indicate, however, that *viroids do not encode any proteins at all*. Cell fractionation schemes indicate that viroids reproduce in the nucleolus, but searches for viroid-specific proteins in the nucleoplasm and cytoplasm of infected cells have yielded nothing. The strongest evidence for the lack of viroid proteins lies in the viroid RNA sequences themselves. Potato spindle tuber viroid, for example, contains no AUG initiation codon in the genomic molecule itself (which is given a "courtesy" designation of "positive-strand RNA" nevertheless) or in its complementary ("negative" strand) RNA. Other viroids have AUGs in either the genome or its complement that might allow the synthesis of short peptides, but the lack of any pattern or consistency between different viroids suggests that these are not, in fact, functional entities. Further evidence that these AUGs are not functional comes from studies that demonstrate that in no instance is any product formed when viroid genomes or their complements are used as templates in *in vitro* translation systems. It is clear from all these studies that viroids must be able to reproduce themselves and cause their characteristic diseases in their plant hosts using only host cell proteins.

WHAT HOST ENZYMES DO VIROIDS USE TO REPLICATE THEMSELVES?

As we have seen, the mechanisms by which RNA viruses replicate their genomes all require virus-encoded proteins. Most single-stranded RNA viruses replicate by creating an intermediate that is complementary to the genome to serve as the template for synthesis of new genomic molecules. The replicative intermediate and new genomes are all synthesized by means of a viral replicase (RNA-dependent RNA polymerase). The retroviruses, in contrast, use viral reverse transcriptase (RNA-dependent DNA polymerase) to create a DNA copy of their genome, insert that copy into a host cell chromosome, and then transcribe new genomes from the inserted cDNA using host cell transcriptase (DNA-dependent RNA polymerase). Since viroids do not appear to encode any proteins at all, how can they replicate themselves?

The first studies seeking to answer this question used various inhibitors of host cell polymerases. These experiments showed that viroid replication was blocked by α amanitin, indicating that viroids may make use of the host cell enzyme RNA polymerase II that is responsible for transcription of host mRNAs. These findings were supported by *in vitro* assays that showed that viroid RNA, perhaps because of its extensive double-stranded helical arrangement, competed effectively with the normal DNA template for that enzyme. More recent work suggests that host cell RNA polymerase I (responsible for host rRNA synthesis) may also be involved in a different step in the replication process.

WHAT IS THE MECHANISM OF VIROID REPLICATION?

Single-stranded viral genomes are replicated by first creating a complementary strand of nucleic acid and then using that template to make numerous new genomic molecules. The size and arrangement of strands in the replicative intermediate provide the clues needed to identify the particular mode of replication being used. Since viroids appear to use cellular enzymes that normally require DNA template molecules, the replication process might somehow involve a reverse transcription step to create a DNA copy of the viroid genome. The only molecules in infected cells complementary to the viroid genome that are detected by nucleic acid hybridization techniques, however, are RNAs, indicating that reverse transcription does not occur.

The size of these complementary RNAs (so-called negative-strand RNA) suggests that the mechanism of replication is not the same as that of the single-stranded RNA viruses. If the negative strands were linear molecules of genome length, then the population of new positive-strand genomes should be of unit length or less. The negative strands found in viroid-infected cells, however, are *longer* than unit length. This suggests that viroids replicate via a *rolling circle* mechanism similar to that seen in the DNA bacteriophage φX174, which also has a single-stranded circular genomic molecule.

Figure 9.2 presents a model for the replication of viroids such as potato spindle tuber viroid. The positive-strand genome serves as template for synthesis of a concatemeric

(1) Rolling circle replication

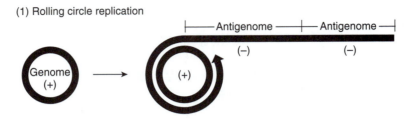

(2) Synthesis of new genomic concatemer

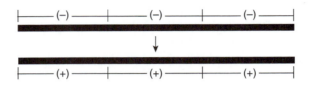

(3) Cleavage of genomic concatemer

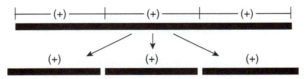

(4) Circularization

FIGURE 9.2 A model for viroid replication. Rolling circle replication of viroids appears to involve formation of a negative-strand concatemer from the positive-strand viroid molecule (step 1). The negative-strand concatemer, a linear molecule, then serves as the template for the synthesis of new genomic molecules (step 2). The mechanism of cleavage of the positive-strand concatemer (step 3) and ligation to form circular genomic molecules (step 4) is unclear.

negative-strand RNA molecule by host cell RNA polymerase II. The negative strand then allows the synthesis of new positive-strand genomes (possibly by RNA polymerase I). These may be synthesized as longer-than-unit-length molecules that are cleaved and circularized while still base-paired to the negative-strand template.

The remaining mystery concerning viroid replication is how the circular genomic molecules are created from a linear template. Cleavage and circularization of positive-strand RNA do not occur *in vitro* unless plant nucleoplasm is also included, so the process would not appear to be autocatalytic like the self-cleavage reactions that remove introns from ribosomal RNA precursor molecules in the protozoan *Tetrahymena*. It may be that cleavage is mediated by one or more specific host cell RNases.

HOW DO VIROIDS CAUSE PLANT DISEASES?

As is the case with most viruses, viroids were first described as disease-causing agents. The diseases associated with viroids include various kinds of stunting, including the potato spindle tuber viroid disease mentioned previously, as well as many forms of mottling and other types of leaf pathologies. The smallest known viroid, the coconut cadang cadang viroid with only 246 bases, is also one of the most virulent, causing the death of hundreds of thousands of coconut palm trees each year in the Philippines. Moreover, different strains of the same viroid produce markedly different diseases in the same type of plant. PSTV variants that differ by single base changes in the sequences of the P domain cause diseases in tomatoes that range from mild to lethal.

How are these diseases produced, given that viroids do not encode any proteins? Obviously the viroid RNA itself must be responsible, presumably through some form of direct interaction with the host cell genome or with some other host cell component. Several candidates for the host factor have been proposed, but none has been positively identified yet.

One interesting candidate is the 7S RNA that, together with a set of plant proteins, is responsible for protein translocation after synthesis. PSTV and other related viroids have runs of bases in their P domains that are complementary to those of 7S RNA, which should permit formation of viroid/7S RNA hybrid molecules. Figure 9.3 presents these possible base pairings. If 7S RNA is bound up in a viroid/7S RNA hybrid, it would not be able to participate effectively in mediating the proper translocation of newly synthesized proteins. This would, in turn, result in formation of abnormal cellular membranes. Viroid infection has been shown to produce alterations in plasma membrane-derived structures (plasmalemmasomes), but how this relates to pathology is uncertain.

Another possible candidate is a 68 kdalton protein that becomes phosphorylated in viroid-infected cells. This protein appears to be a protein kinase similar to that seen in the antiviral defenses of the interferon system in animal cells. Again, the relationship to any pathology is unclear.

WHAT OTHER SUBVIRAL ENTITIES HAVE BEEN FOUND?

Viroids are by no means the only RNA-containing subviral entities that have been described. Other subviral entities may have some characteristics in common with viroids, but are distinct from them in other ways. The entity most like a viroid, although it occurs in animal rather than plant cells, is the *hepatitis delta virus*. The other groups, occurring primarily in plants, are called **satellite viruses** and **satellite RNAs.**

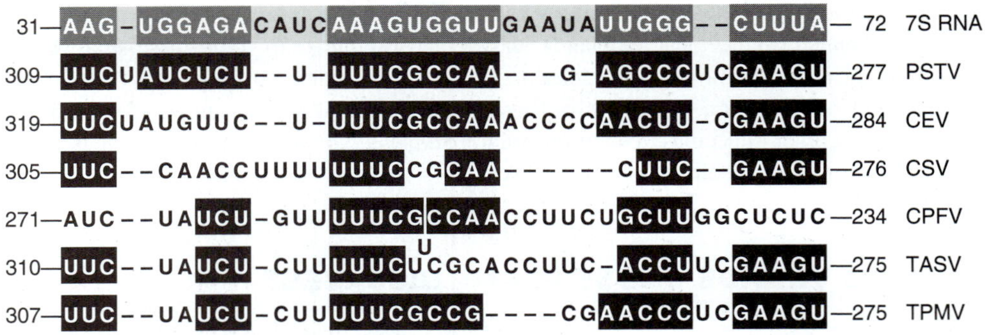

FIGURE 9.3 Possible base pairing between tomato 7S RNA and various viroids. Numbering is from 5′ to 3′. GU base pairs are considered equivalent to GC pairs. The viroids included in this figure are potato spindle tuber viroid (PSTV), citrus exocortis viroid (CEV), chrysanthemum stunt viroid (CSV), cucumber pale fruit viroid (CPFV), tomato apical stunt viroid (TASV), and tomato planto macho viroid (TPMV).

The most important difference between viroids and the other types of subviral RNAs is that only viroids are capable of autonomous replication. Hepatitis delta virus (HDV), satellite viruses, and satellite RNAs all depend on the co-infection of their host cell by a **helper virus** that provides some essential function in their replication and transmission cycles. Although there is a high degree of specificity in each satellite/helper virus relationship, there are usually no significant sequence homologies between their RNAs. This indicates that the helper virus is not the origin of the satellite virus. In the case of the satellite RNAs, the capsids that surround their nucleic acids are composed of helper virus proteins, although they may also contain proteins encoded by their own genomes as well. Satellite viruses and HDV encode their own capsid proteins, but rely on their helper viruses for the RNA-dependent RNA polymerase necessary to replicate their genomes. HDV also relies on its helper for its envelope.

Table 9.2 summarizes some of the important characteristics that distinguish the viroids, HDV, satellite viruses, and satellite RNAs.

HOW DO SATELLITE RNAS AND SATELLITE VIRUSES REPLICATE?

The structure of various satellite RNA molecules and their modes of replication suggest that there are actually two subsets of satellite RNAs. One of these subsets of satellite RNAs has features similar to the viroids, so this group is sometimes referred to as the **virusoids.** Both viroids and virusoids have circular genomes that can form base pairs between the opposite sides of the circle, leading to the formation of rod-shaped structures. They both appear to replicate via a rolling circle mechanism. Although self-cleavage has been proposed to occur in viroids, it has not been demonstrated *in vitro*. *In vitro* self-cleavage has been shown for several virusoids, however. The reaction occurs within a conserved sequence of about 64 bases

Table 9.2 | Comparison of Viroids and Other Infectious RNAs

Characteristic	Viroids	Hepatitis Delta Virus	Satellite RNA	Virusoids	Satellite Viruses
Replication	Autonomous	Helper	Helper	Helper	Helper
Protein encoded	No	Yes	Yes	No	Yes
Capsid source	No capsid	Capsid-self Envelope-helper	Helper	Helper	Self
Circular RNA	Yes	Yes	No	Yes	No
Size (in bases)	250–400	1700	475–1375	325–390	1240
Rolling circle replication	Yes	Yes	No	Yes	No
Host	Plant	Animal	Plant	Plant	Plant
Complementarity to 7S host RNA	Yes	Yes	—	—	—

and produces unusual end products. The cleavage reactions that occur during processing of *Tetrahymena* rRNA or in splicing of hnRNA to form mRNA in eukaryotic cells all create molecules with 5′ phosphates and 3′ hydroxyls. Self-cleavage in virusoids creates the opposite arrangement: a molecule whose 5′ end bears a hydroxyl rather than phosphate group and whose 3′ terminus contains a 2′, 3′ cyclic phosphodiester group rather than a hydroxyl group.

A further feature that relates the virusoid group of satellite RNAs to the viroids is that they do not appear to encode any proteins of their own. The capsid that surrounds them is therefore entirely derived from their helper virus.

Only a few satellite RNAs appear to be related to the viroids, however. Most satellite RNAs have linear rather than circular genomes that are about three to five times larger than viroid genomes. Intrastrand base pairing causes these longer molecules to fold into more complex three-dimensional structures than the rod shape assumed by virion and virusoid RNAs. Moreover, all of these satellite RNAs contain at least one open reading frame encoding a moderate-sized protein. Most evidence indicates that the type of replication of this group is the same as that used by their RNA helper viruses: formation of a unit-length antigenome molecule that then serves as a template for the synthesis of unit-length genomes. The helper virus provides the replicase necessary for these reactions, but it is possible that the satellite RNAs produce a peptide that regulates that activity in some fashion.

Although this discussion has addressed only the characteristics of the RNA satellite viruses, DNA satellite viruses also occur. The replication defective adeno-associated viruses, satellite viruses in the *Parvoviridae* family of single-stranded DNA viruses, were described in Chapter 5.

HOW IS HEPATITIS DELTA VIRUS REPLICATED?

Although hepatitis delta virus (HDV) infects animal rather than plant cells, it appears to be closely related to the viroids. HDV has a circular genome that is about fourfold or fivefold larger than viroid genomes. Nevertheless, HDV and viroid genomes have very similar G + C contents and percentages of bases involved in intrastrand pairs. Sequence analysis has shown that a portion of the HDV genome about 295 bases long has striking homologies with viroid RNA sequences. These bases in the HDV genome are noncoding but probably are essential for replication of the genome.

The remainder of the genome, as well as its complementary antigenomic RNA, contains a number of open reading frames, but only one of these appears to be used. In fact, this open reading frame is used to generate two proteins called *delta antigen S* (195 amino acids) and *delta antigen L* (213 amino acids), that are both required for replication. As Figure 9.4 illustrates, host RNA polymerase II transcribes the open reading frame region of the genome up to a cleavage and polyadenylation signal in the new RNA. This produces a mRNA that is long enough to encode the L protein but contains a UAG termination codon that causes translation to stop after the 195 amino acids of the S protein. In order to form the larger protein, the antigenomic RNA is edited by an RNA adenosine deaminase that results in the conversion of that UAG codon into a UGG codon for tryptophan in the resulting mRNA.

Another fascinating similarity between HDV and viroids is the presence of sequences in the antigenome that are complementary to mammalian 7S RNA. As with the viroids, this suggests that the mechanism of pathogenicity by HDV is disruption of normal protein translocation due to the formation of HDV/7S RNA hybrids. The histological changes seen in HDV-infected hepatocytes are consistent with alterations in internal membrane systems.

HDV replicates by the rolling circle method, similar to that seen in both viroids and virusoids. Despite its other similarities to viroids, however, HDV replication appears to involve the type of self-cleavage seen in virusoids, which results in the formation of a 5′ hydroxyl and a cyclic 2′,3′-monophosphate. This self-cleavage reaction has been demonstrated *in vitro* in both the genomic and antigenomic molecules.

The role of delta antigen S in replication appears to be to block the cleavage of the antigenomic molecule that produces an mRNA, thus permitting synthesis of full-length antigenomes that can serve as template for rolling circle formation of new genomes. These genomes are encapsidated by delta antigen L. Formation of the complete hepatitis delta virus virion requires that the cell be coinfected by hepatitis B virus. The delta antigen L has a lipid attached to it that permits it to interact with the cytoplasmic membrane in the location where hepatitis B virus's surface protein is embedded. The resulting virion consists of a delta virus nucleocapsid surrounded by a hepatitis B virus envelope.

WHAT OTHER SUBVIRAL ENTITIES INFECT ANIMAL CELLS?

HDV is the only viroid-like organism to infect animal cells, but another truly bizarre type of subviral entity is widely distributed among different species of mammals, including humans. These are the agents responsible for the *spongiform encephalopathies,* an array of slowly progressing, inevitably fatal diseases that gradually destroy the central nervous system. The most famous and best studied of these diseases are kuru and Creutzfeldt-Jakob disease in humans, scrapie in sheep, and "mad cow disease" (bovine spongiform encephalopathy) in cattle.

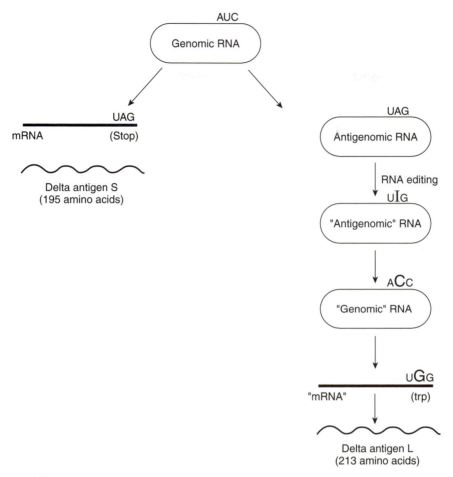

FIGURE 9.4 Two forms of hepatitis delta antigen are produced by RNA editing. Delta antigen S is translated from an mRNA made directly from the viral genome. Delta antigen L is produced when an antigenomic RNA is edited to convert an adenosine in the stop codon site into an inosine (shown as a larger letter). This change produces a tryptophan codon rather than a stop codon in the resulting mRNA, so the protein translated from it is 19 amino acids longer.

Studies on scrapie have provided most of the evidence that the agent responsible for destruction of brain tissue in infected animals is not a conventional virus, even though early experiments showed that the scrapie agent has many characteristics of a virus. After passage through bacterial filters, for example, homogenates of the central nervous system tissue from scrapie sheep injected into susceptible animals produce the disease. Also, dilution experiments clearly demonstrate that the agent replicated within its new host organisms.

Nevertheless, numerous types of analysis indicate that the scrapie agent is not a virus, or even a viroid, *since it does not appear to contain any nucleic acid.* Treatments that alter or inactivate nucleic acids, such as ultraviolet or ionizing radiation, nucleases, hydroxylamine, psoralen, and zinc ions, have no effect on the infectivity of a scrapie preparation. In contrast, agents such as phenol, sodium dodecyl sulfate, or urea that inactivate or denature proteins but have no effect on nucleic acids, or enzymes like trypsin or proteinase K that degrade proteins, do inactivate the scrapie agent. For this reason Prusiner introduced the term **prion** (from *pr*oteinaceous *in*fectious agent plus "on," a suffix originally used to denote particles in physics) to designate these novel entities. Prusiner was awarded the 1997 Nobel prize for his work on prions.

Since the *in vivo* infectivity assays required to detect prions are very time-consuming and expensive to conduct, purification of the scrapie agent and other prions has been extremely difficult. Nevertheless, a protein of 27 to 30 kdalton size was found to be present in the brains of scrapie-infected hamsters but not their uninfected counterparts. This protein was termed *prion protein (PrP) 27–30.* In electronmicrographs (Figure 9.5) PrP 27–30 appears as large macromolecular fibrils similar to (but distinct from) the amyloid fibrils seen in the brains of Alzheimer's disease victims. Several lines of evidence suggest that the PrP 27–30 fibrils are the scrapie infective agent. For example, the fibrils co-purify with scrapie infectivity, their concentration is proportional to infectivity titers, and proteolytic digestion of the fibrils results in a corresponding decrease in infectivity. Similar fibrils have been described in association with Creutzfeldt-Jakob disease and other nonhuman spongiform encephalopathies, also supporting a central role for them in prion diseases.

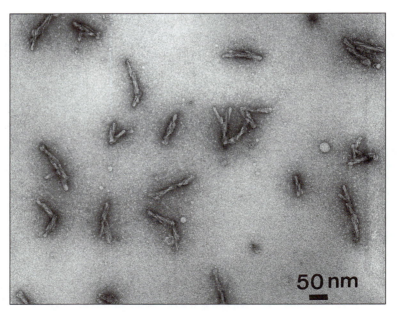

FIGURE 9.5 Electronmicrograph of negatively stained prion fibrils isolated from the brain of a scrapie-infected hamster.

HOW DO PRIONS REPLICATE?

If prions (or PrP 27–30 fibrils) do not in fact contain any nucleic acids, the question of their ability to replicate is indeed vexing. A model for the process has begun to emerge as information about the nature of PrP has accumulated. Determination of the N-terminal amino acid sequence of the scrapie fibrils allowed construction of oligonucleotides to probe for the gene responsible for PrP production. These studies determined that PrP is encoded by a *chromosomal* gene that is expressed at the same level in the brains of both scrapie-infected and normal animals. The normal gene product, designated PrP^c ("c" for cellular), is a glycoprotein of 33 to 35 kdaltons. PrP^c is attached to the plasma membrane via a glycosylphosphatidylinositol anchor and may act as a cell-surface receptor. Although the ligand has not been identified, studies in the 1C11 neuronal cell line indicate that PrP^c cross-linking by antibody reduces the phosphorylation level of the tryosine kinase Fyn, which in turn activates its kinase function. This suggests that PrP^c may play a role in signal transduction in neurites.

The conversion of PrP^c into the amyloid fibrils characteristic of scrapie occurs in a multistep process. PrP^c has four α-helical regions in its native conformation (Figure 9.6). The triggering event in the conversion process is the rearrangement of at least two of these regions into an antiparallel β-pleated sheet conformation. This molecule is termed PrP^{sc} and can be distinguished from PrP^c by its resistance to proteolytic digestion. PrP^{sc} appears to be the "infectious" form of the molecule that catalyzes the conversion of native PrP^c molecules to

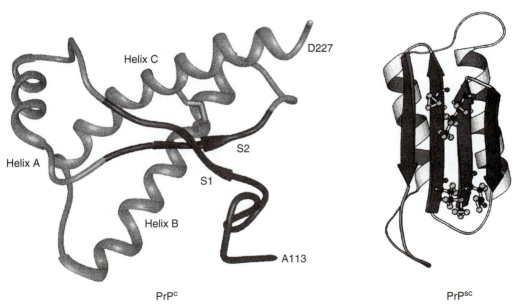

PrP^c PrP^{sc}

FIGURE 9.6 Structures of prion proteins PrP^C and PrP^{SC}. The portion of PrP shown has three α-helical regions that are shifted to β pleated sheet conformation during conversion of the protein into its prion form. These structures are based on cloned proteins comprising amino acids 90 to 231 in PrP and 29–231 in Prp^{SC}.

the PrP^{sc} conformation. Proteolytic removal of the N-terminal 67 amino acids from PrP^{sc} produces a molecule called PrP 27–30, so-named for its molecular weight. PrP 27–30 aggregates into the macromolecular amyloid fibrils associated with the disease scrapie.

Yeast Prion Model Systems

This model, as well as our understanding of how one altered protein catalyzes the conversion of other proteins, has been greatly advanced by the discovery of two different molecules in yeast that behave like prions. Ure2p is a soluble protein that inhibits the production of a plasma membrane transporter protein that brings in nitrogen sources such as ureidosuccinate. When better sources like ammonia or glutamine are available, Ure2p inhibits transcription of the transporter gene. The other protein, Sup35p, is a subunit of the translation release factor. Since both the transport and translation termination events are easy to monitor experimentally, they have provided a powerful tool for studying prion synthesis and activity. Using the Sup35p system, it has been possible to demonstrate *in vitro* the catalytic conversion of the Sup35p protein from its native form to an altered prion form. Introducing the Sup35p prion into yeast cells has also been shown to lead to the self-propagating formation of more prion-like protein.

Some types of prion diseases, like the sporadic form of Creutzfeldt-Jakob disease, appear to arise spontaneously rather than by infection from another individual. Some of these have been shown to arise as the result of a mutation in the gene encoding the PrP protein. Others, however, appear to arise from the *de novo* formation of the prion. The Ure2p system has been used to see how a prion may be generated in yeast. Ure2p is part of a regulatory cascade for nitrogen metabolism and is inhibited by the Mks1p protein. Mks1p is itself regulated by the presence of ammonia or by the Ras pathway (see Figure 8.10). Deletion of the *MKS1* gene or constitutive expression of the Ras pathway greatly reduce the occurrence of prion formation, while overproduction of Msk2p somewhat increases the frequency. These effects are seen only in the formation of the initial prion proteins and do not change the propagation of the prion thereafter. These findings suggest that prion formation may be influenced by other cellular proteins and regulatory cascades.

WHAT IS THE ORIGIN OF THE VARIOUS TYPES SUBCELLULAR ENTITIES?

This question has intrigued scientists for generations, and has led to theories that, at best, contain a good bit of speculation and fail to explain many facets of viral structure and reproduction. This is in part because no viral "fossils" have ever been identified, so there is no historical record of progressive evolutionary changes among the viruses. The only sources of information available to today's researchers are the genomes and proteins found in extant viruses and other subcellular entities like viroids and satellite RNAs. This means that when similar sequences or structures are seen in different viruses, for example, it may be difficult to discern if it is because those viruses share a common evolutionary ancestor (evolutionary homology) or because those sequences or structures represent similar solutions to similar selective pressures by unrelated viruses (convergent evolution).

Three types of theories to explain how viruses arose have been proposed. The first two theories both assume that viruses arose after their host cells had evolved, while the third suggests that viruses have an even more ancient origin than the cells they now infect.

The earliest proposals were *regressive theories* that asserted that viruses are the degenerate progeny of other obligate intracellular parasites. This parallels the endosymbiotic theory for the origin of eukaryotic organelles like the mitochondrion and the chloroplast, which suggests that those organelles began as bacteria that took up residence in the cytoplasm of the eukaryotic progenitor cell and then gradually lost most of their genomes as they became specialized for particular functions within their eukaryotic host. In regressive theories, viruses are entities that have carried this process even further, dispensing with all but a few genes and relying entirely upon their host for their metabolic needs, especially the synthesis of proteins. The progression *Rickettsia* $\longrightarrow$ *Chlamydia* $\longrightarrow$ *Poxviridae* has been suggested as an example of this process (Table 9.3). These theories would predict intermediate forms that have lost most but not all of their metabolic activities, but none have been found. Compared to even the largest and most complicated viruses (the poxviruses), for example, the smallest intracellular parasites (the chlamydia) have vastly more complex genomes and carry on all three of the processes of the Central Dogma using their own cellular machinery. Another difficulty with regressive theories is that they suggest an origin for the DNA viruses but do not explain how RNA viruses might have developed.

The latter difficulty is avoided, at least in part, by the cellular constituent theory, which also proposes that viruses developed after their host cells. In this theory viruses are thought to be descended from *normal cellular DNAs or RNAs* that developed the ability to replicate themselves autonomously. This mechanism requires that a free DNA molecule acquire a replication origin as well as a gene or genes encoding a viral capsid. An RNA molecule would also need to acquire the information encoding an RNA replicase. It has been suggested that some of these combinations may have arisen by "modular evolution" as mobile genetic elements (insertion sequences and transposons) moved or rearranged cellular DNA fragments.

By this theory, bacterial viruses, nearly all of which have DNA genomes, may have evolved from an earlier class of autonomously replicating plasmids that acquired the appropriate genes necessary to encode a capsid. Plasmids, in turn, arose when a circular DNA containing a replication origin became excised from a host cell's genome. Transposons or insertion sequences may have facilitated this process. Scenarios for the formation of the eukaryotic DNA viruses have also been proposed that are based on the structure and activities of transposons. DNA viruses of many types have an organization of terminal repeated

Table 9.3 | Characteristics of Organisms for a Theoretical "Regression" from a Free-Living Bacterium to a Virus

Organism	Free living?	Genome ($\times 10^8$)	Synthesize ATP?	Ribosomes
E. coli	Yes	20	Yes	Present
Rickettsia	No	10–15	Yes	Present
Chlamydia	No	3.6–6.6	No	Present
Poxviruses	No	1.5–2	No	Absent

sequences flanking their structural genes that is similar to the organization of transposons themselves, suggesting a fairly direct pathway.

Cellular constituent theories suggest that RNA viruses may have arisen by two mechanisms related to cellular DNAs and RNAs. One class of RNA viruses may have evolved from retrotransposons, a type of transposon whose movement from one location to another involves both DNA and RNA molecules. Retrotransposons move by means of a three-step process: (a) the retrotransposon DNA sequence is transcribed into an RNA molecule that is (b) reverse-transcribed back into a cDNA molecule that is (c) then reinserted into a new location in the host genome. This process is clearly related to the replication scheme employed by the RNA retroviruses as well as the DNA hepadnaviruses in animals and caulimoviruses in plants.

Other RNA viruses that do not have DNA intermediates in their reproductive cycles may have arisen either as a result of splicing reactions or transpositional events that created combinations capable of independent replication.

An obvious weakness of both the regressive and cellular constituent theories is an explanation for the formation of RNA viruses. The third group of theories on the origin of viruses focuses directly on these viruses. In these models, RNA viruses are the descendants of self-replicating *prebiotic RNA molecules* that became parasites within true cells when they in turn evolved. These theories are based on current ideas concerning the origin of life, which suggest that the first genetic material to develop was RNA. All the classes of reactions required to perpetuate a piece of genetic information; that is, polymerization on a template, cleavage, and ligation, have been demonstrated *in vitro* as reactions catalyzed solely by RNA. "RNA World" theories on the origin of life on earth suggest that prebiotic, self-replicating RNAs gave rise to RNA protein systems, and eventually to DNA-dependent genetic systems. The RNA viruses are therefore extremely ancient in origin, and obviously much modified as they assumed their status as dependents on various types of host cells. This group of theories does not consider how DNA viruses arose, but assumes that numerous other routes are possible since DNA-dependent genetic systems abound.

The Origins of Subviral Entities

The origins of the RNA-containing subviral entities can be fitted into the various schemes proposed for viruses. Satellite viruses very likely followed a form of degeneration pathway, losing their ability to replicate as they evolved from other autonomous RNA viruses. Many of the satellite RNAs may have arisen from satellite viruses as that degenerative process continued even further. The viroids are thought by some to represent a present-day example of self-replicating RNAs derived from some cellular RNA molecule, with the virusoids being derived in turn from them.

This leaves the prions and their origin. If prions are generated solely by posttranslational modification of a normal cellular protein and if they have no independent genetic information of their own, then they clearly must have arisen from some ancestral animal cell that underwent a mutation that created the PrP^{sc} phenotype. When this protein was introduced into another animal (possibly by ingestion), it initiated the replication process associated with pri-

Table 9.4 | Comparison of Theories for the Origin of Different Types of Subcellular Entities

Regressive Model	Cellular Constituent Model	Prebiotic RNA Model
DNA viruses from intracellular parasites	Bacteriophage from plasmids	RNA viruses from self-replicating RNAs
Satellite viruses from autonomous RNA viruses	Eukaryotic DNA viruses via transposons	Viroids from self-replicating RNAs
Satellite RNAs from satellite viruses	Retro-, hepadna-, and caulimoviruses from retrotransposons	
Virusoids from viroids	Prions	
	Viroids?	
	Satellite RNAs?	

ons today. Recent experiments suggest that the prion conversion can occur *de novo* and may be influenced by other cellular proteins or regulatory pathways. This suggests that prion "evolution" may be an ongoing process.

Table 9.4 summarizes ideas regarding the origin of the different classes of subcellular entities. It is very likely that there is no single answer to the question of how the viruses and subviral entities arose and that these entities represent numerous evolutionary lineages.

WHAT FACTORS AFFECT VIRAL EVOLUTION?

While evidence for the events that led to the origin of viruses is lacking, there is abundant evidence that viruses, like all genetic entities, evolve. A classic example of rapid viral evolution involves the myxoma virus of rabbits. Myxoma (rabbitpox) virus was introduced into Australia in 1950 as a means of biological control of wild rabbits that had become agricultural pests. While the program was initially very successful, within 2 years it was evident that the virus was no longer as effective a control agent. When the original virus stock was compared to viruses isolated from infected wild animals in 1952, significant differences were found. More than 99% of laboratory rabbits infected with the original virus died within 2 weeks of exposure, demonstrating that the introduced myxoma virus was extremely virulent and that the Australian rabbit population was very susceptible. In contrast, in more than 80% of the 1952 isolates, the mortality rate had fallen to between 70% and 95%, and the survival time of infected animals had doubled. Clearly the myxoma virus population was rapidly evolving into a less virulent form. Since the virus is transmitted by mosquito vectors, it is likely that strong selective pressure in favor of mutants that allowed the virus to remain in an infected host longer, and therefore available for transmission longer, were responsible for this evolutionary change. Studies on the wild rabbit population also indicated co-evolution for more resistant rabbits was occurring at the same time.

The most important factors in viruses that produce the genetic (and phenotypic) diversity upon which natural selection can operate are *mutation* and *recombination*. Mutation is the more important of these factors, particularly in RNA viruses. Prokaryotic and eukaryotic cells have both proofreading systems associated with their DNA polymerases as well as other postreplicative repair systems that together reduce cellular mutation rates to levels between 10^{-7} and 10^{-11} per base per round of replication. RNA polymerases lack a proofreading activity, so their error rates are on the order of 10^{-4} errors per transcriptional event. The polymerases of RNA viruses and the small DNA viruses do not have error-checking capabilities, so the mutation rates in both types of viruses are many orders of magnitude greater than those of their host cells. This can lead to rapid evolution within a very short time period because even a single virus infecting a host cell may undergo more than a dozen rounds of replication and produce hundreds of thousands of progeny.

Recombination, or the creation of new molecules by combining or substituting pieces of nucleic acid, can be a source of genetic diversity in DNA viruses, particularly when host cell genes are recombined into a viral genome. As we saw in earlier chapters, the transducing bacteriophages in prokaryotes and retroviruses in eukaryotes appear to have acquired cellular genes by this method. An especially intriguing instance of recombination may be the presence of three tRNA genes in the bacteriophage T4. Since these tRNA genes contain introns, and the tRNA genes of their *E. coli* host do not, it appears that the viral genome has undergone recombination with DNA from the eukaryotic host of its bacterial host!

WHAT CONSTRAINS OR COUNTERS THE EFFECTS OF MUTATION AND RECOMBINATION IN THE EVOLUTION OF VIRUSES?

Two obvious constraints on the effects of mutation on the evolution of viruses are the architecture and the lifestyle of viruses themselves. First, the fact that viruses must rely on their host cells for part or all of the functions of the Central Dogma means that their genomes must retain the proper recognition and control sequences for those processes. Second, the capsid of icosahedral viruses imposes an upper limit on the size of nucleic acid molecule that can be encapsulated. Since that limit is usually quite small, virus genomes generally do not contain superfluous or duplicated sequences. Exceptions exist, of course, particularly among the herpesvirus group of large DNA viruses. The assembly process of the capsid also places constraints on the types of changes that are possible. Assembly involves precise interactions between individual capsomer proteins and the viral genome. Mutations that reduce the size of the nucleic acid in the genome or alter capsomer protein–protein or protein-nucleic acid interactions, therefore, often prevent proper assembly. Thus, mutations that would increase or decrease the size of the viral genome, or that significantly alter important sequences within it, are strongly selected against.

Other properties of viruses that may act to counter the effects of mutation and recombination are less obvious. The segmented genomes found in numerous RNA viruses of plants and animals may be one such property. As we noted in Chapter 1, the size of most RNA virus genomes is clustered between about 1×10^6 and 4×10^6 daltons. It has been suggested that this may represent the size of RNA molecules that can be replicated without an excessive accumulation of mutations. Some groups of viruses have created larger genomes than this by distributing their constituent genes among a set of molecules rather than housing them in a single molecule. A mutational event in one segment will inactivate only that segment rather than the entire genome. Thus, a larger number of "hits" may be necessary to accumulate a lethal dose in a segmented genome compared to one with only a single molecule.

A related phenomenon that is capable of generating considerable genetic diversity in segmented viruses is termed *reassortment*. When the same cell is infected concurrently with two different strains (or sometimes even species) of a segmented virus, functional virions may be produced that contain various combinations of the genomic segments drawn from the intracellular pools that accumulate during replication. The resulting viruses can be dramatically different from either of the input viruses.

The rapid evolution of influenza A virus through a combination of the accumulation of small mutations and abrupt reassortments has been extensively documented. Several features of the virus's architecture and reproduction are significant in this evolution. First, the influenza A viral envelope contains two glycoproteins (HA = hemagglutinin and NA = neuraminidase) that evoke protective immune responses in infected persons. These responses are not directed at the entire HA or NA glycoproteins, but rather at various restricted sets of amino acids or sugars within the larger molecules that are termed *antigenic determinants*. The protective response to a particular antigenic determinant of a given strain is lifelong. Second, the HA and NA genes are carried on separate segments of the viral genome. Finally, influenza A virus can infect not only humans, but also several species that live in close association with humans: swine, horses, and wild and domestic birds.

The first type of change in a population of influenza A viruses is called **antigenic drift.** In this process, the antigenic determinants associated with the HA or NA molecules gradually change over a period of years as mutations occur in their encoding RNA segments. As each change occurs, that form of the HA or NA glycoprotein enjoys a selective advantage compared to previous forms that have already evoked the production of specific immune responses in infected animals or persons.

Superimposed on this gradual pattern of evolutionary change are sudden, dramatic alterations in the amino acid sequences of one or both of these glycoproteins. These **antigenic shifts** change the molecular structure of the HA or NA molecules so markedly that they become, in effect, virtually new entities to the immune systems of persons or animals. Antigenic shifts, therefore, are associated with epidemics or pandemics of influenza such as occurred in 1918 (the Spanish flu), 1957 (the Asian flu), 1968 (the Hong Kong flu), and 1977 (the Russian flu). The antigenic shift in 1957 saw changes in both structures (H1N1 $\longrightarrow$ H2N2), that of 1968 only one (H2N2 $\longrightarrow$ H3N2), and that of 1977 both again (H3N2 $\longrightarrow$ H1N1 returned).

Several theories have been advanced to explain how antigenic shifts occur. Nonhuman hosts play key roles in these theories, in part because of the observation that influenza B virus does not infect any nonhuman hosts and also does not exhibit antigenic shifts. The most direct role proposed for nonhuman hosts is that they provide a source of new viral strains that enter the human population with novel antigenicities. The 1957 shift in both HA and NA at the same time suggests such an event. A well-documented and frightening example of direct transfer occurred in 1997, when several people in Hong Kong died after infection with an avian (chicken) influenza A virus of type H5N1. Fortunately, it appears that this particular avian strain can be transferred very efficiently in chickens but not in humans, so no pandemic was produced. Another explanation of antigenic shifts proposes that reassortment creates a new virus type when the same cell is co-infected both by a human and a nonhuman strain of virus. Examples of such animal/human hybrid viruses have been isolated from nature. The usual mechanism appears to involve the simultaneous infection of a pig with both human and avian viruses. Both viruses can reproduce in pig cells, so reassortments are likely to occur. Moreover, the resulting viruses produced in pigs can spread readily to humans to create the possibility of a pandemic. The 1977 return to a combination (H1N1) that had been replaced in 1957 suggests that antigenic shifts may be able to cycle back as the population that is immune to a particular type is replaced by a younger, susceptible group.

HOW DO NONHUMAN VIRUSES GAIN THE ABILITY TO INFECT HUMANS?

Influenza A is actually an avian virus that has left its original host (ducks) to infect first swine, and through them, humans. Since the duck virus cannot infect humans directly, it is evident that some adaptation occurring in swine hosts was required for influenza virus to make its cross-species transfer to become a human virus. Influenza is by no means the only animal virus to make a cross-species transfer to humans. The rapid movement of people around the globe today presents such newly evolved **emerging viruses** with unparalleled opportunities to spread throughout the human species, perhaps with catastrophic consequences. "Emergence" is something of a tricky word. Table 9.5 presents some of the variety of situations in which a virus may be said to have "emerged." The virus may have been present in the human population for a long time, but some change in either the host, or less often in the virus, causes the disease it produces to "emerge" into an epidemic form. For example, poliovirus has been infecting human infants for thousands of years, but epidemic paralytic poliomyelitis emerged as a major health problem only in the last century. This form of polio is a "disease of civilization" that resulted when modern sanitation and water treatment became common. Before that time, the virus was passed to infants and very young children through contaminated water. It appears that when the age at which a person first becomes infected rises above about 5, the risk of developing paralytic poliomyelitis increases markedly. Thus, the good sanitation that saved uncounted

Table 9.5 | Situations in Which New Diseases Emerge

Situation	Virus Examples
1. Change in susceptible population allows disease to emerge from existing human virus	Polio
2. Change in population density allows virus to maintain itself and cause continuing disease	Measles
3. Zoonotic virus with human as dead-end host	Sin Nombre, West Nile
4. Zoonotic virus with some human-to-human transmission	Ebola, Lassa Fever
5. Zoonotic virus mutates to become a human virus	HIV

persons from other bacterial and virus diseases actually contributed to the emergence of paralytic poliomyelitis.

Another factor that may produce episodic "emergence" of a virus is population size. A virus like measles virus requires a sizable vulnerable population (about 500,000 people) in order to be maintained in a given area. If a relatively isolated population is smaller than that limit, each new measles virus introduction produces an epidemic. It is likely that measles virus actually entered the human population from horses (rinderpest virus is the likely source) in the Middle East when that region began to contain large urban centers.

The powerful tools of recombinant DNA technology and PCR are increasingly being used to identify new viruses. For example, these tools have been used to discover a number of viruses that cause hepatitis, including the elusive hepatitis C virus that was long called the "non-A, non-B virus" since its only identity was that it was not either of those other hepatitis viruses. Human herpesvirus 8, which is implicated in Kaposi's sarcoma, was first found by a form of competitive PCR that compared DNA sequences in normal and cancer cells. The possibility also exists that new viruses may be introduced via xeno-transplantation, or transplantation of animal tissues into humans. The use of pig cells in experimental therapies to treat Parkinson's disease, diabetes, and renal dialysis patients raises the fear that porcine endogenous retrovirus (PERV) may be transferred to the human receiving the pig cells. At least 50 copies of the PERV provirus are carried in the pig genome, and some of these have been shown to produce viruses that can infect human cells in culture.

Zoonotic Viruses

True emergence requires crossing species boundaries. Many members of the Flaviviridae and the Bunyaviridae appear to be **zoonotic** viruses, or animal viruses that can infect humans. Humans are a dead-end host, however, since the virus is not spread from one person to another. The Sin Nombre hantavirus that caused an outbreak of a swiftly lethal form of pneumonia in the Four Corners area of New Mexico in 1993 is an example of such a zoonotic virus. In this case, deer mice are thought to be the natural host and pass the virus in their urine and feces. The good growing conditions following an especially wet spring in 1993 allowed an increase in the deer mouse population, which in turn resulted in more humans coming into contact with infected materials.

Modern ease of movement appears to play a major role in zoonotic outbreaks. In 1999, the West Nile flavivirus, an African virus as its name indicates, emerged in the United States. The virus's normal host is birds but humans may become infected when a mosquito feeds first on an infected bird and then a person. The virus is spreading very rapidly because it appears that numerous species of birds can serve as hosts and that the virus can be transmitted by a wide variety of different mosquitos. In 1999 outbreaks occurred only in New York, New Jersey, and Delaware; a year later eight more states on the Eastern seaboard reported cases.

Some zoonotic viruses can be transmitted from human to human under special circumstances. The filoviruses that cause Ebola fever are zoonotic viruses whose natural host and mode of transfer to humans have not yet been identified. Like other zoonotic viruses, Ebola virus causes a severe disease with very high mortality. Unlike most other zoonotic viruses, however, Ebola virus can be transmitted via blood, tissues, or other body fluids from one person to another. The horrific nature of the hemorrhagic fever it produces, coupled with the possibility of direct transmission, has made Ebola virus the subject of popular books and films focused on emerging pathogens.

Lest we become too anthrocentric, it should be noted that human pathogens can also be passed in the other direction and infect animals. Gorillas, chimpanzees, baboons, and monkeys in Africa have all been shown to have become infected with organisms brought to them by scientists and tourists. Along with a variety of bacterial and protozoan pathogens, measles, polio, and influenza viruses have been identified.

The Origin of HIV

The present AIDS pandemic is the result of a cross-species transfer of a zoonotic virus that mutates rapidly into forms that can be passed from human to human. When human immunodeficiency virus (HIV) was identified as the causative agent of AIDS, an important question was "Where did it come from?" Since another human retrovirus that also infects cells of the immune systems, human T-lymphotropic virus I (HTLV-I), had been shown to have a close relative infecting monkeys, the simian T-lymphotropic virus (STLV), the search focused on monkey and ape viruses. Although the search was first successful in captive Asian macaque monkeys, the wild population found to harbor a monkey virus similar to HIV was the African green monkey. This simian immunodeficiency virus (SIV) appears to have made its evolutionary accommodation with African green monkeys, since it does not produce any noticeable suppression of the infected animals' immune systems. When SIV infects Asian monkeys, however, it produces a disease similar to AIDS, suggesting that SIV could be quite virulent in different hosts. Four groups of SIV resident in nonhuman primates have now been described based on nucleotide sequences and host organisms: $SIV_{mac/sm}$ in macaque/mangabey monkeys, SIV_{agm} in African green monkeys, SIV_{mnd} in mandrill monkeys, and SIV_{cpz} in chimpanzees.

HIV and SIV appear to share the same core proteins but to have different envelope proteins. A second, less virulent, HIV has been identified in West Africa, which is called HIV-2 to distinguish it from the original virus (HIV-1). HIV-2 shares both core and envelope antigens with SIV and is also genetically more closely related to $SIV_{mac/sm}$ than is HIV-1. The only difference in the genetic organization of HIV-1 compared to HIV-2 is in two regulatory genes. HIV-1 encodes a 15 kdalton protein VPU overlapping with the 5′ end of the *env* gene; HIV-2 lacks this gene but instead encodes a 15 kdalton protein VPX overlapping the 3′ end of its *pol* gene. The function of both of these proteins, and therefore any possible relationship to virulence, is unknown.

HIV-1 has a variety of different subtypes whose relationships to each other and to HIV-2 are diagrammed in Figure 9.7 HIV-1 has three distinct clusters called M (major), O (outlier), and a single type called YBF30 that fits with neither the M nor O groups. Each of these appears to have evolved from a strain of SIV_{cmz} that infects the chimpanzee subspecies (*Pan troglodytes troglodytes*) that lives in Gabon, Equatorial Guinea, and southern Cameroon in West Africa. Beatrice Hahn and her coworkers speculate that the transfer from ape to human occurred when chimpanzees were butchered for food. Sequence analysis of the group M viruses suggests that the members of that cluster began to diverge about 1931, indicating that HIV has been present in the human population for at least 70 years. The epidemic spread of the virus in the 1970s appears to be the result of increased travel to and from Africa at a time when large cities were developing there and the sexual revolution was taking place in Europe and the United States.

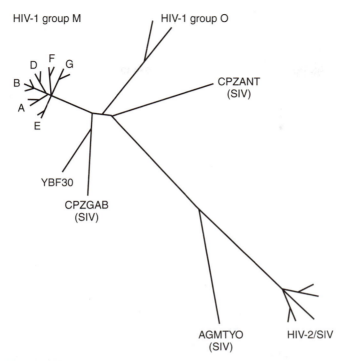

FIGURE 9.7 Genetic relationships between groups of HIV and SIV. The M and O groups and strain YBF30 of HIV-1 appear closely related to a Gabon chimpanzee SIV (CPAGAB) and less so to other chimpanzee SIVs. HIV-1 is only more distantly related to HIV-2 and the sooty mangabey SIV and the African green monkey SIV (AGMTYO).

SUMMARY

A variety of subviral entities with RNA genomes have been found in eukaryotic cells. The plant viroids are autonomously replicating infectious RNA molecules that appear to encode no proteins at all. The other RNA-containing subviral entities (hepatitis delta virus in animal cells, satellite RNAs, and satellite viruses in plant cells) all form encapsidated structures but require the assistance of a helper virus to replicate their genomes (all three) and to provide capsid proteins (HDV and satellite RNAs). Another subviral entity found in animal cells, the prion, appears to be an infectious protein since it contains no nucleic acid at all.

Three types of theories have been advanced to explain how viruses and subviral entities may have arisen. Regressive theories propose that viruses are the degenerate progeny of their obligate intracellular parasites. Cellular constituent theories suggest that viruses are descended from normal cellular RNAs or DNAs that acquired the ability to replicate themselves autonomously. The third group of theories focuses on the RNA viruses and proposes that they are the extant descendants of prebiotic RNA molecules that became self-replicating. Whatever their origins, viral genetic diversity upon which natural selection can operate is maintained through mutation and recombination.

Viral "emergence" may take several forms. Zoonotic viruses cross the species barrier between their usual animal host and humans, but do not continue to spread in their new host population. Other viruses undergo mutations that allow them to proliferate in humans. HIV-1 has evolved from a chimpanzee virus (SIV) to become a true human pathogen.

SUGGESTED READINGS

Viroids

1. "Potato spindle tuber virus: A plant virus with properties of a free nucleic acid" by T.O. Diener and W.B. Raymer, *Science* 158:378–381 (1967), presents experiments that suggested that PSTV was not a conventional virus.

2. "Nucleotide sequence and secondary structure of potato spindle tuber viroid" by H.J. Gross, H. Domdey, C. Lossow, P. Jank, M. Raba, H. Alberty, and H.L. Sanger, *Nature* 273:203–208 (1978), presents the first complete molecular structure for a eukaryotic pathogen.

3. "Domains in viroids: Evidence of intermolecular RNA rearrangement and their contribution to viroid evolution" by P. Keese and R.H Symons, *Proc Natl Acad Sci USA* 82:4582–4586 (1985), uses sequence homologies among various viroids to present a model for five structural domains in viroids.

4. "Viroid replication is inhibited by α-amanitin" H-P Muhlbach and H.L. Sanger, *Nature* 278:185–187 (1979), and "DNA-dependent RNA polymerase II of plant origin transcribes viroid RNA into full-length copies" by H-R. Rackwitz, W. Rohde, and H.L. Sanger, *Nature* 291:297–301 (1981), give evidence that viroids use RNA polymerase II to replicate their genomes.

5. "A replication cycle for viroids and other small infectious RNAs" by A.D. Branch and H.D. Robertson, *Science* 223:450–455 (1984), presents a model for rolling circle replication of viroids.

6. "The 7S RNA from tomato leaf tissue resembles a signal recognition particle RNA and exhibits a remarkable sequence complementarity to viroids" by B. Haas, A. Klanner, K. Ramm, and H.L. Sanger, *EMBO J* 7:4063–4074 (1988), and "Pathogenesis by antisense" by R. Symons, *Nature* 338:542–543 (1989), discuss the possible involvement of 7S RNA/viroid RNA hybrids in viroid pathogenesis.

Satellite RNAs and Viruses

1. "Plant virus satellites" by R.I.B. Francki, *Ann Rev Microbiol,* 39:151–174 (1985), presents a fine overview of these related subviral entities.

2. "Non-enzymatic cleavage and ligation of RNAs complementary to a plant virus satellite RNA" by J.M. Buzayan, W.L. Gerlach, and G. Bruening, *Nature* 323:349–352 (1986), presents evidence for an unusual form of self-cleavage followed by ligation to generate satellite RNA genomes.

3. "Self-cleavage of virusoid RNA is performed by the proposed 55-nucleotide active site" by A.C. Foster and R.H. Symons, *Cell* 50:9–16, (1987), discusses additional evidence in support of a model for self-cleavage that results in unusual 5′ and 3′ termini.

Hepatitis Delta Virus

1. "Structure and replication of the genome of the hepatitis δ virus" by P-J. Chen, G. Kalpana, J. Golberg, W. Mason, B. Werner, J. Gerin, and J. Taylor, *Proc Natl Acad Sci USA* 83:8774–8778 (1986), and "Structure, sequence and expression of the hepatitis delta (δ) viral genome" by K.-S. Wang, Q.-L. Choo, A.J. Weiner, J.-H. Ou, R.C. Najarian, R.M. Thayer, G.T. Mullenback, K.J. Denniston, J.L. Gerin, and M. Houghton, *Nature* 323:508–514 (1986), analyze HDV and compare it to plant viroids, virusoids, and satellite RNAs.

2. "An ultraviolet-sensitive RNA structural element in a viroid-like domain of the hepatitis delta virus" by A.D. Branch, B.J. Benenfeld, B.M. Batoudy, F.V. Wells, G.J. Gerin, and H.D. Robertson, *Science* 243:649–652 (1989), describes the evidence that HDV has a viroid-like region separate from a protein-encoding region in its genome.

3. "Antigenomic RNA of human hepatitis delta virus can undergo self-cleavage" by L. Sharmeen, M.Y.P. Kuo, G. Dinter-Gottlieb, and J. Taylor, *J Virol* 62:2674–2679 (1988), "Characterization of self-cleaving RNA sequences on the genome and antigenome of human hepatitis delta virus" by M.Y.P. Kuo, L., Sharmeen, G. Dinter-Gottlieb, and J. Taylor, *J Virol* 62:4439–4444 (1988), refine the self-cleavage model for HDV.

4. "Regulation of polyadenylation of hepatitis delta virus antigenomic RNA" by S.-Y. Hsieh and J. Taylor, *J Virol* 65:6438–6446 (1991), "Hepatitis delta virus RNA editing is highly specific for the amber/W site and is suppressed by hepatitis delta antigen" by A.G. Polson, H.L. Ley, B.L. Bass and J.L. Casey, *Mol Cell Biol* 18: 1919–1926 (1998), and "Transcription of hepatitis delta antigen mRNA continues throughout hepatitis delta virus (HDV) replication: A new model of HDV RNA transcription and replication" by L.E. Modahl and M.M. Lai, *J Virol* 72: 5449–5456 (1998), present models of how hepatitis delta virus produces two forms of the delta antigen during its replication.

Prions

1. "Novel proteinaceous infectious particles cause scrapie" by S.B. Prusiner, *Science* 216:136–144 (1982), presents the evidence that led Prusiner to coin the term *prion* to describe the scrapie agent.

2. "Identification of a protein that purifies with the scrapie prion" by D.C. Bolton, M.P. McKinley, and S.B. Prusiner, *Science* 218:1309–1310 (1982), "Further purification and characterization of scrapie prions" by S.B. Prusiner, D.C. Bolton, D.F. Groth, K.A. Bowman, S.P. Cochran, and M.P. McKinley, *Biochem* 21:6942–6950 (1982), and "A protease-resistant protein is a structural component of the scrapie prion" by M.P. McKinley, D.C. Bolton, and S.B. Prusiner, *Cell* 35:57–62 (1983), characterize the PrP 27–30 molecule and its possible role as the scrapie prion itself.

3. "Scrapie infectivity, fibrils and low molecular weight protein" by H. Diringer, H. Gelderblom, H. Hilmert, M. Ozel, C. Edelbluth, and R.H. Kimberlin, *Nature* 306:476–478 (1983), and "Scrapie-associated fibrils in Creutzfeldt-Jakob disease" by P.A. Merz, R.A. Somerville, H.M. Wisniewski, L. Manuelidis, and E.E. Manuelidis, *Nature* 306:474–476 (1983), are a pair of back-to-back articles describing the amyloid-like fibrils associated with prion infections.

4. "A cellular gene encodes scrapie PrP 27–30 protein" by B. Oesch, D. Westaway, M. Walchli, M.P. McKinley, S.B.H. Kent, R. Aebersold, R.A. Barry, P. Tempst, D.B. Teplow, L.E. Hood, S.B. Prusiner, and C. Weissmann, *Cell* 40:735–746 (1986), and "Scrapie and cellular PrP isoforms are encoded by the same chromosomal gene" by K. Basler, B. Oesch, M. Scott, M. Walchli, D.F. Groth, M.P. McKinley, S.B. Prusiner, and C. Weissmann, *Cell* 46:417–428 (1986), and "Molecular cloning and complete sequence of prion protein cDNA from mouse brain infected with the scrapie agent" by C. Locht, B. Chesebro, R. Race, and J.M. Keith, *Proc Natl Acad Sci USA* 83:6372–6376 (1986), present evidence that the scrapie PrP is the product of a normal cellular gene.

5. "Molecular biology of prion diseases" by S.B. Prusiner, *Science* 252:1515–1522 (1991), reviews the structural and molecular biological evidence that prions are formed from a normal cellular protein, and suggests mechanisms for that reaction.

6. "Prions" by S.B. Prusiner, *Proc Natl Acad Sci USA* 95: 13363–13383 (1998), is an abbreviated version of Prusiner's Nobel lecture.

7. "Prion domain initiation of amyloid formation in vitro from native Ure2p" by K.L. Taylor, N. Cheng, R.W. Williams, A.C. Steven, and R.B Wickner, *Science* 283: 1339–1343 (2000), and "A protein required for prion generation: [URE3] induction requires the Ras-regulated Mks1 protein"

by H.K. Edskes and R.B Wickner, *Proc Natl Acad Sci USA* 97: 6625–6629 (2000), describe control of the formation and propagation of a yeast prion.

8. "Evidence for the prion hypothesis: Induction of the yeast [PSI+] factor by in vitro converted Sup35 protein" by H.E. Sparrer, A. Santoso, F.C. Szoka Jr., and J.S. Weissman, *Science* 289: 595–599 (2000), shows that prions can be generated *in vitro*.

9. "Signal transduction through prion protein" by S. Mouillet-Richard, M. Ermonval, C. Chebassier, J.L. Laplanche, S. Lehmann, J.M. Launay, and O. Kellerman, *Science* 289: 1925–1928 (2000), presents evidence that mammalian PrP^c may be part of a signaling cascade system in neuronal cells.

Viral Origins and Evolution

1. "Origin of retroviruses from cellular moveable genetic elements" by H.M. Temin, *Cell* 21:599–600 (1980), reviews evidence in support of evolution of retroviruses from cellular components.

2. "Speculations of the early course of evolution" by J.E. Darnell and W.F. Doolittle, *Proc Natl Acad Sci USA* 83:1271–1275 (1986), "Modern metabolism as a palimpsest of the RNA world" by S.A. Benner, A.D. Ellington, and A. Tauer, *Proc Natl Acad Sci USA* 86:7054–7058 (1989), and "Circular RNAs: Relics of precellular evolution?" by T.O. Diener, *Proc Natl Acad Sci USA* 86:9370–9374 (1989), discuss prebiotic RNAs as the origin of viruses and subviral entities.

3. "The evolution of RNA viruses" by D.C. Reanney, *Ann Rev Microbiol* 36:47–73 (1982), "Rapid evolution of RNA viruses" by D.A. Steinhauer and J.J. Holland, *Ann Rev Microbiol* 41:409–433 (1987), and "Evolution of RNA viruses" by J.H. Strauss and E.G. Strauss, *Ann Rev Microbiol* 42:657–683 (1988), discuss the features leading to rapid evolution in RNA viruses, as well as probable evolutionary relationships among the RNA virus groups.

4. "Molecular mechanisms of variation in influenza viruses" by R.G. Webster, W.G. Laver, G.M. Air, and G.C. Schild, *Nature* 296:115–121 (1982), reviews the processes of antigenic drift and antigenic shift in influenza A virus.

5. "Molecular basis for the generation in pigs of influenza A viruses with pandemic potential" by T. Ito, J.N.S. Couceiro, S. Kelm, L.G. Baum, S. Krauss, M.R. Castrucci, I. Donatelli, H. Kida, J.C. Paulson, R.G. Webster, and Y. Kawaoka, *J Virol* 72: 7367–7373 (1998), provides support for the reassortment model of new influenza A virus types.

Emerging Viruses

1. "Xenografts and retroviruses" by R.A. Weiss, *Science* 285: 1221–1222 (1999), and "Search for cross-species transmission of porcine endogenous retrovirus in patients treated with living pig tissue" by K. Paradis, G. Lanford, Z. Long, W. Heneine, P. Sandstrom, W.M. Switzer, L.E. Chapman, C. Lockey, D. Onions, the Xen 111 Study Group, and E. Otto, *Science* 285: 1236–1241 (1999), discuss the possibility of transfer of a pig virus to humans when pig cells are used therapeutically.

2. "Origin of HIV-1 in the chimpanzee *Pan troglodytes troglodytes*" by F. Gao, E. Bailes, D.L. Robertson, Y. Chen, C.M. Rodenburg, S.F. Michael, L.B. Cummins, L.O. Arthur, M. Peeters, G.M. Shaw, P.M. Sharp, and B.H. Hahn, *Nature* 397: 436–441 (1999), and "Timing the ancestor of the HIV-1 pandemic strains" by B. Korber, M. Muldoon, J. Theiler, F. Gao, R. Gupta, A. Papedes, B.H. Hahn, S. Wolinsky, and T. Bhattacharya, *Science* 288: 1789–1796 (2000), present evidence that HIV-1 emerged from chimpanzees about 70 years ago.

The Virologist's Toolkit

Virologists ask a wide variety of questions about the viruses they study. What is their composition? How large are they? How do they interact with their host cells? How do they express and replicate their genomes? To answer these and many other questions, virologists draw on experimental techniques from a number of disciplines. This appendix is designed as an aid to refresh the reader's memory by providing brief descriptions of some of the major techniques commonly used in virology.

I. ANIMAL CELL CULTURE

Viruses must have living cells as hosts to support their replication. Before the development of techniques for the *in vitro* cultivation of animal cells, laboratory animals and embryonated eggs were employed for virus isolation. These expensive methods have largely been supplanted now by the use of *cell cultures* derived from a wide variety of human or animal tissues. Cell cultures are initiated by dissociating small pieces of tissue into single cells. This can be done either by treatment with a proteolytic enzyme such as trypsin and a chelating agent like ethylenediaminetetraacetic acid (EDTA), or by pressing it through a fine mesh sieve. The cells are then introduced into a culture flask, tube, or plate containing cell culture medium. The basic cell culture medium is a very complex mixture of a buffered salt solution, glucose, vitamins, amino acids, and a pH indicator. This medium is usually supplemented by the addition of serum and antibiotics, and for certain cell types, hormones or growth factors. The cells attach to the surface of the culture vessel and replicate until all the available surface of the vessel has been occupied.

Primary cultures are derived directly from a tissue. Primary cultures usually contain a mixture of both fibroblastic and epithelioid cells. Figure A1.1 presents a scheme for establishing a primary culture of mouse mammary epithelial cells that uses the enzyme collagenase to dissociate the minced mammary gland tissue. Primary cultures can often be passaged by treating the cells with trypsin and EDTA to remove them from the culture vessel surface, diluting them, and then replating in new vessels. The cells in the mixed populations of primary cultures rarely have the same rate of division, so one cell type (usually fibroblastic) comes to dominate as they reproduce or are subpassaged. The *cell strains* thus produced are usually diploid and divide for only a limited number of generations before they die out.

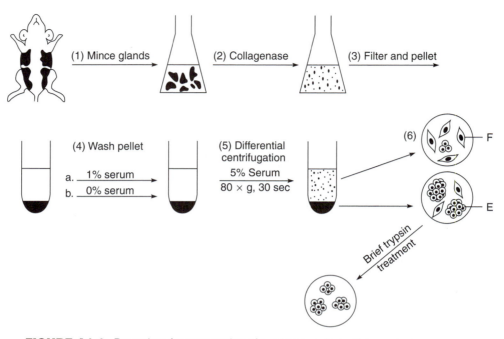

FIGURE A1.1 Procedure for preparation of a primary culture of normal mouse mammary epithelial cells. Mammary glands are removed and minced (1), then digested for 60 to 90 minutes in collagenase (2), and the resulting suspension passed through cheesecloth to remove any undigested lumps of tissue (3). After washing the cells in medium with 1% serum and then without serum (4), the cells are subjected to a gentle centrifugation in medium containing 5% serum. Epithelial cells tend to clump in this medium and so are pelleted, while fibroblastic cells do not clump and remain largely in suspension (5). 24 hours after plating the cells from the pellet, a gentle trypsin treatment removes the small number of fibroblastic cells from the culture (6). F = fibroblastic cells; E = epithelial cells.

Occasionally a mutation (sometimes called the "immortality event") will occur in one of the cells in the strain, leading to the establishment of a *continuous cell line* that is aneuploid and has no limit on the number of generations that can be subpassaged. Figure A1.2 shows the appearance of the fibroblastic cell line BHK-21 (baby hamster kidney) and the epithelioid cell line MCF-8 (mouse mammary tumor).

II. MICROSCOPY

The human eye is able to resolve objects down to about 100,000 nm in size, but both viruses and their host cells are smaller than this limit (mammalian epithelial cells ≈ 30,000 nm diameter, bacteria ≈ 2000 nm diameter, viruses ≈ 10 to 250 nm diameter). Microscopes of various types are therefore needed to *magnify* the image of the material of interest. *Light microscopes* are used for materials larger than about 200 nm, while *electron microscopes*

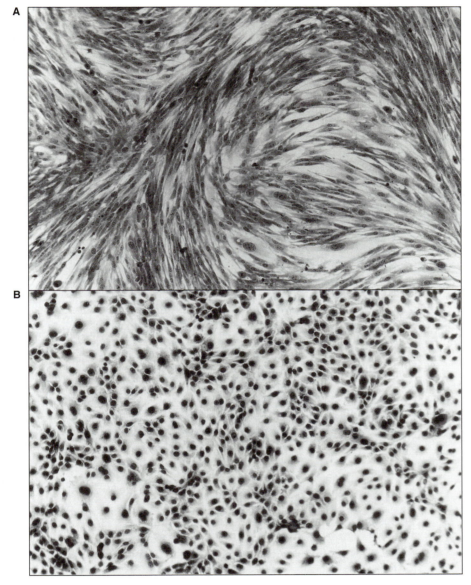

FIGURE A1.2 The appearance of fibroblastic and epithelial cells in culture. **A,** The fibroblast cell line BHK-21 produces sheets of thin cells in swirls and parallel arrays. **B,** The epithelial cell line MCF-8 produces a sheet with a cobblestone appearance.

are used for smaller specimens like the viruses. The basic organizations of light and electron microscopes are comparable (Figure A1.3), except that electron microscopes use electrons rather than visible light as their illumination, and electromagnetic coils rather than glass lenses to focus that illumination.

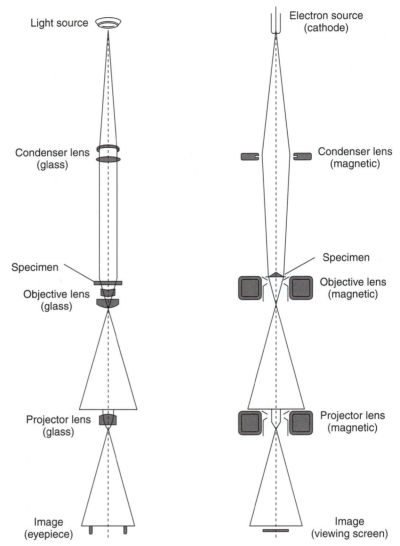

FIGURE A1.3 Comparison of the optical systems of the compound light microscope *(left)* and the transmission electron microscope *(right)*.

Specimens for examination in the electron microscope must be stained with electron-dense heavy metals in order to improve image contrast. *Whole mounts* of viruses can be prepared by three different methods: (a) positive staining, or coating the specimen with heavy metals, (b) negative staining, or coating the background with heavy metals so that the specimen is seen in contrast to that dark background, and (c) shadow casting, or evaporating heavy metals from one side of the specimen to create shadows behind the particles.

III. RADIOISOTOPES

Isotopes are forms of a particular element that have the same atomic number but different atomic masses. Radioisotopes differ with respect to the *type of radiation* that is emitted (alpha particles = helium nucleus, beta particles = electron, or gamma radiation = electromagnetic radiation or photons), the *energy level* of the emitted radiation, and the *half-life* of the isotope or the length of time required for half the atoms in a sample to decay. Table A1.1 presents the characteristics of radioactive isotopes most commonly used in biological research. The fact that they have different energy levels allows each isotope in a mixture to be detected independently.

Radioactive isotopes are widely used in biological studies as *tracers* that allow the researcher to *label* compounds of interest and follow them through an experiment. Since the isotopes of a given element all have identical chemical properties (will form the same chemical bonds, etc.), they can participate interchangeably in the metabolic processes of the cell. For example, a glucose molecule containing ^{14}C is metabolized in exactly the same manner as one containing the usual, nonradioactive isotope ^{12}C.

Types of Labeling Experiments

There are two basic experimental designs using isotopic tracers, although there are hundreds of variations on these basic themes. In a *continuous labeling* experiment, cells are exposed to an isotopically labeled compound that becomes incorporated into the cells' constituents by the usual metabolic pathways of the cell. Although labeling periods can be any length, many experiments use short labeling periods called *pulses*. Consider, for example, an experiment to determine how long after infection a virus stimulates cellular DNA synthesis. In such an experiment (Figure A1.4), a set of dishes of cultured cells are infected by the virus at the same time. At various lengths of time after infection, ^{3}H-thymidine is added to one dish of the set for a 15-minute pulse-labeling period. During each of the pulse-labeling periods, ^{3}H-thymidine will become incorporated into new DNA molecules as they are synthesized. At the end of each pulse-labeling period, the amount of radioactive DNA in the cells in that dish is then determined as a measure of DNA synthesis in those cultures.

The second basic experimental design for the use of isotopic tracers, called *pulse-chase,* allows the researcher to follow what happens to a set of biological molecules that were synthesized during a brief pulse interval. For example, in the experiment to determine how polio viral proteins are processed after their synthesis, described in Chapter 4 (page 122),

Table A1.1 | Properties of the Radioisotopes Most Commonly Used in Biological Research

Radioisotope	Half-life	Emission Type	Energy*	Decays Into
$^{3}H_1$	12.3 y	Beta	0.018	$^{3}He_2$
$^{14}C_6$	5700.0 y	Beta	0.156	$^{14}N_7$
$^{32}P_{15}$	14.3 d	Beta	1.710	$^{32}S_{16}$
$^{35}S_{16}$	87.1 d	Beta	0.167	$^{35}Cl_{17}$
$^{131}I_{53}$	8.04 d	Beta	0.606	$^{131}Xe_{54}$
		Gamma	0.364	$^{131}Te_{52}$

*In million eV.

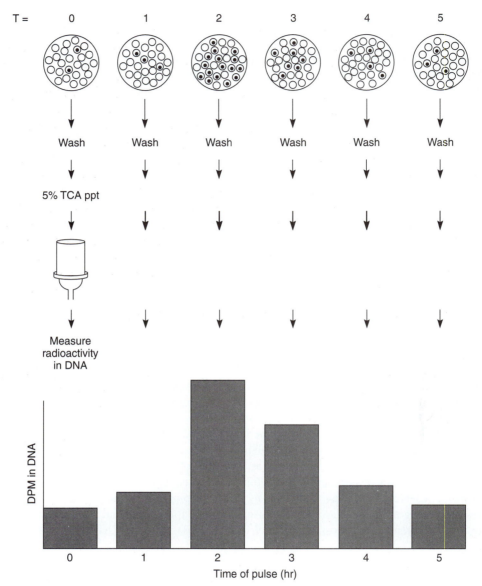

FIGURE A1.4 A pulse-labeling experiment. At time = 0, a set of six cultures was infected with a virus. At T = 0 and at hourly intervals thereafter, ^{3}H-thymidine was added to one dish of the set for a 15-minute pulse period. After the pulse period, the medium on the labeled dish was removed, the cells washed several times in buffer, and then ice-cold 5% trichloroacetic acid (TCA) was added to precipitate macromolecules such as DNA. The precipitated macromolecules were captured on filters and the amount of radioactivity on each filter determined. The results indicate that viral infection stimulates DNA synthesis about 2 hours after infection.

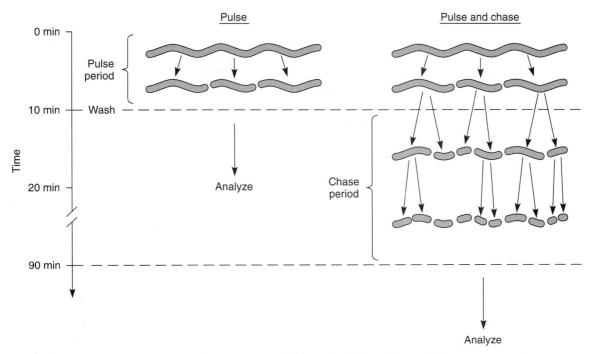

FIGURE A1.5 A pulse-chase experiment. A pair of dishes of poliovirus-infected cells were given a 10-minute ^{35}S-methionine pulse during which a very large poliovirus-encoded protein was synthesized, incorporating the radioactive amino acid, and a series of cleavage reactions begun. At the end of the pulse period the culture fluids containing the ^{35}S-methionine were removed from each dish, the cultures washed, and the contents of one of the dishes processed for determination of the size of the viral proteins. Fresh medium containing nonradioactively labeled methionine was added to the other dish to begin the 90-minute chase period. During this chase period, additional cleavages converted the original very large protein into 10 smaller proteins. At the end of the chase period, the contents of this dish were analyzed. Figure 4.4 shows the actual results of this experiment.

^{35}S-labeled methionine was given to infected cell cultures for a *pulse* of 10 minutes (Figure A1.5). All the proteins synthesized in the cell during this 10-minute pulse will incorporate the radioactive methionine along with unlabeled methionine. At the end of the pulse period, the culture fluid containing ^{35}S-labeled methionine was removed and replaced with fluid containing a large amount of unlabeled (nonradioactive) methionine. This diluted any remaining radioactive methionine so much that the likelihood of the cell incorporating a significant amount of it was very low. The cells were then allowed a 90-minute period of normal metabolism (the *chase*). During this period new proteins were synthesized but they were *not* radioactively labeled; the only radioactive proteins in the cells during the chase are those synthesized during the pulse period. The viral proteins were then separated by electrophoresis and the amount of radioactivity in each determined. The results of this experiment, shown in Figures 4.4 and 4.5, indicate that the large proteins synthesized during the pulse period were cleaved into smaller proteins during the chase.

Detection of Radioactivity

In biological research the decay of radioactive isotopes is usually detected in one of two ways, depending on whether the objective is to *quantify* the amount of decay that occurs in a sample (as in a pulse-chase experiment), or is to *localize* the presence of isotope in a preparation (as in an experiment to determine which cells in a monolayer are synthesizing DNA). *Liquid scintillation spectrometry* is usually the method of choice for determining the amount of radioisotope in samples (Figure A1.6). The biological sample (usually in liquid form) is added to a solution (called "cocktail" in lab slang) that contains *phosphors* or *scintillants* that have the property of absorbing the energy emitted from radioisotopes and releasing that energy again in the form of light that can be detected by the photocells in a liquid scintillation spectrometer. Each

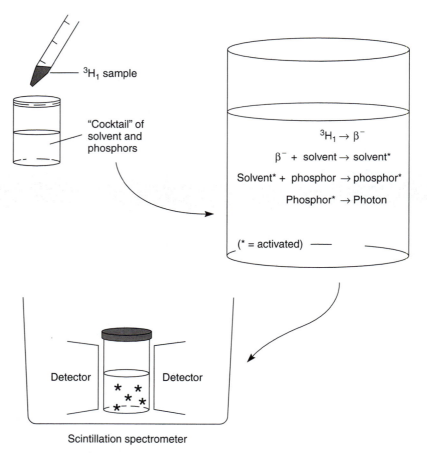

$^3H_1$ sample

"Cocktail" of solvent and phosphors

$^3H_1 \rightarrow \beta^-$

$\beta^- + \text{solvent} \rightarrow \text{solvent*}$

$\text{Solvent*} + \text{phosphor} \rightarrow \text{phosphor*}$

$\text{Phosphor*} \rightarrow \text{Photon}$

(* = activated)

Detector   Detector

Scintillation spectrometer

FIGURE A1.6 Detection of radioactivity by liquid scintillation spectrometry. Energy from radioactive disintegrations raises the energy level of solvent molecules and then phosphor molecules. The flashes of light emitted by phosphors are counted by detectors in the liquid scintillation spectrometer.

radioactive decay event thus results in a flash of light that can be detected and recorded by the counter. Scintillation counters can distinguish the flashes produced by the emission of beta particles or gamma rays by their different energy levels, so the components of two- or three-part mixtures of isotopes in the same sample can be counted separately.

Autoradiography is used to determine the location of radioactive compounds within a solid sample such as a tissue section or an electrophoresis gel. The process takes advantage of the fact that the particles released during radioactive decay can "expose" a photographic emulsion in exactly the same manner that photons of light normally do. In autoradiography for use with light or electron microscopy, a photographic emulsion is poured on tissue sections (in the dark, of course) and then developed after a suitable exposure time (Figure A1.7).

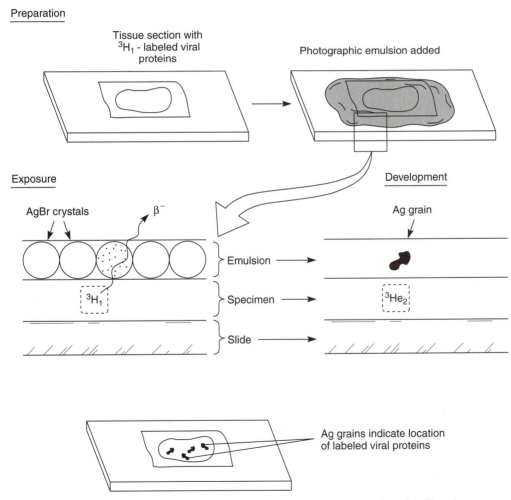

FIGURE A1.7 Detection of radioactivity by autoradiography. The location of radioactive isotope molecules within a specimen is given by the position of Ag grains in a developed photographic emulsion overlying the specimen.

Each radioactive decay event releases a particle that may travel through the emulsion and strike a particle of AgBr, creating a "latent" image that will result in the deposition of a silver grain directly above the site of the radioactive element in the specimen after development of the emulsion. Since the developed emulsion is transparent, the section can be examined under the microscope and the location of the emitting isotope determined. Autoradiography is also used in association with electrophoretic techniques, as will be described in a later section.

IV. TECHNIQUES FOR THE SEPARATION AND ANALYSIS OF MACROMOLECULES

Macromolecules can be separated from one another by a variety of techniques based on differences in the molecules' sizes, charges, or physical/chemical properties. Two basic techniques, *molecular sieve chromatography* and *centrifugation,* separate macromolecules that differ in size. Two other techniques, *electrophoresis* and *affinity chromatography* separate molecules that differ in their electrical charges or other physical/chemical properties.

Molecular Sieve Chromatography

Molecular sieve chromatography and electrophoresis are similar in that the separation takes place as the mixture of molecules moves through a matrix of some inert material. In *molecular sieve chromatography,* also called *gel filtration,* the matrix, in the form of small beads with holes of controlled sizes in them, is placed in a column and the mixture of macromolecules is percolated down the column. Large molecules slip around the beads because they are too large to enter the holes in them and thus are washed (eluted) from the column quickly; all that can be determined about these molecules is that they are all larger than some particular size (called the *exclusion limit*) of that matrix. Smaller molecules, which can enter the holes in the beads, move more slowly down the column since they spend part of their time within beads rather than moving between beads. The smaller the molecule, the more often it will enter a bead and therefore the slower it will move down the column (Figure A1.8, *A*).

An example of this type of analysis would be determining the number and sizes of the proteins in a virion. The virion, dissociated into its component proteins by addition of a detergent (sodium dodecyl sulfate, Triton X-100, or cetylpyridinium chloride, for example), to the solution, is layered onto a column containing beads of sepharose (Sephadex), the fluid coming through the column collected in aliquots (often called "fractions"), the concentration of protein in each aliquot determined chemically or by measuring A_{280}, and those values graphed against the volume of fluid that has passed through the column. The position of the peaks on the resulting graph relative to those of proteins of known sizes allows an estimate of molecular weight to be made (Figure A1.8, *B*). Molecular sieve chromatography is frequently used as a preparative as well as an analytical tool since the materials in the aliquots are available for further use if desired.

Centrifugation

Centrifugation is another technique for separating macromolecules by size that can be used either analytically or preparatively. It is particularly useful in purifying virions from cell culture fluids or cell homogenates. The technique separates particles or macromolecules in an artificial gravitational field (which can be as great as $700,000 \times g$). The actual basis of

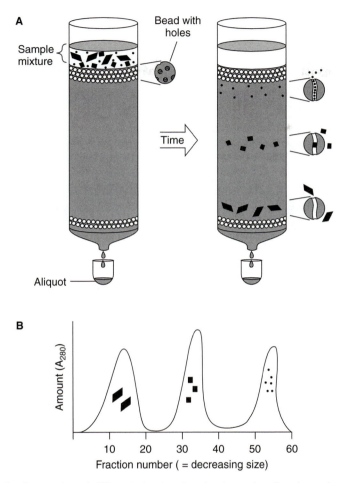

FIGURE A1.8 Separation of different sized molecules by molecular sieve chromatography.
A, As a mixture of molecules passes down the column, small molecules are slowed as they enter the holes in the matrix beads, while large molecules simply pass around the beads. The largest molecules therefore leave the column first. **B,** The amount of protein in each aliquot (or fraction) is determined from its A_{280}. These values are plotted against the volume of liquid that has passed through the column or against the fraction number.

separation of macromolecules by centrifugation is complex since both the molecular weight of the molecules and their three-dimensional configuration in fluid play roles in affecting the separation. For this reason separation is often referred to as dependent on the particles' *buoyant densities.* Sizes determined by centrifugation are termed *sedimentation coefficients,* measured in units called Svedbergs (S); the size of the prokaryotic ribosome is 70S, for example.

Two forms of centrifugation are commonly used in virology. In the first the objective is to remove particles (often virions) or macromolecules from a suspension. In this simple technique the sample is spun at the speed necessary to create a gravitation force sufficient to *pellet* (force to the bottom of a centrifuge tube) the desired materials in a reasonable length

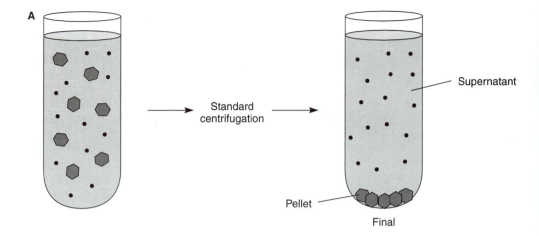

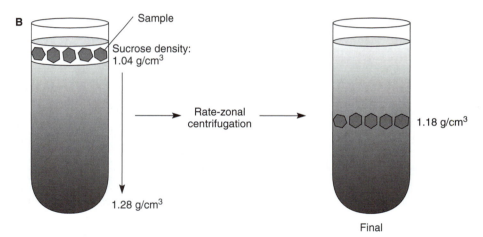

FIGURE A1.9 Two types of centrifugation techniques. **A,** Centrifugation to remove particles or macromolecules from a homogeneous suspension. **B,** Purification of avian retroviruses on a continuous sucrose density gradient. The virions band at a sucrose density of 1.18 g/cm³.

of time (Figure A1.9, *A*). The fluid (supernatant) remaining above the pellet is removed and the sample recovered. The pelleted materials may not be pure, however, since every material with an S value greater than or equal to the material of interest will also be pelleted.

A second use of centrifugation allows high resolution separations of molecules. *Rate-zonal centrifugation* separates molecules based on their mass. In this method a density gradient of some inert material such as sucrose is created in the centrifuge tube and the materials to be separated are then layered on the top of this preformed gradient. During centrifugation the materials are forced down the gradient until they encounter a concentration of sucrose that "buoys" them up so much that their further migration down the tube is greatly slowed. The individual components therefore form bands in the gradient (Figure A1.9, *B*).

An example of how these two techniques are combined is in the purification of avian retroviruses. The culture fluids from infected cells are first centrifuged at 10,000 × g for 10 minutes to remove cell debris. The supernatant is then centrifuged at 80,000 × g for 30 minutes to pellet the viruses. The viruses are then layered onto a 15%–60% sucrose density gradient (density from 1.04 to 1.28 g/cm³) and centrifuged at 250,000 × g for 60 minutes. The viruses form a sharp band at a sucrose density of 1.18 g/cm³. After centrifugation the material in the gradients is recovered by dripping the gradient through a hole in the bottom of the tube or by inserting a needle through the side of the tube at the appropriate level.

Equilibrium density-gradient centrifugation separates molecules based on their densities. In this method the materials to be separated (such as DNA and RNA) are mixed with an inert material (such as $CsCl_2$) that will form a gradient during the actual centrifugation. As the gradient is created, the materials to be separated form bands at their equivalent densities in the gradient.

Electrophoresis

In molecular sieve chromatography the macromolecules are carried through the matrix by the flow of fluid. In *electrophoresis,* charged macromolecules are forced through a meshwork of matrix material by the flow of an electric current (*electro* + *phoresis* = "carrying" in Greek). The most common matrix materials (referred to as *gels* since they are semisolid aqueous suspensions) are agarose (for nucleic acids) and polyacrylamide (for proteins as well as nucleic acids). The rate at which a molecule is drawn through the gel depends upon *charge-mass ratio;* two molecules with the same mass and shape but different total electrical charges will migrate at rates dependent on their charges. For example, at a given pH the various amino acids of two different proteins of the same mass may have different charges, so the proteins will migrate at different rates in a polyacrylamide gel.

Although many applications of electrophoresis use the technique to separate molecules based on their charge differences, two very important applications of electrophoresis separate molecules based on their sizes also. If the subunits of the macromolecule all have the same charge-mass ratio, as is the case with nucleic acids since each nucleotide in a chain has about the same unit mass and charge, small molecules will be carried through the gel quickly while larger molecules will be "hung up" and will be moved more slowly. For example, to determine the sizes of genomic molecules in a virus with a segmented genome, the collection of nucleic acids is loaded into a well at one end of an agarose gel, and a uniform electric current is applied for a set period of time (Figure A1.10). The location of the various molecules in the gel is then usually determined by staining techniques or by autoradiography if the nucleic acids were radioactively labeled.

Proteins can also be separated based on their sizes if they are first denatured by a detergent like sodium dodecylsulfate (SDS). One molecule of SDS (with a negative charge at neutral pH) binds to each amino acid in the peptide chain. Since the charged SDS molecules repel each other, each peptide assumes a rodlike shape with each subunit having the same mass-charge ratio. Separation therefore is based on the length of the peptide. Figure 4.5 shows an autoradiogram of ^{35}S-labeled poliovirus proteins separated on a 10% polyacrylamide gel containing SDS.

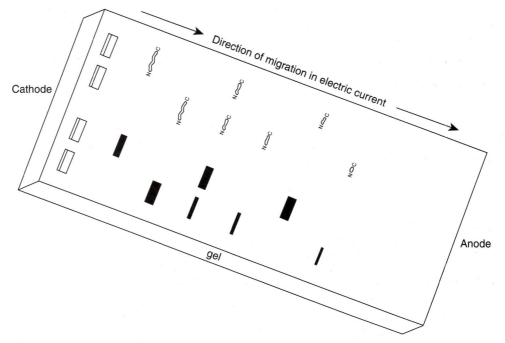

FIGURE A1.10 Separation of different sized protein molecules by electrophoresis. Mixture A was placed in wells 1 and 3, and mixture B in wells 2 and 4. The smaller a protein is, the more rapidly it is moved through the gel by the electric current. The relative sizes of the protein molecules in mixtures A and B are diagrammed in wells 1 and 2. The bands that would actually be seen in a stained gel are diagrammed in wells 3 and 4.

As Figure 4.5 illustrates, an advantage of electrophoresis compared to molecular sieve chromatography is that a number of samples can usually be analyzed and compared to standards at the same time. Other advantages of electrophoresis are that electrophoresis can usually resolve molecules that are small or close in size to each other and that the sample analyzed can be quite small (often in the order of μgs of material).

Affinity Chromatography

A powerful method of purifying proteins that takes advantage of their individual physical/chemical properties is *affinity chromatography.* This technique takes advantage of the highly specific kinds of binding interactions that are a hallmark of biological systems. One component of a binding system (molecules like the peptide hormone insulin, for example) is attached to an inert carrier (often a gel filtration bead) and a solution containing the other component of the binding system (the insulin receptor protein, for example) is added. As the solution percolates around the beads, a specific binding reaction occurs between the insulin receptor in solution and the insulin molecules linked to the beads. When the beads are washed with buffer, all the unattached proteins are removed and the desired insulin receptor protein is retained. It is then removed from the affinity column by adjusting the pH or

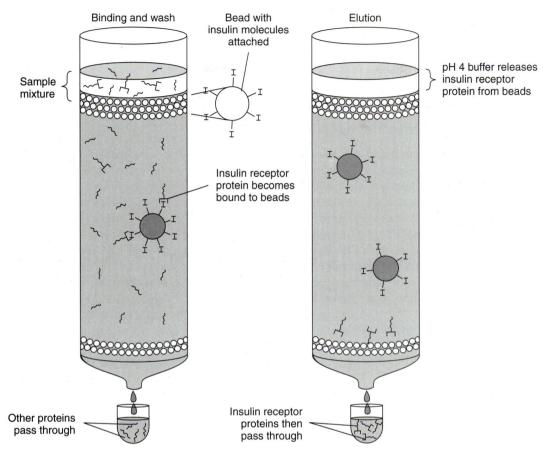

Binding and wash

Bead with insulin molecules attached

Elution

Sample mixture

pH 4 buffer releases insulin receptor protein from beads

Insulin receptor protein becomes bound to beads

Other proteins pass through

Insulin receptor proteins then pass through

FIGURE A1.11 Affinity chromatography. A pH 7 solution containing proteins solubilized from plasma membranes is run through an affinity column containing beads with insulin molecules attached to them. The insulin receptor proteins become bound to the beads while all other proteins are washed from the column. The insulin receptor proteins are then released from the column by washing with a pH 4 buffer.

ionic strength of the buffer such that the specific binding reaction is neutralized (Figure A1.11). This type of technique can also be applied in its reciprocal form: The specific binding protein is attached to the beads (antibodies and lectins [plant proteins that bind specific carbohydrates] are commonly used in this manner) and their ligands are percolated through the column.

IV. ANALYSIS OF NUCLEIC ACIDS

DNA sequencing and nucleic acid hybridization techniques provide powerful tools for the analysis of viral genomes and their transcription products. *Nucleic acid hybridization* is the term applied to a variety of techniques based on the observation that when two single-stranded nucleic acid molecules with complementary base sequences are brought together under suitable conditions, base pairing will lead to the formation of a double-stranded molecule. The complementary sequences can both be in DNA molecules, both in RNA molecules, or one in DNA and the other in RNA. There are three different types of experimental questions that can be answered by hybridization studies: (1) what is the sequence complexity of the nucleic acid, (2) what is the degree of relatedness of two nucleic acid sequences to each other, and (3) what is the location of particular target sequences?

Reassociation Kinetics Hybridization

Sequence complexity, that is, the number and length of particular sequences within a larger molecule, is determined by performing *reassociation kinetics hybridizations.* In a typical experiment, double-stranded DNA is denatured by heat or ionic conditions to separate the individual strands so they can diffuse apart. The temperature is then lowered or the ionic conditions rendered favorable for base pairing, so that renaturation can occur. The amount of double-stranded DNA is determined over various time intervals. The rate of renaturation (also called reannealing) depends upon how readily complementary sequences find each other in the solution. If there are numerous copies of complementary sequences in the solution, they can find each other and base pair quickly. Sequences that occur only once, on the other hand, may require a very long time to come into contact with each and form base pairs. Figure A1.12 compares the reassociation kinetics of DNAs from the bacteriophages T4, *E. coli,* and calf thymus. Nearly the entire *E. coli* genome consists of single copy sequences of moderate length, so the reassociation kinetics curve is sigmoidal around a $C_0t = 9$ (C_0t is the product of C_0, the concentration of DNA at time = 0, and the time the reaction is run in seconds). The curve for the bacteriophage T4, which has a much smaller genome than *E. coli,* centers around $C_0t = 0.3$, while the eukaryotic DNA with its larger size and repeated sequences produces a more complex curve.

Saturation Hybridization

Saturation hybridization is used to compare the sequences of two different nucleic acids. For example, it has been used to determine what portion of the RNA sequences of various members of the picornaviruses are the same. To test the relatedness of two serotypes of cox-sackievirus, A7 and A10, radiolabeled A7 viral RNA was generated by infecting cells in the presence of ^{32}P. This RNA, called the *probe,* was then mixed with different amounts of nonradiolabeled A7 RNA or with A10 RNA. Sufficient time was allowed for all possible complementary sequences to find each other, and then an enzyme was added to digest away all single-stranded RNA that had not formed base pairs with its complement. The amount of radioactivity in the remaining double-stranded RNA was then determined. As the graph in Figure A1.13 shows, the amount of double-stranded RNA formed when A7 and A10 RNAs were mixed is about one-third that formed when the same experiment was performed using A7 RNA as both probe and target, indicating that A7 and A10 share about a third of their RNA sequences.

In Situ Hybridization

In situ (*situs* is Latin for "local place") *hybridization* is used to determine where target sequences are located within a specimen. *In situ* hybridization was used, for example, to determine the location at which mouse mammary tumor virus (MMTV) provirus DNA becomes

A

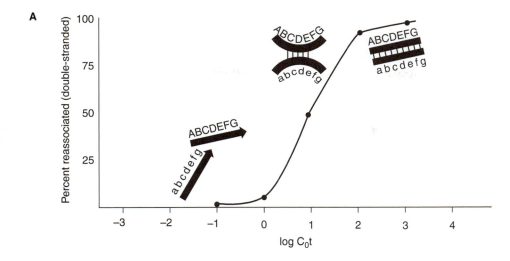

B

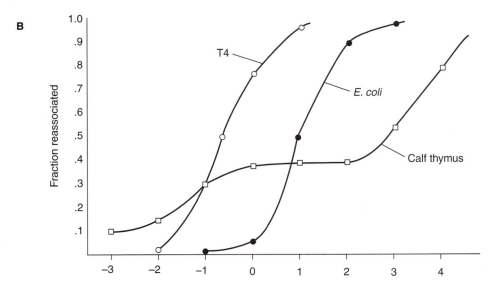

FIGURE A1.12 Reassociation kinetic hybridization. **A,** The proportion of molecules that have reannealed or hybridized increases as C_0t increases. The uppercase and lowercase letters represent DNA sequences that are complementary. **B,** The reassociation kinetics of T4, *E. coli,* and calf thymus DNAs. T4 DNA is 50% reassociated at $C_0t = 3 \times 10^{-1}$ and *E. coli* DNA at $C_0t = 9 \times 10^{1}$. Calf thymus has a series of phases rather than a single sigmoid curve.

integrated into the host cell chromosomal DNA. A karyotyping chromosome squash is prepared from mammary epithelial cells from a GR strain mouse and then treated to denature the chromosomal DNAs into single strands. Radioactive probe made by labeling the MMTV

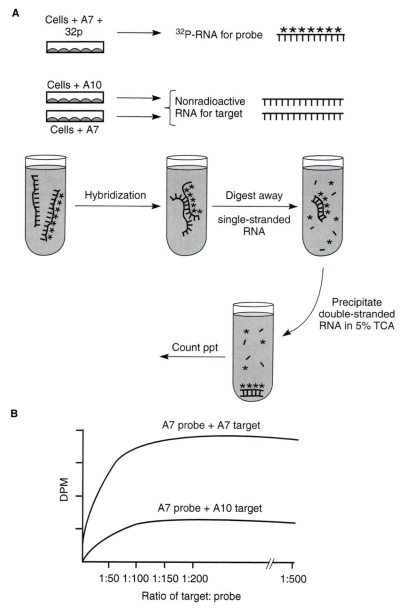

A, Cells + A7 + 32p → ^{32}P-RNA for probe

Cells + A10 → Nonradioactive RNA for target

Cells + A7

Hybridization → Digest away single-stranded RNA → Precipitate double-stranded RNA in 5% TCA → Count ppt

B,

A7 probe + A7 target

A7 probe + A10 target

DPM

1:50 1:100 1:150 1:200   1:500

Ratio of target: probe

FIGURE A1.13 Saturation hybridization. **A,** ^{32}P-labeled coxsackievirus strain A7 RNA (probe) and nonlabeled coxsackievirus strain A10 and strain A7 RNAs (targets) were isolated. Target and probe RNAs were mixed at various ratios and allowed to hybridize. After hybridization, single-stranded RNA was digested away by ribonuclease, the double-stranded RNA hybrids precipitated in trichloroacetic acid, and the amount of radioactivity in the precipitate determined. **B,** The degree of relatedness of the A7 and A10 genomes is shown by the amount of A7 RNA probe that binds to A10 RNA relative to the binding of the A7 probe to A7 RNA. A7 and A10 appear to share about a third of their sequences.

genome is added and allowed to hybridize with the chromosomal DNAs of the squash. Unbound probe is washed off and the specimen then analyzed by autoradiography. The presence of grains over a small region of chromosome 18 indicates that this is the site of the provirus DNA in these cells.

Blotting Techniques

Two very powerful techniques for analyzing nucleic acid sequences are related in concept to *in situ* hybridization. The experimental object in both *Southern blotting* and *northern blotting* is to determine the location of particular sequences within a set of nucleic acid fragments. A Southern (with a capital "S" since it is named for Edward Southern who earned a Nobel prize for developing this procedure) blotting experiment is presented in Figure A1.14. DNA is first cleaved by restriction endonucleases, enzymes that

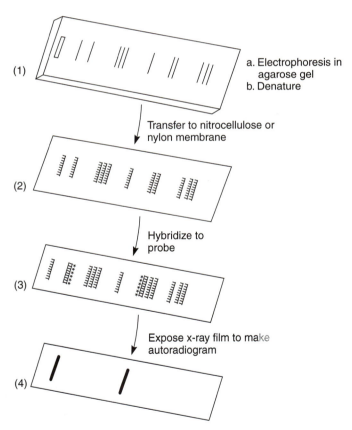

FIGURE A1.14 Southern blotting. After electrophoresis and denaturation, DNA is transferred to a nitrocellulose or nylon membrane. Sequences of interest are detected by hybridization followed by autoradiography.

recognize specific base sequences and cut at defined locations each time. The resulting fragments are then separated by size by agarose or polyacrylamide gel electrophoresis. Rather than visualizing the fragments at this time, however, the fragments are denatured into single strands and then transferred by capillary action to a sheet of nitrocellulose or nylon. The arrangement of the fragments relative to each other is not altered by the denaturation and transfer processes. A form of *in situ* hybridization is then performed using a radioactive probe for the desired sequence. The probe will bind only to those fragments bearing the target sequences. The location of those fragments is then determined by autoradiography.

A procedure in which RNA rather than DNA is separated by electrophoresis before transfer and probing is termed *northern blotting* (RNA being the "opposite" of DNA). *Western blotting,* described on page 371, is an analogous technique for analyzing proteins. The scientific community breathlessly awaits the advent of eastern blotting.

Nucleic Acid Sequencing

The ultimate analysis of a nucleic acid is the determination of its base sequence. Since *DNA sequencing* is relatively simple, RNAs are sequenced by converting them into their complementary DNA and analyzing that molecule. Several methods are available for DNA sequencing that differ only in the initial steps that generate collections of DNA fragments that differ in length. In the *dideoxy* method developed by Sanger, the DNA to be sequenced serves as the template in four separate *in vitro* DNA synthesis reactions (Figure A1.15). Each reaction mixture contains the template DNA, primer DNA that is radiolabeled (often with ^{35}S), polymerizing enzymes, and three of the four deoxynucleotide bases. The fourth base is present in two forms: the usual deoxynucleotide form (dNTP) and in the form of a *di*deoxynucleotide (ddNTP) that lacks a hydroxyl group at the 3′ position where formation of the phosphodiester bond would occur when that base has been added to a chain that is serving as the growing point. Incorporation of the dideoxynucleotide therefore causes elongation of that molecule to cease. To produce a collection of fragments that end in a guanine, for example, both dGTP and ddGTP are used. The ratio of dGTP to ddGTP is such that the products of the reaction will include some chains that end at each position where a guanine occurs. Since there are reactions for each of the four bases, there should be chains ending at every position possible. The four reaction mixtures are then separated on adjoining lanes of a polyacrylamide gel and the results visualized by autoradiography. The sequence is read up from the bottom of the gel.

Polymerase Chain Reaction

The *polymerase chain reaction (PCR)* is a powerful new tool for the detection or cloning of rare DNA sequences. The technique is based on amplification of target sequences that lie between two selected sites in a genomic DNA molecule by repeated cycles of DNA synthesis using specific primer molecules. The process begins with the synthesis of primers, oligodeoxynucleotide molecules about 15 to 30 bases long that are complementary to the two selected sites (Figure A1.16). A cycle begins with heating to denature the DNA molecule containing the target sequences and addition of the primer DNAs. Cooling the mixture allows the primer molecules to anneal to the target DNA followed by a round of DNA synthesis

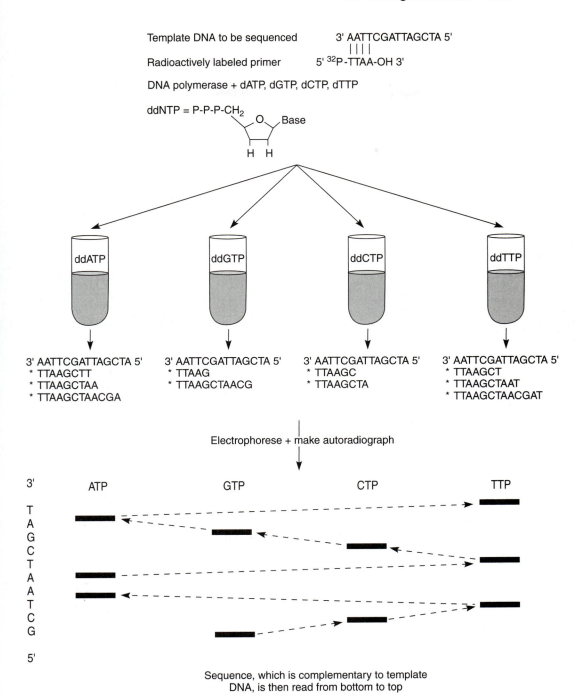

FIGURE A1.15 DNA sequencing by the Sanger dideoxynucleotide method. The sequence of interest serves as template in four separate DNA synthesis reactions, which will terminate when a dideoxynucleotide is incorporated into the nascent strand. After the reaction mixtures containing a series of incompletely elongated fragments are separated by gel electrophoresis, the bands are detected by autoradiography. The sequence is then read up the gel from the shortest fragment.

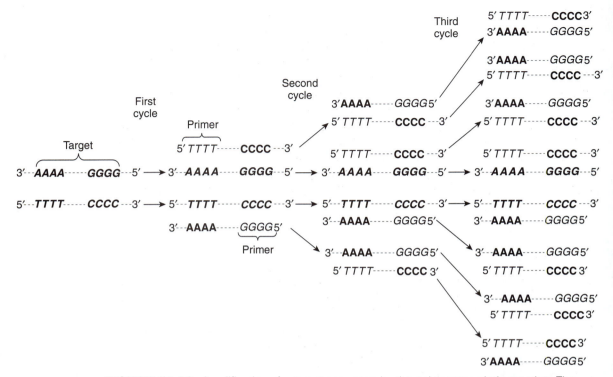

FIGURE A1.16 Amplification of a target sequence by the polymerase chain reaction. The sequence of bases at the ends of the target sequence is used to make primer oligonucleotides. Denaturation of the DNA strands containing the target sequences permits the oligonucleotides to bind and prime a round of DNA synthesis by *Taq* polymerase. A second cycle of denaturation, primer binding, and DNA synthesis produces two DNA strands consisting only of the target sequences. Successive cycles quickly enrich the reaction mixture for those sequences.

catalyzed by a heat-stable enzyme called *Taq* polymerase. This first cycle produces two new DNA strands, each of which begins with a primer sequence at its 5′ end and extends past the target sequences at its 3′ end. The reaction mixture is then heated to begin a second cycle of denaturing to create single-stranded DNA, annealing of primers to that DNA, and then synthesis of new strands. Two of the eight strands resulting from the second cycle contain only the target sequences that are flanked by the primer sequences. Each succeeding cycle enriches the fraction of strands that contain only the target, until after 60 cycles there are 2.3×10^{18} copies of the target sequence (Table A1.2).

PCR amplification is used in many research and clinical settings. For example, PCR is the basis of a very sensitive test for HIV that can detect a single infected lymphocyte in a population of 10^6 uninfected cells. The primer oligodeoxynucleotides for this assay are designed to bind only to conserved regions of the HIV proviral DNA. The test is particularly useful in determining the status of infants too young to have produced the HIV-specific antibodies that are the basis of other HIV detection methods.

Table A1.2 | Amplification of Target Sequences by the Polymerase Chain Reaction

Number of Cycles (n)	Total Strands (2^{n+1})	Strands with Extensions ($2n + 2$)	Target-Only Strands	(%) Target-Only Strands
0	2	2	0	0
1	4	4	0	0
2	8	6	2	25
3	16	8	8	50
4	32	10	22	69
5	64	12	52	81
10	2048	22	2026	99
20	2.1×10^6	42	2.1×10^6	99 +
40	2.2×10^{12}	82	2.2×10^{12}	99 +
60	2.3×10^{18}	122	2.3×10^{18}	99 +

VI. IMMUNOLOGIC TECHNIQUES

For many years immunological techniques have been a mainstay for clinical virology as well as having many uses in research settings. Many of these methods employ *antibodies* (also called *immunoglobulins*) found in the blood serum, and are therefore called *serological* methods. The antibodies produced in an infected organism are directed against various small biochemical groupings (*antigenic determinants*) exposed on a virion or on the surfaces or in the interior of infected cells. The serum is *polyclonal* since it contains a mixture of specificities of antibodies to the same virus. *Monoclonal* antibody, or antibody of a single specificity, can be produced *in vitro* by fusing a single antibody-producing cell to a myeloma cell, creating a hybridoma cell line that continues to synthesize its specific antibody.

Indirect immunofluorescence assay (IFA), radioimmunoassay (RIA), and enzyme-linked immunosorbent assay (ELISA) techniques are the most broadly used methods for detection of viruses in clinical specimens. *Immunoblotting,* or as it is more commonly called, *western blotting,* is principally a research method at present but has important clinical uses as well (the detection of HIV infection is one). All of these methods are called "indirect" antibody assays because the binding of specific antibodies (the primary antibody) to viral antigenic determinants is detected by a second reaction also involving antibodies. The detection system employs antibodies (the secondary antibody) from a different species raised against the antibodies of the species that supplied the serum being assayed. For example, the primary antibody may be human and the secondary antibody rabbit antihuman antibody. The secondary antibody has attached to it (conjugated) a fluorescent dye (fluorescein isothiocyanate or rhodamine), a radionuclide (often ^{125}I), or an enzyme (horseradish peroxidase, alkaline phosphatase, and others). Figure A1.17, *A* shows the general scheme for immunoassays using conjugated antibodies.

Immunofluorescence Assays

IFA typically detects viral antigens in infected cells. The test can be for the viral antigens themselves, either in cultured cells or clinical specimens, or for the presence of antibodies

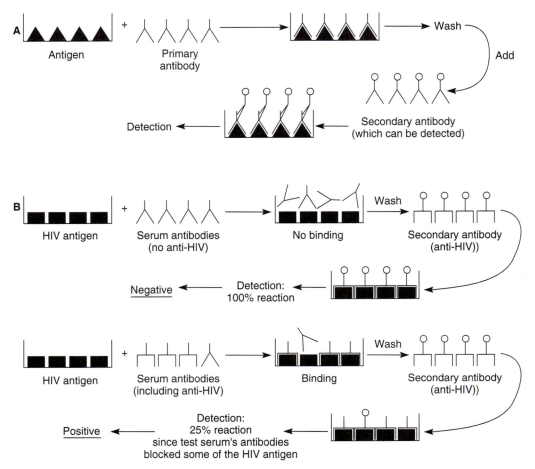

FIGURE A1.17 Immunoassays using conjugated antibodies. **A,** The basic scheme uses antigens immobilized on a surface. Primary antibodies bind to antigen molecules so that they are not removed during a wash step. Conjugated secondary antibodies then bind to the primary antibodies. Detection of bound secondary antibodies depends on the nature of the conjugate. **B,** Competitive immunoassays are used to determine the presence of specific antibodies in a serum sample. An ELISA test for anti-HIV antibodies compares the amount of binding of a conjugated anti-HIV antibody to immobilized HIV antigens in a control well that contains no other anti-HIV antibody to that occuring when a test serum sample is added. If the test sample contains anti-HIV antibodies, they will block binding of the conjugated antibody and reduce the amount of signal detected.

in serum that are specific for a particular virus constituent. An example of the second type of use of IFA is the detection of antibodies to varicella-zoster virus. VZV-infected cells are fixed as a source of antigen, incubated with the human serum being tested, and then conjugated antibodies against human antibody are added. Fluorescent staining of the VZV-infected cells indicates that the serum sample contained antiVZV antibodies.

RIA and ELISA Assays

RIA and ELISA assays are often done as competitive immunoassays in which the ability of a test serum to inhibit the binding of a standard antibody to the target antigen is determined. For example, the RIA test blood banks use to screen for hepatitis B virus uses competition between standard antiHBsAg antibodies attached to a solid surface and any antiHBsAg antibodies that may be present in a test serum sample for binding soluble radiolabeled HBsAg. If the serum contains antiHBsAg antibodies, binding of the radiolabeled antigen to the immobilized standard antibodies will be inhibited. An ELISA test for anti-HIV antibodies is performed in the wells of microtiter dishes. The dish is coated with HIV antigen, the test serum added, followed by addition of the enzyme-conjugated standard antibody. Competition between any anti-HIV antibodies in the test serum and the standard antibody will reduce the amount of binding of the standard antibody, and consequently the amount of product produced when the enzyme's substrate is added (Figure A1.17, *B*).

Immunoblot or Western Assays

Immunoblots or western blots are used to detect the presence of a specific protein in a mixture. The proteins to be analyzed are separated by electrophoresis on polyacrylamide gels and then transferred ("blotted") to nitrocellulose or nylon filters for assay. The filter is then incubated with antibody to the protein of interest. The location of antibody binding to its specific protein is detected by autoradiography if the antibody was radiolabeled or by a chemical reaction if enzyme-conjugated antibody was used.

A western blot assay is used to confirm the results of a positive ELISA test for the presence of anti-HIV antibodies in a person's serum. Virion proteins are separated by electrophoresis and transferred to a nitrocellulose filter that is then incubated with the human serum. The presence of human immunoglobulins bound to specific HIV proteins is detected by the addition of conjugated antihuman immunoglobulin antibody and development of the sample.

Characteristics of Selected Viruses

The following outline briefly describes the features of the major viral groups discussed in this book. Page references direct the reader to more detailed discussions of certain features. The Baltimore classification scheme is presented inside the front cover.

I. Class I Viruses: Double-stranded DNA
    A. Bacteriophages
        1. Myoviridae (T-even phage group)
           a. genome: linear with terminal redundancies, 172 Kbp, 120×10^6 daltons (page 89)
           b. architecture: head 110×80 nm, "contractile" tail 110 nm (page 42)
           c. attachment/penetration: via tail fibers; penetration by rearrangement of sheath proteins (page 43)
           d. genome expression: host RNA polymerase; sequential modification of recognition factors (page 92)
           e. genome replication: formation of linear concatemers via terminal redundancies (page 94)
           f. maturation/release: lysis (page 219)
        2. Siphoviridae (λ-phage group)
           a. genome: linear with cohesive ends, 48 Kbp, 33×10^6 daltons (page 95)
           b. architecture: head 60 nm diameter, "noncontractile" tail up to 500 nm (page 44)
           c. attachment/penetration: via tail (pages 45)
           d. genome expression: host RNA polymerase; sequential antitermination (page 97)
           e. genome replication: circularization via cohesive ends, theta replication followed by rolling circle (page 97)
           f. maturation/release: lysis
           g. other: may become lysogenic (page 219)

3. Podoviridae (T7-phage group)
    a. genome: linear with terminal redundancy, 40 Kbp, 25×10^6 daltons (page 85)
    b. architecture: head 65 nm diameter, tail 20 nm
    c. attachment/penetration: via tail (page 85)
    d. genome expression: host, then viral, RNA polymerases (page 85)
    e. genome replication: formation of linear concatemers via terminal redundancies (page 89)
    f. maturation/release: lysis

B. Animal Viruses
  1. Papovaviridae
      a. genome: circular, 5–8 Kbp, 3 to 5×10^6 daltons (page 177)
      b. architecture: naked icosahedral capsid, 40 to 55 nm diameter (page 172)
      c. attachment/penetration: endocytosis, uncoating in nucleus
      d. genome expression: host RNA polymerase, differential splicing of transcripts (page 178)
      e. genome replication: host DNA polymerase, theta replication (page 178)
      f. maturation/release: assembly in nucleus, lysis (page 223)
      g. other: may transform nonpermissive host cells (page 305)
  2. Adenoviridae
      a. genome: linear, 36–38 Kbp, 20 to 30×10^6 daltons, with inverted terminal redundancies and protein attached to 5′ terminus (page 181)
      b. architecture: naked icosahedral capsid with spikes, 70–90 nm diameter (page 174)
      c. attachment/penetration: endocytosis, uncoating in nucleus (pages 52, 68)
      d. genome expression: host RNA polymerase, differential splicing of transcripts (page 181)
      e. genome replication: viral DNA polymerase using protein rather than RNA primers (page 182)
      f. maturation/release: assembly in nucleus, lysis (page 223)
      g. other: may transform nonpermissive host cells (page 308)
  3. Herpesviridae
      a. genome: linear, 124–235 Kbp, 80 to 150×10^6 daltons, may have single-stranded nicks (page 174)
      b. architecture: icosahedral nucleocapsid (130 nm diameter) with envelope (page 174)
      c. attachment/penetration: direct fusion of envelope with plasma membrane, uncoating in nucleus (pages 55, 62)
      d. genome expression: host RNA polymerase regulated by viral factors (page 184)
      e. genome replication: viral DNA polymerase, circularization, and then perhaps rolling circle replication (page 185)

      f. maturation/release: assembly in nucleus, budding through nuclear membrane into ER (page 232)

      g. other: may transform permissive or nonpermissive cells (page 310)

  4. Poxviridae

      a. genome: linear with inverted terminal repeats and covalently closed ends, 130–375 Kbp, 85 to 240×10^6 daltons (page 174)

      b. architecture: complex capsid (several layers and often envelope), oval or brick-shaped, 200–400 nm long (page 174)

      c. attachment/penetration: direct fusion of envelope with plasma membrane or by endocytosis, uncoating in cytoplasm

      d. genome expression: viral RNA polymerase (page 186)

      e. genome replication: viral DNA polymerase, displacement synthesis to form concatemers (page 187)

      f. maturation/release: assembly in cytoplasm, budding (page 232)

  5. Hepadnaviridae

      a. genome: circular but neither strand closed; negative strand, 3.2 Kbp, 1.6×10^6 daltons, circularized by incomplete positive strand of variable length (page 193)

      b. architecture: icosahedral nucleocapsid with envelope, 42 nm diameter (page 193)

      c. attachment/penetration: not known

      d. genome expression: after genome converted to double-stranded covalently closed circular molecule, host RNA polymerase makes nested set of three mRNAs (3′ co-terminal), longest of which is about 200 bases longer than the genome itself (page 194)

      e. genome replication: reverse transcription of longest mRNA in cytoplasm, second strand synthesis in virion (page 195)

      f. maturation/release: assembly in cytoplasm, release mechanism not known

      g. other: may transform permissive host cells

C. Plant Viruses

  1. Caulimovirus

      a. genome: circular but neither strand closed, 8 Kbp, 4 to 5×10^6 daltons; negative strand has one discontinuity, positive strand has two discontinuities (page 197)

      b. architecture: naked icosahedral capsid, 50 nm diameter

      c. attachment/penetration: aphid vector (page 70)

      d. genome expression: host RNA polymerase makes two nested (3′ co-terminal) mRNAs, the longest of which is longer than the genome (page 197)

      e. genome replication: reverse transcription of longer mRNA molecule (page 198)

      f. maturation/release: assembly in cytoplasm (viroplasm), aphid vector

II. Class II: Single-stranded DNA Viruses

  A. Bacteriophages

    1. Microviridae (φX174 phage group)

      a.   genome: circular, 5.4 Kb, 1.7×10^6 daltons; positive strand (page 100)

      b.   architecture: icosahedral capsid with short spikes, 27 nm diameter

      c.   attachment/penetration: spike attaches to cell wall lipoprotein, penetration aided by pilot protein on DNA

      d.   genome expression: host RNA polymerase makes three polycistronic mRNAs; nonstructural proteins are translated from overlapping reading frames or from different initation sites in same frame (page 100)

      e.   genome replication: synthesis of antigenomic strand, followed by first theta and then rolling circle replication (page 102)

      f.   maturation/release: lysis (page 216)

  B. Animal Viruses

    1. Parvoviridae

      a.   genome: linear with self-complementary ends, 5 Kb, 1.5 to 2.2×10^6 daltons; negative strand in one genus, either in two genera (page 174)

      b.   architecture: naked icosahedral capsid, 18–26 nm diameter (page 174)

      c.   attachment/penetration: endocytosis

      d.   genome expression: host RNA polymerase, alternative splicing of transcripts, cleavage of polyprotein, overlapping reading frames all used (page 188)

      e.   genome replication: self-primed displacement synthesis to form concatemers (page 190)

      f.   maturation/release: assembly in cytoplasm; lysis

      g.   other: two genera are autonomous, one is replication defective (page 188)

  C. Plant Viruses

    1. Geminiviruses

      a.   genome: circular, 2.7–2.9 Kb, 0.7 to 1×10^6 daltons; one or two molecules per complete virion; positive strand (page 175)

      b.   architecture: fused pair of icosahedral capsids (page 175)

      c.   attachment/penetration: leafhopper and whitefly vector (page 70)

      d.   genome expression: host RNA polymerase, transcription from genome and its complement (page 191)

      e.   genome replication: possibly rolling circle via a double-stranded intermediate

      f.   maturation/release: nuclear assembly, leafhopper, and whitefly vectors

III. Class III: Double-stranded RNA Viruses

  A. Animal Viruses

    1. Reoviridae

      a.   genome: 10 to 12 segments; total 16–23 Kbp, 12 to 20×10^6 daltons (page 151)

    b.   architecture: naked icosahedral outer capsid (60–80 nm diameter) with short spikes containing isometric nucleocapsid also with spikes (page 151)

    c.   attachment/penetration: endocytosis followed by removal of outer capsid; inner nucleocapsid remains intact in cytoplasm (page 151)

    d.   genome expression: integral viral RNA polymerase transcribes negative strands of genome into monocistronic mRNAs that are extruded from nucleocapsid (page 151)

    e.   genome replication: mRNAs are encapsidated in nucleocapsid and negative-strand RNAs synthesized (page 154)

    f.   maturation/release: outer capsid assembled around nucleocapsid in cytoplasm; lysis (page 232)

IV.   Class IV: Positive-strand RNA Viruses

    A.   Bacteriophages

        1.   Leviviridae

            a.   genome: linear, 3.5–4.7 Kb, 1.2×10^6 daltons (page 107)

            b.   architecture: quasi-icosahedral capsid, 23 nm diameter (page 107)

            c.   attachment/penetration: attachment to F pilus (page 45)

            d.   genome expression: proteins are translated from overlapping reading frames or from same reading frame using read-through of weak terminator (page 107)

            e.   genome replication: viral replicase synthesizes antigenome as template for new genomes (page 110)

            f.   maturation/release: lysis

    B.   Animal Viruses

        1.   Picornaviridae

            a.   genome: linear, 7.2–8.4 Kb, 2.4×10^6 daltons; VPg at 5′ and 3′ poly-A tail (page 119)

            b.   architecture: naked icosahedral capsid, 27–30 nm diameter (page 119)

            c.   attachment/penetration: endocytosis; uncoating in cytoplasm (page 51)

            d.   genome expression: genome translated into single polyprotein that is cleaved into individual proteins (page 121)

            e.   genome replication: viral RNA polymerase synthesizes antigenome as template for new genomes (page 126)

            f.   maturation/release: assembly in cytoplasm; lysis (page 213)

        2.   Togaviridae

            a.   genome: linear, 9.7–11.8 Kb, 4×10^6 daltons; 5′ cap and 3′ poly-A tail (page 120)

            b.   architecture: icosahedral nucleocapsid (30–42 nm diameter) with envelope with spikes (60–65 nm diameter) (page 120)

            c.   attachment/penetration: endocytosis and fusion with vesicle membrane; uncoating in cytoplasm (page 62)

      d.   genome expression: one polyprotein translated directly from 5′ end of genome and a second from an mRNA corresponding to the 3′ end of the genome transcribed from the antigenome (page 129)

      e.   genome replication: viral RNA polymerase synthesizes antigenome as template for new genomes (page 131)

      f.   maturation/release: assembly in cytoplasm; budding from cytoplasmic membrane

   3. Coronaviridae

      a.   genome: linear, 20–30 Kb, 9 to 11 × 106 daltons; 5′ cap and 3′ poly-A tail (page 120)

      b.   architecture: helical nucleocapsid, envelope with spikes, 60 to 220 nm diameter (page 120)

      c.   attachment/penetration: direct fusion of envelope with plasma membrane or by endocytosis and fusion with vesicle membrane; uncoating in cytoplasm

      d.   genome expression: initial translation of 5′ end of genome into a polyprotein, followed by translation of a nested set of mRNA's (3′ co-terminal) synthesized from the antigenome (page 133)

      e.   genome replication: viral RNA polymerase synthesizes antigenome as template for new genomes

      f.   maturation/release: bud into rough ER; release by lysis or fusion of vesicles with plasma membrane

 C. Plant Viruses

   1. Potyvirus (potato Y virus group)

      a.   genome: linear, 9.5 Kb, 3 to 3.5 × 10^6 daltons; VPg at 5′ and 3′ poly-A tail (page 138)

      b.   architecture: flexible helical capsid, 11 nm diameter, 700 nm long

      c.   attachment/penetration: aphid vector (page 70)

      d.   genome expression: genome translated into a single polyprotein that is cleaved into individual proteins (page 138)

      e.   genome replication: viral RNA polymerase synthesizes antigenome as template for new genomes

      f.   maturation/release: assembly in cytoplasm, aphid vector

   2. Tymovirus (turnip yellow mosaic virus group)

      a.   genome: linear, 6.3 Kb, 2 × 10^6 daltons; 5′ cap and 3′ tRNA rather than poly-A tail (page 138)

      b.   architecture: icosahedral capsid, 29 nm diameter

      c.   attachment/penetration: beetle vector (page 70)

      d.   genome expression: one polyprotein translated directly from 5′ end of genome and a second from an mRNA corresponding to the 3′ end of the genome transcribed from the antigenome (page 138)

      e.   genome replication: viral RNA polymerase synthesizes antigenome as template for new genomes

      f.   maturation/release: assembly in cytoplasm, beetle vector

3. Tobamovirus (tobacco mosaic virus group)
   a. genome: linear, 4.8 Kb, 2×10^6 daltons; 5' cap and 3' tRNA instead of poly-A tail (page 138)
   b. architecture: rigid helical capsid, 18 nm diameter, 300 nm long (page 203)
   c. attachment/penetration: mechanical or seed transmission (page 70)
   d. genome expression: initial translation of 5' end of genome into two polyproteins via read-through of weak terminator, followed by translation of a nested set of mRNA's (3' co-terminal) synthesized from the antigenome (page 138)
   e. genome replication: viral RNA polymerase synthesizes antigenome as template for new genomes
   f. maturation/release: assembly in cytoplasm, mechanical or seed transmission (page 203)
4. Comovirus (cowpea mosaic virus group)
   a. genome: two linear molecules, 5.9–3.5 Kb, 2.4×10^6 daltons and 1.4×10^6 daltons; both 5' VPg and 3' poly-A tail (page 138)
   b. architecture: bipartite icosahedral capsids, 28 nm diameter (page 138)
   c. attachment/penetration: beetle vector (page 70)
   d. genome expression: each RNA translated into a polyprotein (page 138)
   e. genome replication: viral RNA polymerase synthesizes antigenome as template for new genomes
   f. maturation/release: beetle vector

V. Class V: Negative-strand RNA
   A. Animal Viruses
      1. Rhabdoviridae
         a. genome: linear, 13–16 Kb, 3.5 to 4.6×10^6 daltons; termini are complementary (page 143)
         b. architecture: flexible helical nucleocapsid coiled into bullet shape (45–100 nm diameter, 100–430 nm long), envelope with spikes (page 141)
         c. attachment/penetration: endocytosis followed by fusion of envelope with vesicle membrane; uncoating in cytoplasm (page 55)
         d. genome expression: monocistronic mRNAs transcribed from genome by RNA polymerase that is part of virion (page 143)
         e. genome replication: viral RNA polymerase synthesizes antigenome and then new genomes (page 144)
         f. maturation/release: sites of assembly and budding are variable (page 232)
         g. other: two genera can infect both vertebrate and insect host; another genus infects plant hosts

2. Paramyxoviridae
   a. genome: linear, 16–20 Kb, 5 to 7×10^6 daltons; termini are complementary (page 141)
   b. architecture: flexible nucleocapsid; envelope with spikes, 150 nm diameter but very pleomorphic (page 141)
   c. attachment/penetration: fusion of envelope with cytoplasmic membrane; uncoating in cytoplasm (pages 55, 62)
   d. genome expression: monocistronic mRNAs transcribed from genome by RNA polymerase that is part of virion, RNA editing (page 145)
   e. genome replication: viral RNA polymerase synthesizes antigenome and then new genomes
   f. maturation/release: assembly in cytoplasm; budding from cytoplasmic membrane

3. Filoviridae
   a. genome: linear, 19.1 Kb, 4.2×10^6 daltons; complementary termini (page 141)
   b. architecture: helical nucleocapsid (50 nm diameter); envelope with spikes; very pleomorphic; varies in length greatly and may have branches (page 141)
   c. attachment/penetration: unknown
   d. genome expression: mRNAs synthesized by virion RNA polymerase; RNA editing and translational frameshifting occurs in GP synthesis (page 145)
   e. genome replication: viral RNA polymerase synthesizes antigenome and then new genomes
   f. maturation/release: assembly in cytoplasm; budding from cytoplasmic membrane

4. Orthomyxoviridae
   a. genome: eight segments of linear RNA, total 10–13.6 Kb, 4.5×10^6 daltons; all segments have same termini (page 147)
   b. architecture: flexible helical nucleocapsid, envelope with spikes, virion 100 nm in diameter (page 141)
   c. attachment/penetration: endocytosis followed by fusion of envelope with vesicle membrane; uncoating in cytoplasm (pages 54, 66)
   d. genome expression: 5′ cap and 8 to 15 bases of host mRNAs used as primer for viral RNA polymerase to synthesize mRNAs from each segment; smallest two mRNAs are differentially spliced (page 147)
   e. genome replication: viral RNA polymerase synthesizes antigenome and then new genomes (page 148)
   f. maturation/release: assembly in cytoplasm; budding from cytoplasmic membrane (pages 223, 227)

5. Bunyaviridae
   a. genome: three segments of linear RNA, total 11–21 Kb, 6×10^6 daltons; smallest segment in one genus is ambisense (page 150)
   b. architecture: flexible helical nucleocapsids, envelope with spikes; virion about 100 nm diameter

      c. attachment/penetration: endocytosis followed by fusion of envelope with vesicle membrane; uncoating in cytoplasm

      d. genome expression: mRNAs synthesized by elongation of host primers by viral RNA polymerase

      e. genome replication: viral RNA polymerase synthesizes antigenome and then new genomes

      f. maturation/release: assembly in cytoplasm; budding from golgi

      g. other: all but one genus transmitted by arthropod vectors

   6. Arenaviridae

      a. genome: two linear RNA molecules, 10–14 Kb, 3×10^6 daltons and 1.3×10^6 daltons; both are ambisense (page 141)

      b. architecture: flexible helical nucleocapsids, envelope with spikes, virion 60 to 300 nm diameter (page 141)

      c. attachment/penetration: endocytosis followed by fusion with vesicle membrane or by direct fusion with cytoplasmic membrane; uncoating in the cytoplasm

      d. genome expression: one mRNA is synthesized from each genome segment by viral RNA polymerase and a second from the antigenome of that segment (page 150)

      e. genome replication: viral RNA polymerase synthesizes antigenome and then new genomes

      f. maturation/release: assembly in cytoplasm; buds from cytoplasmic membrane

      g. other: virion contains ribosomes

VI. Class VI: Retroviruses

  A. Animal Viruses

   1. Retroviridae

      a. genome: linear, 7.2–11 Kb, 2 to 3.5×10^6 daltons, 5′ cap and 3′ poly-A tail; two identical strands per virion (page 155)

      b. architecture: icosahedral nucleocapsid, envelope with spikes, 100 nm diameter (page 155)

      c. attachment/penetration: endocytosis followed by fusion of envelope with vesicle membrane; HIV by direct fusion with cytoplasmic membrane (pages 56, 64)

      d. genome expression: RNA genome is copied into a double-stranded DNA provirus by viral reverse transcriptase (page 157), the provirus is inserted into the host cell's genome (page 160) and then transcribed by host RNA polymerase into a polycistronic RNA that is differentially spliced for mRNAs encoding polyproteins (page 162)

      e. genome replication: provirus transcribed to form genome (page 163)

      f. maturation/release: assembly in cytoplasm, budding from cytoplasmic membrane

      g. other: may transform permissive host cells (page 297)

2,5-oligo   chains of adenylic acid residues linked by 2′-5′ phosphodiester bonds rather than the usual 3′-5′; product of 2,5-oligoA synthase that is induced by interferon and activates RNase L.

A

A protein   protein of RNA bacteriophages involved in attachment.

abortive transformation   reversion to a normal phenotype following transient neoplastic transformation of cultured cells by DNA viruses.

ambi-sense   a single strand of RNA that is both positive- and negative-strand RNA.

antigenic drift   gradual change in antigenic determinants of a viral component as a result of the accumulation of point mutations.

antigenic shift   abrupt significant change in antigenic structure of a viral component, possibly caused by reassortment or recombination.

antigenome   a nucleic acid molecule that is complementary to a genomic molecule.

antiterminator   viral protein that causes readthrough of termination codons.

antiviral state   cell contains the interferon-induced proenzymes 2,5-oligoA synthase and 67 kd protein kinase.

apoptosis   programmed cell death.

attachment   specific binding of a virus to its host cell's surface.

autonomous virus   a virus that can replicate itself in a suitable host cell.

B

Baltimore classification   an organizational scheme that relates viruses' genomes to the mRNA used to synthesize viral proteins.

budding   final step in formation of enveloped virus, in which the nucleocapsid evaginates from a host membrane system.

C

cap   a 7-methyguanosine added 5′ to 5′ in eukaryotic mRNA.

capsid   protein coat that surrounds the viral nucleic acid.

capsomer   the morphological unit of the capsid that is composed of one or several structural subunit proteins.

cell culture   growth of dispersed eukaryotic cells *in vitro*.

cell cycle   the stages through which a cell passes from one division to the next (M $\longrightarrow$ G1 $\longrightarrow$ S $\longrightarrow$ G2 $\longrightarrow$ M).

cell wall   a rigid structure external to the plasma membrane in plant or bacterial cells.

Central Dogma   (of molecular biology) term designating the processes of expression of genetic information (DNA is replicated into DNA and transcribed into RNA, which is translated into protein).

chronic infections   a type of viral infection in which the virus persists over a period of time and infectious virus can be recovered by conventional methods.

circular permutation   property of a population of viruses with terminally redundant genomes in which each individual genome begins with a different base within the overall sequence; produces a circular genetic map even though the genome is a linear molecule.

coat protein   the capsid protein of RNA bacteriophages.

cohesive ends   (also called "sticky" ends) complementary unpaired bases at the termini of linear genomes in certain viruses that permit circularization of the molecule.

coliphage   a virus that infects *E. coli*.

concatemer   a nucleic acid molecule in which the sequence of a genome is repeated numerous times.

coreceptor   molecule required for penetration after virus has attached to host cell.

core polysaccharide   a chain of sugars attached to lipid A in lipopolysaccharide.

***cos* site**   sequence at which λ virus genomic concatemer is cut to produce "cohesive" ends.

cosmids   cloning vectors constructed by inserting *cos* sequences into plasmids, enabling them to be packaged into λ particles *in vitro*.

cytocidal   killing a cell.

cytolytic   causing a cell to disintegrate.

cytopathic effects   morphological indications of viral infection in animal cells.

D

defective interfering particle   virus that cannot multiply due to a mutation in an essential gene but that can interfere with the multiplication of normal viruses infecting the same host cell.

defective virus   a virus that requires coinfection of its host cell by another virus in order to replicate.

DNA gyrase   enzyme that creates negative supercoils in DNA to permit replication.

DNA ligase   enzyme that closes gaps in the phosphodiester backbone of DNA molecule.

DNA polymerase   enzyme that synthesizes, and often proofreads, DNA.

DNA translocating vertex   structure at junction of head and tail of complex bacteriophage virion.

E

early genes   viral genes expressed at beginning of infective process, usually before and during replication of the genome (cf. *late genes*).

eclipse   period of viral reproductive cycle between penetration and maturation.

efficiency of plating (EOP)   the proportion of virions that actually produce infected cells.

endocytosis   uptake of extracellular material into a vesicle formed by invagination of the plasma membrane.

endogenous   carried in the form of a provirus and therefore transmitted vertically (cf. *vertical transmission*).

enhancer sequence   a set of bases that regulate the expression of a nearby promoter on the same nucleic acid molecule (cf. *transactivator*).

envelope   a membrane that surrounds the nucleocapsid in some viruses.

ether-sensitive virus   a virus with an envelope.

excisionase   enzyme responsible for site-specific recombination to remove lysogenic viral genome from host genome.

exogenous   virus acquired by infection and therefore transmitted horizontally (cf. *horizontal transmission*).

exon   bases in a eukaryotic transcript that encode the information for a protein.

F

flagellum   whip-like structure in bacteria that permits motility.

fluid-mosaic membrane model   model of biological membranes as phospholipid bilayers in which proteins "float."

fusion peptide   series of hydrophobic amino acids in fusion protein thought to mediate fusion of viral envelope with a cellular membrane.

fusion protein   viral envelope protein responsible for fusion of the envelope and host cell membrane during penetration and/or uncoating.

G

gene activator protein   protein that facilitates RNA polymerase binding to particular promoter sequences.

gram-negative bacteria   bacteria with a thin layer of peptidoglycan and an external membrane system.

gram-positive bacteria   bacteria with a thick layer of peptidoglycan and no external membrane system.

H

helper virus   an autonomous virus that provides some essential function or product for the replication of another virus or subviral entity infecting the same cell.

hemagglutination   clumping of erythrocytes (red blood cells) as a result of cross-linking cells by simultaneous attachment to the same virus particles.

hexamer   one of six protein subunits that meet on adjoining triangular faces of icosahedral virions.

hexon   a hexamer.

horizontal transmission   passed from cell to cell by infection.

host-induced modification   change in methylation pattern of progeny viruses that occurs when a bacteriophage infects new cell with restriction/modification systems different from the cell that produced the infecting virus.

host range   the types of different cells or organisms that a given virus can infect.

I

icosahedron   a solid figure having 20 triangular faces.

inclusion body   microscopically visible site of viral synthesis, assembly, or damage.

initiation codon   the triplet AUG to which a ribosome binds to begin translation of a protein from mRNA.

integral proteins   in the fluid-mosaic membrane model, proteins that span the lipid bilayer.

integrase   viral enzyme responsible for recombination during formation of a prophage or provirus.

interferon   a class of small proteins induced by viruses and other agents that is secreted and acts to induce an antiviral state in cells to which it binds.

intron   bases in a eukaryotic transcript that do not code for protein and are therefore spliced out during posttranscriptional modification of the transcript.

in vitro   outside the organism (literally "in glass").

in vivo   in the organism (literally "in the living [state]").

L

late genes   viral genes expressed toward the end of the infective process, usually after replication of the genome (cf. *early genes*).

latent infections   a type of viral infection in which the viral genome persists in the infected cell but infectious virus is present only during "reactivation" episodes.

lipid A   constituent of the outer membrane of gram-negative bacteria, consisting of N-acetyl-glucosamines, phosphates, and long-chain fatty acids.

lipopolysaccharide (LPS)   a constituent of the outer membrane of gram-negative bacteria, consisting of lipid A, core polysaccharide, and O antigen.

long terminal repeats   identical base sequences found at the termini of a linear nucleic acid molecule.

lyse (lysis)   cause dissolution or destruction.

lysogenic virus   virus that has become integrated into its host's chromosome (cf. *lytic virus*).

lytic cycle   viral reproduction that causes host cell to lyse.

lytic virus   virus that reproduces itself and then disrupts its host cell.

M

maturation   process of assembly of virions during virus replication.

monocistronic mRNA   RNA encoding only a single protein.

monoclonal antibody   antibody of a single specificity produced by the progeny of a single original antibody-producing cell.

monopartite   having a single genomic nucleic acid molecule.

multipartite   having more than one nucleic acid molecule, with each molecule being separately encapsidated.

multiplicity of infection   the average number of viruses that enter each infected cell.

N

naked virus   virus without an envelope.

negative strand   nucleic acid (DNA or RNA) that is complementary to mRNA (cf. *positive strand*).

neoplastic transformation   conversion to the phenotype of a tumor cell, either benign (noninvasive) or malignant (invasive).

nested set   a group of progressively shorter mRNA molecules that all have the same 3′ terminus.

nonpermissive host   cell that lacks some factor critical for viral reproduction.

nonsense codon   one of three termination triplets (UGA, UAA, UAG).

nucleation   interaction of nucleic acid with a few capsid subunits to form an assembly initiation complex.

nucleocapsid   nucleic acid and its surrounding capsid proteins.

O

O antigen   constituent of the outer membrane of gram-negative bacteria, consisting of a chain of repeating sequences of sugars.

oncogene   viral or cellular gene capable of neoplastically transforming an animal cell.

one-step growth experiment   classic experiment demonstrating that virus replication involves assembly rather than division.

operon   in prokaryotes, a group of contiguous structural genes that are transcribed into a polycistronic mRNA, together with their controlling elements.

ORF   an "open reading frame" or DNA sequence containing in-phase initiation and termination codons, and therefore presumably encoding a protein.

origin   site of initiation of DNA synthesis.

overlapping genes   genes translated in different reading frames on the same sequence of bases.

P

p53 protein that blocks cell cycle in G$_1$ if DNA damage has occurred.

pac **site** DNA or RNA sequence that interacts with capsid proteins during nucleation.

penetration introduction of viral nucleic acid into a host cell's cytoplasm.

pentamer one of the protein subunits that surrounds each vertex in an icosahedral virion.

penton a pentamer.

peripheral proteins in the fluid-mosaic membrane model, proteins that are lightly embedded in only one side of the lipid bilayer.

peptidoglycan constituent of bacterial cell walls, consisting of chains of *N*-acetyl-glucosamine and *N*-acetyl-muramic acid cross-linked by short chains of amino acids.

permissive host cell that has all the factors necessary to support a productive viral infection.

persistent infection infection that results in the continued presence of virus within the infected cells or organism.

pilot protein a protein that provides assistance in transferring viral DNA into the cytoplasm and in initiating DNA replication.

pilus hair-like appendage in bacteria permitting attachment to surfaces or transmission of genetic information during conjugation.

plaque assay method of titering viruses based on counting the cleared areas (plaques) on a continuous cell sheet produced by viral destruction of their host cells.

plasmadesma hole through plant cell walls permitting formation of cytoplasmic connections between adjacent cells.

plasmid small, autonomously replicating, circular, double-stranded DNA molecules in prokaryotes.

poly-A tail chain of adenosines added to the 3′ end of a eukaryotic transcript during posttranscriptional modification.

polycistronic mRNA single prokaryotic mRNA, which encodes several proteins to be separately translated.

polyprotein a single peptide containing the sequences of several individual proteins.

portal protein protein with hexadecimal symmetry that mediates functions of DNA translocating vertex.

positive strand nucleic acids with the same sequence as mRNA (thymine replacing uracil in DNA) (cf. *negative strand*).

posttranscriptional modifications three changes made to eukaryotic transcripts to form mRNA: capping, adding a poly-A tail, and splicing.

primer a short RNA molecule that is elongated by DNA polymerase to form an Okazaki fragment.

prion term given to infectious agent causing certain subacute spongiform encephalopathies, possibly a protein of cellular origin.

procapsid an empty icosahedral capsid during the assembly process.

productive infection infection leading to the manufacture of new virions.

promoter sequence in DNA to which RNA polymerase binds to initiate transcription.

protomer one of the individual proteins that form a viral capsid; also called a "structural subunit."

proto-oncogene a normal cellular gene homologous to a viral oncogene.

provirus viral genome that has become incorporated into its host cell's genome.

R

Rb protein that regulates entry of cell into S phase of the cell cycle.

reading frame the codon sequence determined by grouping bases in threes from a fixed starting point.

release process by which virions leave their host cells.

replicative form or RF a double-stranded intermediate in the replication of a single-stranded DNA or RNA genome (cf. *replicative intermediate*).

replicative intermediate or RI during replication of a single-stranded DNA or RNA genome, a template molecule with a number of daughter molecules being copied from it simultaneously (cf. *replicative form*).

repressor protein a molecule that blocks transcription or translation by binding to a specific location on the template molecule.

restriction the condition in which a bacteriophage is unable to infect a host cell strain different from the strain that produced it.

restriction endonucleases enzymes that cleave DNA following the recognition of specific sequences in a DNA chain.

reverse transcriptase   RNA-dependent DNA polymerase.

reverse transcription   DNA synthesis from an RNA template molecule.

rho   a peptide that enhances the recognition of some transcription termination sequences by RNA polymerase in prokaryotes.

RNA replicase   RNA-dependent RNA polymerase.

RNA-dependent RNA-polymerase   enzyme that uses RNA as a template to synthesize a complementary RNA molecule.

rolling circle replication   asymmetric replication of a circular double-stranded molecule in which one strand is nicked and peeled away from the template molecule at its 5′ end while its 3′ end is continuously elongated (cf. *sigma intermediate*).

S

satellite RNA   small RNA molecules requiring a helper virus for replication.

satellite virus   small virus requiring a helper virus for replication.

scaffolding proteins   proteins that assist in assembly of an icosahedral procapsid but do not become part of the mature capsid.

sigma intermediate   the appearance of a molecule during rolling circle replication.

single-stranded binding proteins (SSBs)   proteins that hold template strands of DNA open during replication.

specialized transduction   transfer of one or more cellular genes as part of an infecting virus's genome.

spike   protein/glycoprotein structure that projects from the surface of nucleocapsids of some naked viruses or occur on viral envelopes; involved in attachment and penetration.

splicing   removal of introns and joining of exons during posttranscriptional modification of eukaryotic transcripts.

sticky ends   see *cohesive ends.*

structural gene   a gene that encodes a protein.

structural subunit   individual proteins of the capsomer; a protomer.

superinfection exclusion   the resistence to infection of a cell already infected by the same virus, often because the incoming viral genome is degraded.

superinfection immunity   the resistence to infection of a lysogenic cell by other viruses of the same type.

T

T antigens   "tumor-specific" antigens; nonstructural proteins encoded by the papovaviruses and adenoviruses that are associated with tumor production and transformation.

target tissues   the cells in which viruses produce their particular diseases.

tegument   a thin structure occurring between the core and envelope in herpesviruses.

teichoic acid   constituent of gram-positive bacterial cell surface, consisting of a long chain of glycerols or ribitols with substitutions.

temperate   another term for "lysogenic."

terminal redundancy   the same sequence occurs at both ends of a linear genome.

terminase   complex of proteins that both cuts DNA concatemers and aids in translocation of genome-sized molecules into capsids.

theta intermediate   the form resembling a Θ seen during bidirectional replication of a circular double-stranded DNA genome.

transactivator   a genetic regulatory protein that can act on sequences physically distant from its own gene, even on different molecules of nucleic acid (cf. *enhancer sequence*).

transfection   uptake and incorporation of naked DNA by eukaryotic cells.

transformation   an infective process that causes the host cell to take on the characteristics of a tumor cell. See *neoplastic transformation.*

tumor suppressor   protein that prevents growth of cells with damaged DNA or that have entered G_0.

tumor viruses   viruses capable of causing neoplastic transformation.

U

uncoating   process (after or coincident with penetration) that makes the viral nucleic acid within the capsid available for expression.

V

vector   an organism that transmits a virus from one host to another; often an insect.

vertical transmission   passage of a provirus from one generation of cells to the next via the germ line.

virion   the complete virus particle, which may be found extracellularly.

viroid   very small circular RNA molecule that replicates autonomously.

virus particle   a virion.

virusoid   a group of satellite RNAs with properties similar to those of viroids.

Z

zoonotic   animal virus that infects humans.

CREDITS

PHOTOGRAPHS

Chapter 1

Figure 1.4A: © Omikron/Photo Researchers, Inc.; 1.4B: Courtesy of Centers for Disease Control; 1.5C: © SPL/Photo Researchers, Inc.; 1.7A: © Bruce A. Voyles; 1.7B: © Lee D. Simon/Photo Researchers, Inc.; 1.7C: Courtesy of NASA; 1.8A: Courtesy of the Centers for Disease Control; 1.8B: © J. Gennaro Jr./Photo Researchers, Inc.; 1.11A: From Terry J. Beveridge, "Thin-section envelope profiles of conventionally fixed and embedded bacteria" in *Microbiological Reviews,* 55:684–705, figure 1B, 1991. Reprinted by permission of American Society for Microbiology; 1.12A: From Terry J. Beveridge, "Thin-section envelope profiles of conventionally fixed and embedded bacteria" in *Microbiological Reviews*, 55:684-705, figure 2, 1991. Reprinted by permission of American Society for Microbiology; 1.14B (1-2): © Bruce A. Voyles; 1.15: From Marie Chow, "Plaque phenotypes of poliovirus VP2074E mutants" in *Journal of Virology* 66:1641–1648, figure 2 p. 164, 1992. Reprinted by permission of American Society for Microbiology; 1.16: From B.W. Falk, *Phytopathology* No. 69, p. 617, 1979, American Phytopathological Society.

Chapter 2

Figure 2.1: © Bruce A. Voyles; 2.6: Courtesy of Lenore T. Durkee; 2.7: Courtesy of Lucien Caro, University of Geneva, Switzerland; 2.9 A,B: © Lee D. Simon/Photo Researchers, Inc.; 2.10B: Courtesy of F. Eiserling, D. Caspar, and M. Moody; 2.13: Courtesy of Frederick Murphy, Centers for Disease Control; 2.14 A, B: Courtesy of Dr. Martin Lochelt, from *Journal of Virology,* 74:2885–2887, 2000. Reprinted by permission of the American Society for Microbiology; 2.17A, B: From C. Morgan and C. Howe, "Viruses as Observed in the Electron Microscope" in *Journal of Virology* 2, 1968. Reprinted by permission of American Society for Microbiology; 2.19 A, B: Courtesy of K. Simons from Simons, Gardoff, and Karenbeck, "First and last stages of infection" from *Scientific American*, Vol. 246 (2); 1982.

Chapter 3

Figure 3.8: From F. William Studier, "Time course of protein synthesis during T7 infection" from *Journal of Molecular Biology*, 166:477–535. Copyright © 1983 Academic Press, Ltd.; 4.2A: Courtesy of the Centers for Disease Control; 4.2B: From Dennis T. Brown and Jeffrey G. Gliedman, "Morphological variants of Sindbis virus obtained from infected mosquito tissue culture cells" in *Journal of Virology*, 12:1534–1539, 1973, fig. 1 page 1536. Reprinted by permission of American Society for Microbiology.

Chapter 4

Figure 4.2C: Courtesy of the Centers for Disease Control; 4.5 Modified from Shih, et al., *Proceedings of the National Academy of Sciences*, 75:5808, 1978. Courtesy of Roland Rueckert; 4.17A: From E.L. Palmer and M.L. Martin, *An Atlas of Mammalian Viruses*, 1982, Courtesy of the Centers for Disease Control; 4.17B: From E.L. Palmer and M.L. Martin, *Electron Microscopy in Viral Diagnosis,* 1988, Courtesy of the Centers for Disease Control; 4.17C: Courtesy of Dr. Frederick A. Murphy, Centers for Disease Control and Prevention; 4.17D: Courtesy of the Centers for Disease Control; 4.17E: Courtesy of Kathy Kuehl and J.E. White, USAMRID; 4.17F, 4.24A: From E.L. Palmer and M.L. Martin, *Electron Microscopy in Viral Diagnosis*, 1988, Courtesy of the Centers for Disease Control; 4.26 Courtesy of A.R. Bellamy, from "Reovirus Reaction Cores" in *Journal of Virology* 14:324, 1974. Reprinted by permission of American Society for Microbiology; 4.27A: Reprinted with permission from Gonda, "Sequence Homology and Morphologic Similarity of HTLV-III and Visna Virus, a Pathogenic Lentivirus" in *Science* 227: 173–177. Copyright © 1985 by the American Association for the Advancement of Science.

Chapter 5

Figure 5.1A, B: Courtesy of the Centers for Disease Control; 5.1C, D: From E.L. Palmer and M.L. Martin, *Electron Microscopy in Viral Diagnosis*, 1988, Courtesy of the Centers for Disease Control; 5.1E: Courtesy of the Centers for Disease Control; 5.1F: © Photo Researchers, Inc.; 5.2A, B: Courtesy of Robert G. Milne.

Chapter 6

Figure 6.2A: From P.J. Butler and A. Klug, "The Assembly of a Virus" in *Scientific American*, Vol. 239, pp. 62–69, November 1978, page 63. Reprinted by permission of the author; 6.15: Courtesy of K. Simons from Simons, Gardoff, and Karenbeck, "First and last stages of infection" from *Scientific American*, Vol. 246 (2),1982.

Chapter 7

Figure 7.1A: Courtesy of J.H.M. Willison, Dalhousie University, Nova Scotia; 7.1B: From A. Takeuchi and K. Hashimoto, *Infection and Immunity*, Vol. 13, page 569, 1976, American Society for Microbiology; 7.2: Courtesy of the Centers for Disease Control; 7.3A (left & right): Courtesy of H.B. Greenberg, *Journal of Virology*, 65(8):4190–4197, 1991. Reprinted by permission of American Society for Microbiology; 7.3B (left & right): © Bruce A. Voyles; 7.4: © Dr. Gopal Murti/Science Photo Library/Photo Researchers, Inc.; 7.5: Courtesy of Jean Lucas-Lenard from "Polyacrylamide gel analysis of proteins synthesized in mengovirus-infected L cells during the course of infection" in *Journal of Virology*, 56(1):161–171, 1985. Reprinted by permission of American Society for Microbiology.

Chapter 8

Figure 8.1A, B, 8.2 A-C: © Bruce A. Voyles; 8.8: From Bernard Davis, Renato Dulbecco, Herman Eisen, Herol Ginsberg, and Wood, *Microbiology,* © Lippincott Williams & Wilkins.

Chapter 9

Figure 9.5: Courtesy of H. Diringer and M. Ozel, Robert Koch Institute, Berlin, from *Nature*, Vol. 306, 1983; 9.6 A, B: From Stanley Prusiner, "Prions" in *Proceedings of the National Academy of Sciences*, November 10, 1998 Vol. 95 Issue 23, pp. 13363–13383, p. 13370. Reprinted by permission.

Appendix A

A1.2 (A, B): © Bruce A. Voyles

ILLUSTRATIONS AND TEXT

Chapter 1

Figure 1.2: From Ricki Lewis *Life,* 3rd edition. Copyright 1998 WCB McGraw-Hill, Dubuque IA. Reprinted by permission. Table 1.1: Host cell values are modified from data given in *Growth of the Bacterial Cell* by Ingraham, Maaloe, and Neidhardt, 1983; virus values are from *The Genetics of Bacteria and Their Viruses* by Hayes, 1964.

Chapter 2

Figure 2.4: Courtesy of F. Eiserling and the American Society for Microbiology. Figure 2.10A: Courtesy of T. Crowther. Table 2.3: Modified from Mathews, et al., *Bacteriophage T4,* 1983.

Chapter 4

Figure 4.4: Modified from Pallansch et al., *J Virol* 49:874, 1984. Courtesy of R. Rueckert. Figure 4.6: Adapted from S. R. Stewart and B. L Semler *Semin. Virol.* 8:242–255, 1997. Figure 4:27B: Courtesy of M. A. Gonda. *Science* 227: 173–177, copyright 1985 by the AAAS.

Chapter 5

Table 5.1: Data are taken from Field, et al., *Virology,* Tables 56-2 and 56-3, page 1596.

Chapter 6

Figure 6.1: Courtesy of W. B. Wood. Figure 6.2B: From The Assembly of a Virus: Butler JG, Klug A. Copyright 1978 by Scientific American, Inc. All rights reserved. Figure 6.3: From The Assembly of a Virus: Butler JG, Klug A. Copyright 1978 by Scientific American, Inc. All rights reserved. Figure 6.5: From The Assembly of a Virus: Butler JG, Klug A. Copyright 1978 by Scientific American, Inc. All rights reserved. Figure 6.7A: Courtesy of M. G. Rossmannaud and V. E. Johnson (Ann. Rev. Biochem. 58: 533–578, 1989). Figure 6.7B: Courtesy of M. Chow, *Science* 229:1360 (1985). Figure 6.7C: Courtesy of G. Harrison. Figure 6.9: Courtesy of J. E. Johnson.

Chapter 7

Table 7.4: Data are taken from W. Arber and D. Dussoix, J Mol Biol 5:18–36, 1962.

Chapter 9

Figure 9.1: Courtesy of R. N. Symons. Figure 9.3: Courtesy of R. N. Symons. Figure 9.6: Michael Balter, *Science* 281:1425 copyright 1998. Figure 9.7: From S. B. Prusiner *PNAS* 95:13370 copyright 1998.

Genome Architecture

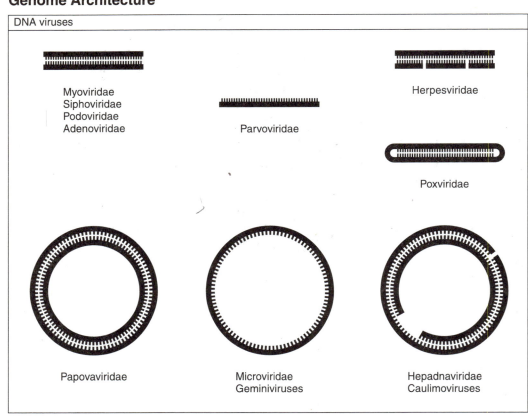

DNA viruses

Myoviridae
Siphoviridae
Podoviridae
Adenoviridae

Parvoviridae

Herpesviridae

Poxviridae

Papovaviridae

Microviridae
Geminiviruses

Hepadnaviridae
Caulimoviruses

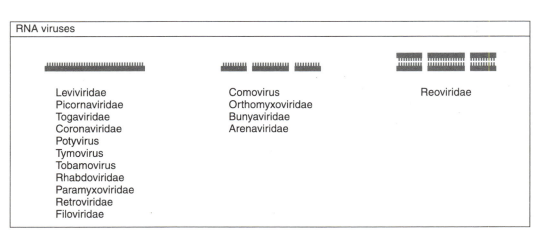

RNA viruses

Leviviridae
Picornaviridae
Togaviridae
Coronaviridae
Potyvirus
Tymovirus
Tobamovirus
Rhabdoviridae
Paramyxoviridae
Retroviridae
Filoviridae

Comovirus
Orthomyxoviridae
Bunyaviridae
Arenaviridae

Reoviridae